Advances in Mine Backfilling Technology and Materials

Advances in Mine Backfilling Technology and Materials

Guest Editors

Yuye Tan
Xun Chen
Yuan Li

MDPI

Basel • Beijing • Wuhan • Barcelona • Belgrade • Novi Sad • Cluj • Manchester

Guest Editors
Yuye Tan
Resources Engineering Department
The University of Science and Technology Beijing
Beijing
China

Xun Chen
Department of Mining Engineering
Shandong University of Technology
Zibo
China

Yuan Li
Department of Mining Engineering
University of Utah
Salt Lake City
United States

Editorial Office
MDPI AG
Grosspeteranlage 5
4052 Basel, Switzerland

This is a reprint of the Special Issue, published open access by the journal *Minerals* (ISSN 2075-163X), freely accessible at: www.mdpi.com/journal/minerals/special_issues/9A22SX627F.

For citation purposes, cite each article independently as indicated on the article page online and using the guide below:

Lastname, A.A.; Lastname, B.B. Article Title. *Journal Name* **Year**, *Volume Number*, Page Range.

ISBN 978-3-7258-4042-7 (Hbk)
ISBN 978-3-7258-4041-0 (PDF)
https://doi.org/10.3390/books978-3-7258-4041-0

Contents

About the Editors

Yuye Tan

Yuye Tan, a Professor and Doctoral Supervisor at the University of Science and Technology Beijing (USTB), specializes in green mining technologies, backfill materials, and mining economics. With extensive expertise in sustainable mining practices, her research focuses on developing innovative solutions for metal deposit exploitation.

Her scientific contributions span three key areas: (1) Development of ultra-fine tailing backfill materials; (2) Damage mechanisms in rock–backfill composite structures; (3) High-stage backfill mining techniques.

Notable honors include the following: (1) First Prize in Metallurgical Science and Technology Awards (2022 and 2025); (2) 13th Metallurgical Youth Science and Technology Award (2023); (3) Four additional provincial/ministerial first prizes; (4) Yonggang Excellent Teacher Award at USTB.

Nationally Funded Research Projects (Past Five Years):

(1) National Natural Science Foundation of China (NSFC), "Evolution of Hydration Temperature Field and Time-Dependent Strength Mechanism of Backfill Materials under Deep Mine High-Temperature Conditions".

(2) National Natural Science Foundation of China (NSFC), Young Scientists Fund, No. 52004019, "Macro-Meso Damage and Fracture Evolution of Backfill-Rock Composite Structures".

(3) Ministry of Science and Technology, National Key R&D Program (Key Special Project), No. 2022YFC2905003, "Development and Application of Green Mining and Beneficiation Technologies for Copper-Molybdenum-Rhenium Polymetallic Resources in Tibet".

Xun Chen

Xun Chen, Ph.D., is a lecturer and a master's student supervisor at Shandong University of Technology. In 2022, he received a Ph.D. degree in Mmining Engineering from the University of Science and Technology Beijing. His research interests include cemented tailing backfill, solution mining, and deep ore deposit mining. He has hosted and participated in more than 10 projects, including the National Natural Science Foundation of China (NSFC) Program, the Shandong Provincial Natural Science Foundation, and so on. He has authored/co-authored more than 40 refereed journal papers and more than 10 patents. Moreover, he has received eight science and technology awards, including the Second Prize of the Ministry of Education Natural Science Award, the First Prize of Science and Technology Award of China Gold Association, and others.

Yuan Li

Dr. Yuan (Stuart) Li is an Assistant Professor in the Department of Mining Engineering at the University of Utah, with a research focus on mine tailings management. He holds a Ph.D. in Mining Engineering from the University of British Columbia, where he conducted extensive research on the sedimentation behavior of tailings suspensions. His work integrates theoretical modeling, laboratory experiments, and field applications to promote sustainable mine waste practices. Dr. Li has authored peer-reviewed publications in high-impact journals and contributed to international mining and tailings conferences through service on technical and editorial committees. He also brings teaching expertise in hydraulics and mine waste management. Prior to his academic appointment, he gained valuable industry experience as a mine waste specialist with leading engineering firms such as AECOM and Hatch. His research continues to bridge the gap between academia and industry, addressing critical challenges in tailings management.

Preface

Backfill mining technology has emerged as a sustainable solution for mine waste management and ground subsidence control in underground mining operations. Recognized for its efficiency, environmental benefits, and potential for mine environment rehabilitation, this technology has gained widespread adoption in modern mining practices.

Recent decades have witnessed significant advancements in backfill mining research, encompassing mining methods, material development, and performance evaluation of backfill systems. These developments have substantially promoted the application of backfill technology in underground mines.

This Special Issue presents a collection of innovative studies that advance backfill technology across multiple dimensions. The featured research covers fundamental investigations into material properties, including mechanical behavior under dynamic loading and environmental cycles, as well as practical applications such as novel dewatering techniques and optimized mining methods. Several contributions focus on material enhancement through fiber reinforcement and thermal treatment, while others address operational challenges through scheduling optimization and curing agent development.

We extend our sincere gratitude to all authors for their valuable contributions, to the reviewers for their constructive feedback, and to the editorial board for their dedicated efforts in preparing this Special Issue. Their collective work has significantly enriched the field of sustainable mining technologies.

This Special Issue demonstrates the remarkable progress in backfill technology, bridging theoretical understanding with practical engineering solutions to address contemporary challenges in mine waste management and sustainable resource extraction.

Yuye Tan, Xun Chen, and Yuan Li
Guest Editors

minerals

MDPI

Article

Research on the Long-Term Mechanical Behavior and Constitutive Model of Cemented Tailings Backfill Under Dynamic Triaxial Loading

Yuye Tan [1,2], Jinshuo Yang [1,2,*], Yuchao Deng [1,2], Yunpeng Kou [2,3], Yiding Li [4] and Weidong Song [1,2]

1 State Key Laboratory of High-Efficient Mining and Safety of Metal Mines of Ministry of Education, University of Science and Technology Beijing, Beijing 100083, China; tanyuye@ustb.edu.cn (Y.T.)
2 School of Resources and Safety Engineering, University of Science and Technology Beijing, Beijing 100083, China
3 Filling Engineering Laboratory, Shandong Gold Group Co., Ltd., Laizhou 261441, China
4 Beijing Guoxin Anke Technology Co., Ltd., Beijing 100160, China
* Correspondence: yangjinshuo@xs.ustb.edu.cn

Abstract: Cemented tailings backfill (CTB) plays an important role in mine filling operations. In order to study the long-term stability of CTB under the dynamic disturbance of deep wells, ultrafine cemented tailings backfill was taken as the research object, and the true triaxial hydraulic fracturing antireflection-wetting dynamic experimental system of coal and rock was used to carry out a static true triaxial compression test, a true triaxial compression test under unidirectional disturbance, and a true triaxial compression test under bidirectional disturbance. At the same time, the acoustic emission monitoring and positioning tests of the CTB were carried out during the compression test. The evolution law of the mechanical parameters and deformation and failure characteristics of CTB under different confining pressures is analyzed, and the damage constitutive model of the filling body is established using stochastic statistical theory. The results show that the compressive strength of CTB increases with an increase in intermediate principal stress. According to the change process of the acoustic emission ringing count over time, the triaxial compression test can be divided into four stages: the initial active stage, initial calm stage, pre-peak active stage, and post-peak calm stage. When the intermediate principal stress is small, the specimen is dominated by shear failure. With an increase in the intermediate principal stress, the specimen changes from brittle failure to plastic failure. The deformation and failure strength of CTB are closely related to its loading and unloading methods. Under a certain stress intensity, compared with unidirectional unloading, bidirectional unloading produces a greater deformation of the rock mass, and the failure strength of the rock mass is higher. This study only considers the confining pressure within the compressive limit of the specimen. Future research can be directed at a wider range of stresses to improve the applicability and reliability of the research results.

Keywords: ultrafine cemented tailings backfill; true triaxial; acoustic emission; failure characteristics; constitutive model

Academic Editors: Elsabe Kearsley, Abbas Taheri and Andrey G. Kalinichev

Received: 7 January 2025
Revised: 10 February 2025
Accepted: 6 March 2025
Published: 8 March 2025

Citation: Tan, Y.; Yang, J.; Deng, Y.; Kou, Y.; Li, Y.; Song, W. Research on the Long-Term Mechanical Behavior and Constitutive Model of Cemented Tailings Backfill Under Dynamic Triaxial Loading. *Minerals* **2025**, *15*, 276. https://doi.org/10.3390/min15030276

1. Introduction

In recent years, with the gradual depletion of surface and shallow mineral resources, deep mining has become the main trend of mining development [1–3]. The process of underground metal mining, with increasing depth, faces a series of challenges, such as high ground pressure, high temperature, and the poor stability of ore rock, and disasters such as rock burst are prominent [4–6]. On 20 October 2018, a major rock burst accident occurred in the drainage roadway and No.3 contact roadway of the 1303 working face of Shandong Longyun Coal Industry Co., Ltd., Shandong Energy Longkuang Group, Shandong, China, resulting in 21 deaths and 4 injuries, with a direct economic loss of 56.398 million yuan. The buried depth of the accident area was 1027 m to 1067 m, and the high self-weight stress of coal and rock mass was one of the main causes of the accident. The filling mining method has been widely used in the deep metal mining process due to its advantages of high production capacity, high recovery strength, and high safety [7–9]. As a key part or even permanent part of the underground stope, the mechanical properties and overall stability of the CTB are very important for controlling the pressure of the mining site during the mining process and maintaining the long-term stability of the underground stope structure [10–12]. For example, CTB may be affected by dynamic loads (such as blasting vibration) during mining, resulting in damage to its internal structure. The results show that after dynamic loading, the microstructure of CTB degrades, the porosity increases, the wave velocity decreases, and the uniaxial compressive strength and elastic modulus decrease significantly. This damage reduces the bearing capacity of CTB and affects the long-term stability of the mine [13]. Therefore, it is of great importance to study the mechanical properties of CTB to ensure the safety of underground engineering.

In order to study the failure mechanism of deep CTB, scholars at home and abroad have carried out a lot of research and achieved fruitful results. Wang et al. studied the strength characteristics and failure characteristics of rock and CTB through triaxial loading and unloading experiments [14–18]. Rong et al. analyzed the deformation and failure characteristics and mechanical parameter evolution of deep rock under different stress paths by carrying out true triaxial experiments on rock under different stress paths. The study found that under triaxial compression conditions, rock deformation mainly manifests as axial compression and expansion along the direction of minimum principal stress, and that its failure mode is shear failure. Under the unloading conditions, the rock deformation mainly manifests as strong dilatancy along the unloading direction, and the failure mode is tensile–shear composite failure [19–22]. Miao et al. studied the energy evolution law of rock under cyclic loading and unloading, discussed the relationship between energy dissipation and rock strength, and proposed that according to the evolution characteristics of the energy dissipation ratio curve and stress–strain curve, the cyclic loading and unloading process can be divided into five stages: the initial compaction and elastic stage, crack stable development stage, crack unstable development stage, post-peak unstable rupture stage, and residual strength stage. The corresponding energy dissipation ratio shows 'spoon'-shaped evolution characteristics of slight decrease, stable development, slow rise, and gentleness after significant rise. [23,24]. Lu et al. conducted a numerical study on the mechanical behavior of backfill–rock composite structure (BRCS) under triaxial compression, including deformation, failure mode, strength characteristics, and acoustic emission evolution. The influence factors of deformation, strength characteristics, and failure mode of BRCS under triaxial compression were obtained, and the characteristics of each stage of the stress–strain curve of BRCS under triaxial compression were analyzed [25]. Fu et al. studied the influence of structural parameters on the creep mechanical properties of surrounding rock and filling body composite specimens by carrying out conventional triaxial tests and triaxial creep tests, and revealed the variation law of creep energy and damage evolution characteristics [26]. Yang et al. studied the fracture mode and hydraulic fracture propagation law of limestone under the combined action of multiple factors by means of a true triaxial hydraulic fracturing test and numerical simulation [27].

In the process of loading, rock releases internal stored energy and produces new cracks, accompanied by strong acoustic emission signals. Acoustic emission monitoring is used to capture and characterize the micro-damage inside the CTB, which can reasonably characterize the internal variation characteristics of the CTB during the loading process [28,29]. Cheng et al. conducted an acoustic emission test on the backfill-surrounding rock combination under uniaxial compression and found that the cumulative count of acoustic emission ringing increased in an 'S' shape with time and constructed a coupling relationship between the cumulative acoustic emission ringing count and the damage constitutive equation [30]. Gao et al. conducted triaxial compression and acoustic emission tests on different types of reservoir rocks and found that the failure types and acoustic emission characteristics of different types of reservoir rocks under triaxial stress are significantly different. [31]. Zhao et al. studied the acoustic emission characteristics of rock burst under different unloading rates and found that when the unloading rate is high, the rock is prone to strain-type rock burst [32]. Some scholars have also studied the relationship between rock damage theory and the acoustic emission parameters of rocks. Chen et al. carried out acoustic emission monitoring tests of sandstone under true triaxial graded loading and proposed a damage variable and damage evolution model based on the cumulative window waveform number [33].

In this paper, the ultrafine cemented tailings backfill is taken as the research object. Through the development of a coal–rock true triaxial hydraulic fracturing antireflection-wetting dynamic experimental system and acoustic emission monitoring technology, the mechanical properties of the CTB under dynamic true triaxial experimental conditions are analyzed. The temporal and spatial evolution law of damage, ringing count, and energy in the compression process is studied, and the fracture location is carried out. The long-term strength evolution law of CTB under the dynamic disturbance of deep wells is summarized, which is helpful to provide the basis for the strength design of CTB in deep mining.

2. Materials and Schemes

2.1. Experimental Materials

In this experiment, the ultrafine tailings of Jiaojia gold mine, Shandong, China, a laboratory-made cementitious material, and tap water were selected as the experimental materials, and the physical and chemical properties of the filling material were studied via particle size analysis, X-ray diffraction analysis, and X-ray fluorescence spectrometer (XRF) analysis. The particle size composition, mineral composition, and main chemical composition of the whole tailings were obtained.

2.1.1. Ultrafine Tailing

Ultrafine tailings were selected in the experiment. The particle size composition is shown in Figure 1, and the main chemical composition is shown in Figure 2. The experimental results show that the average weighted particle size of all the tailings is 43.88 μm. The chemical composition of ultrafine tailings is mainly made up of SiO_2 and Al_2O_3, and its mineral composition is relatively simple, including quartz, feldspar, and zeolite, so it is a good inert filling material.

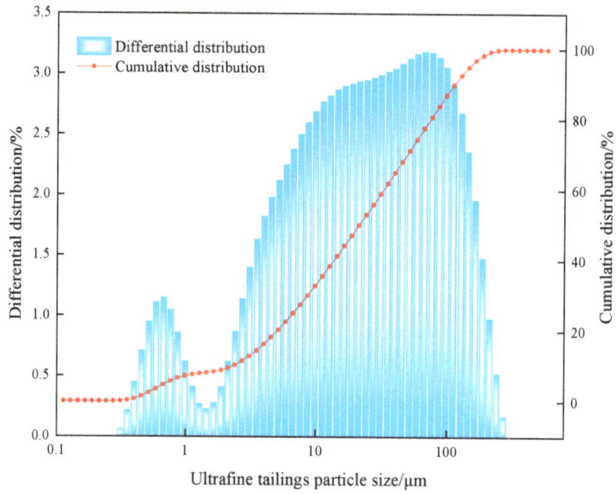

Figure 1. Particle size composition of ultrafine tailings.

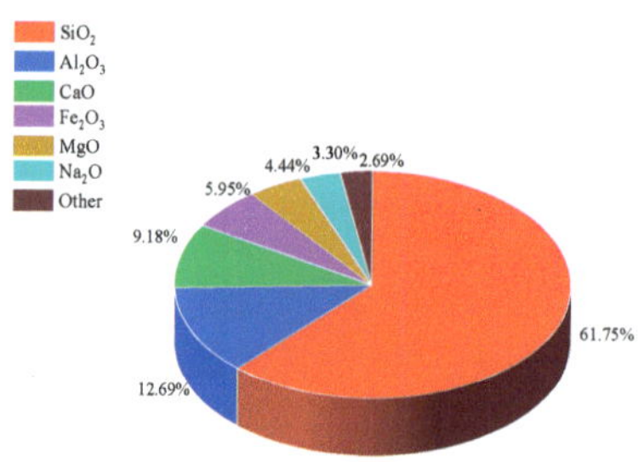

Figure 2. Chemical composition of ultrafine tailings.

2.1.2. Cementitious Material

In this experiment, a laboratory's self-developed cementitious material, 'Filling C Material' was used for the research. Its mineral composition is mainly made up of slag and gypsum. The chemical composition of filling C material is shown in Figure 3; its chemical composition is mainly made up of CaO, SiO_2, and Al_2O_3, and contains a small amount of MgO, Fe_2O_3, and TiO_2, indicating that it is an active cementitious material and suitable for underground filling raw materials.

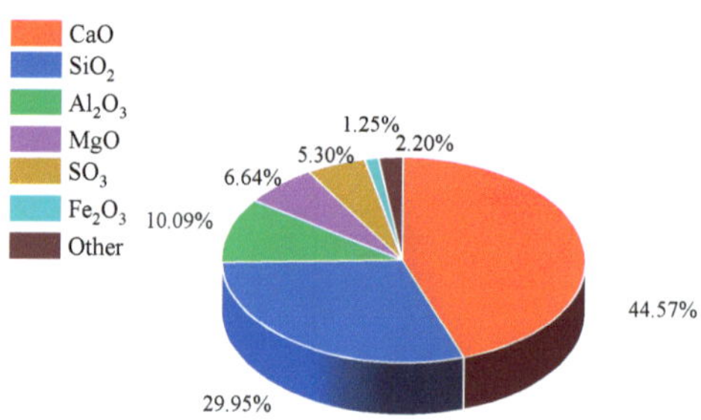

Figure 3. Chemical composition of cementitious materials.

2.2. *True Triaxial Loading and Unloading Experimental Scheme*

2.2.1. Specimen Preparation

According to the cement–sand ratio of 1:4, the appropriate amounts of cementitious materials and ultrafine tailings were weighed, and the appropriate amount of tap water was added for stirring. First, the mixture was manually stirred evenly, and then the NRJ-411 A cement mortar mixer was used for strong stirring to ensure that the stirred slurry had no obvious particles or agglomerates. The filling slurry, with a mass concentration of 64%, was prepared. The shape of the specimen was a cube, and the size was 100 mm × 100 mm × 100 mm. After 48 h of preparation, the specimen was demolded and placed in a standard curing box with a constant temperature (temperature 20 ± 1 °C) and constant humidity (humidity 95 ± 1%) for 60 days to obtain a CTB specimen under long-term curing conditions. In each of the three experiments, three specimens were prepared for each different intermediate principal stress condition, amounting to a total of twenty-seven specimens. The surface of the obtained sample was carefully ground using a grinding stone machine, and the surface roughness was controlled within 3 mm to ensure an as smooth and flat surface as possible. At the same time, each pair of opposite surface samples should be parallel to each other, and the angle deviation should be controlled within 0.25° to meet the flatness requirements of the true triaxial test samples.

2.2.2. Experimental Scheme

In the experiment, the triaxial compression experiment of the CTB was carried out using the true triaxial hydraulic fracturing antireflection-wetting dynamic experimental system of coal and rock. In the compression process, the acoustic emission monitoring and positioning test were carried out with the DS5 full information acoustic emission analyzer. The acoustic emission sensor is fixed on the indenter to monitor the failure characteristics of acoustic emission during the failure process of the specimen, Vaseline is applied between the probe and the indenter to reduce the attenuation of the acoustic emission signal by enhancing the coupling, the noise threshold of the experimental instrument is 30 mv, and the data acquisition interval is 0.1 s. Set the initial stress, where σ_1 represents the maximum principal stress, along the Z direction. σ_2 represents the intermediate principal stress along the Y direction and σ_3 represents the minimum principal stress along the X direction, as shown in Figure 4. Figure 5 shows the process of the whole experiment.

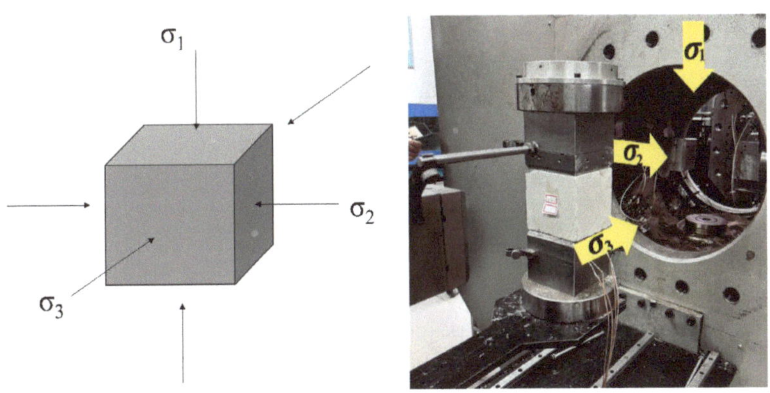

Figure 4. Loading diagram.

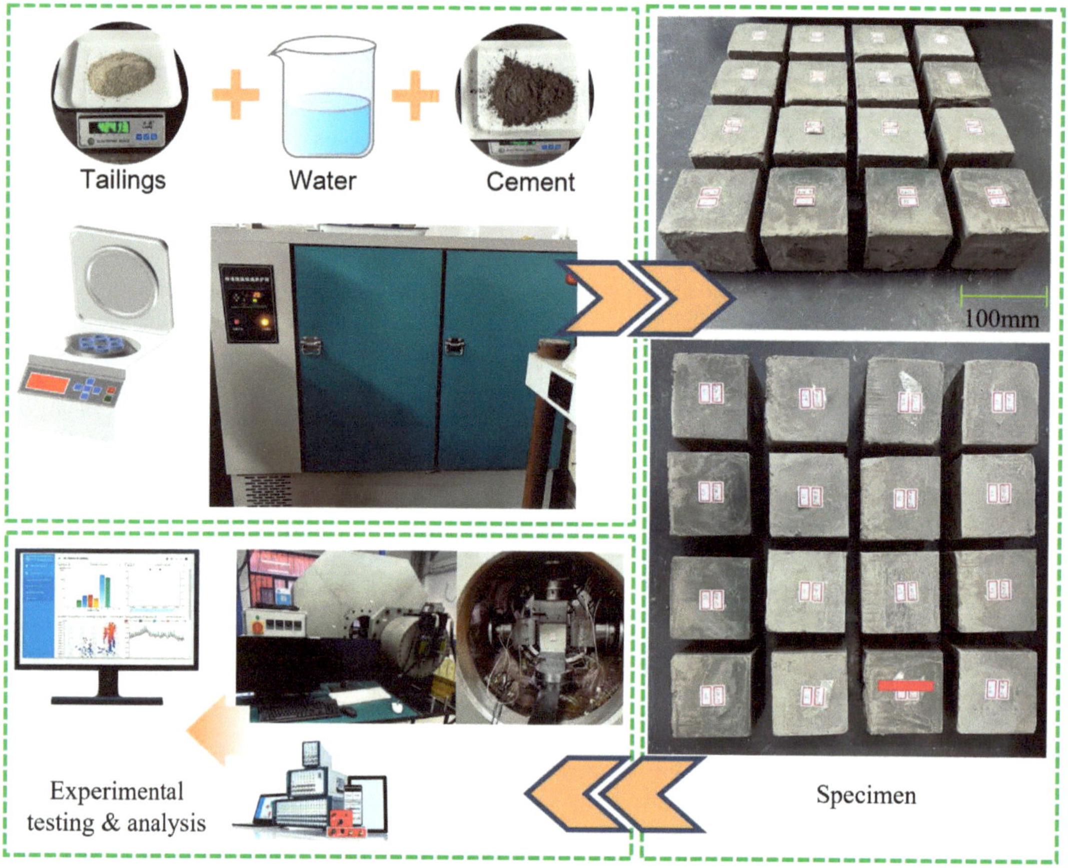

Figure 5. Experimental process.

Scheme 1: Static true triaxial compression experiment. ① Two methods of displacement control and load control were used in the test loading. The displacement control method loaded the specimen σ1 to 0.5 MPa at a loading rate of 0.01 mm/s, so that the end face of the specimen was in contact with the indenter. Then, keeping the principal stress unchanged, the intermediate principal stress σ_2 was loaded to 0.7 MPa using the same loading method. Finally, keeping the intermediate principal stress unchanged, σ_3 was loaded to 0.5 MPa. ② After the initial stress state was loaded, the specimen was stable for about 60 s under this stress state, and the σ_2 and σ_3 stresses remained unchanged. ③ The load control mode was changed to a loading rate of 5×10^{-4} m/min until the specimen was destroyed;

Scheme 2: Constant axial pressure single-sided unloading test. ①, ② the same as Scheme 1. ③ The σ1 was loaded to 80% of the triaxial strength, and on the basis of this static stress, it was maintained for 3–5 min, and then the disturbance load was applied. ④ σ2 was unloaded at 0.002 MPa/s until the specimen failed;

Scheme 3: Constant axial pressure double-sided unloading test. ①, ②, and ③ the same as Scheme 2. ④ The σ2 and σ3 were unloaded at 0.002 MPa/s until the specimen failed.

3. Results and Analyses

3.1. Study on Mechanical Properties of Backfill Under Static True Triaxial Compression

3.1.1. Stress–Strain Curve Analysis

Figure 6 shows the stress–strain curve of the triaxial compression test of the CTB under different loading conditions. It was be found that, similar to the results of the true triaxial compression experiments conducted by Rong et al. [19], the corresponding axial stress increases gradually with an increase in confining pressure. At the initial stage, the CTB specimen is in the compaction stage, the pores and cracks inside the specimen are compacted, and the curve is concave. After that, the specimen is in the elastic stage, and the stress–strain curve is approximately linear. After reaching peak strength, the slope of the curve becomes negative, the CTB specimen is damaged from the compression state to the expansion state, and the bearing capacity decreases rapidly. Under the same curing time, with an increase in confining pressure, the peak strength of the CTB specimen increases continuously, and the bearing capacity of the CTB specimen increases gradually.

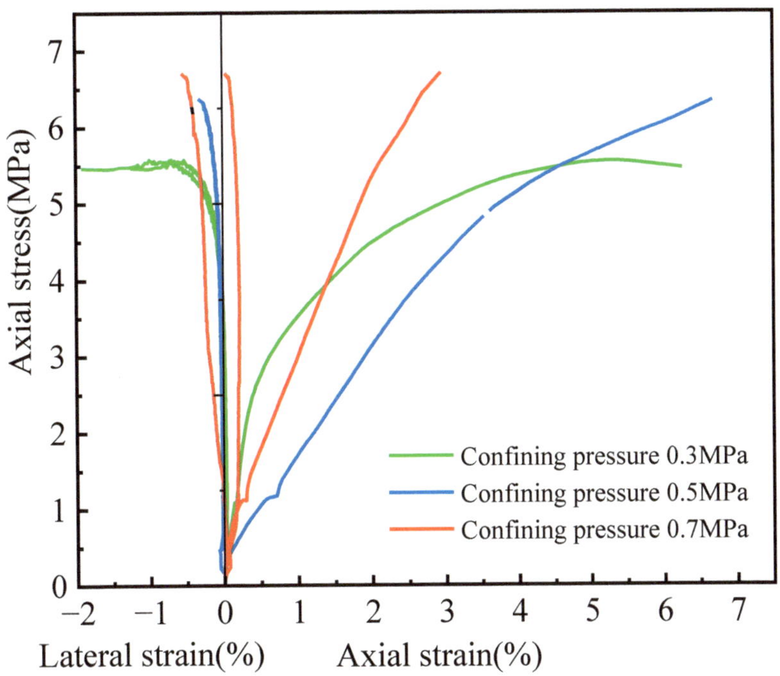

Figure 6. Stress–strain curves of CTB specimens under different loading conditions.

3.1.2. Analysis of Acoustic Emission Characteristics

The acoustic emission ringing energy is the number of oscillations of the transducer crossing the threshold signal. It is a piece of monitoring data related to the signal amplitude in the acoustic emission experiment. It is a basic measurement parameter that can reflect the fracture of the CTB, and is widely used in the analysis of the internal crack evolution of the rock mass. Figure 7 shows the relationship diagram of the time–ringing energy–cumulative ringing energy of CTB under static true triaxial compression under different intermediate principal stresses.

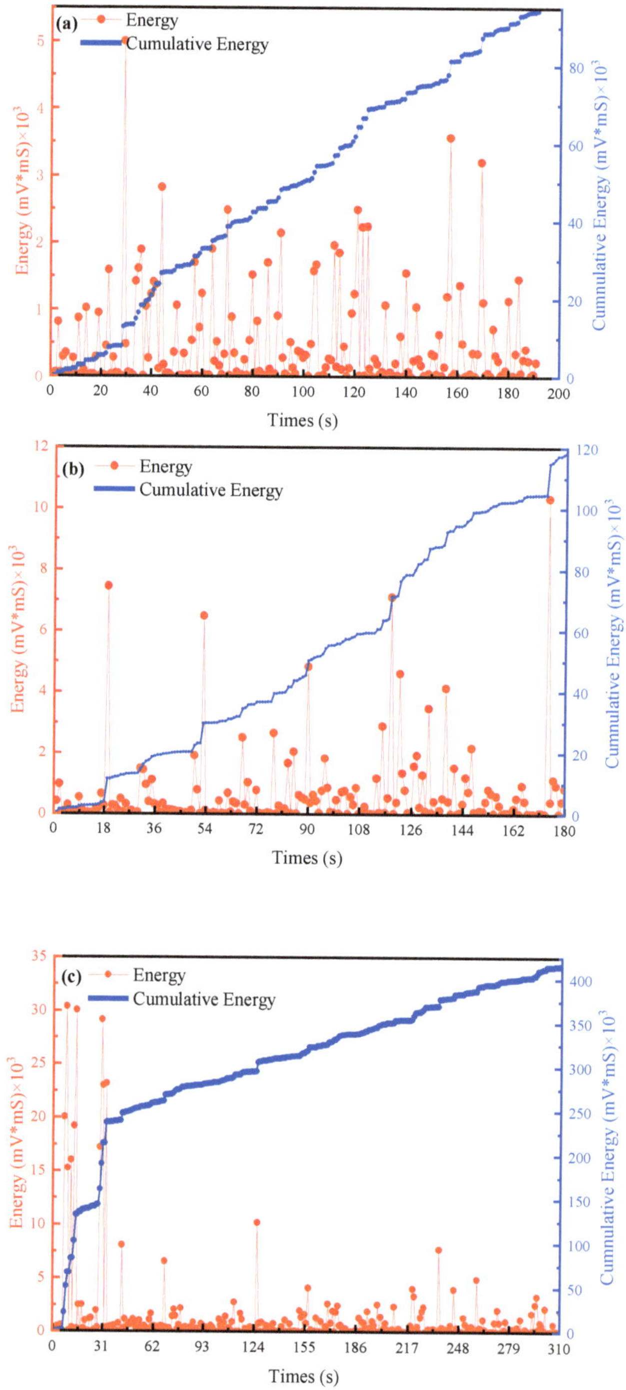

Figure 7. Loading time–energy acoustic emission curve of CTB under different loading conditions: (**a**) σ_2 = 0.3 MPa; (**b**) σ_2 = 0.5 MPa; and (**c**) σ_2 = 0.7 MPa.

It can be seen from Figure 7 that in the triaxial static compression test of the CTB under different test conditions, the signal energy is mainly manifests in four stages with the loading time. In the initial active stage, corresponding to the pore compaction stage of the CTB, the ringing energy signal shows a concentrated and disorderly distribution. This phenomenon is due to the joint action of the surface structure of the CTB to complete the compaction, while the internal structure is still in the compaction process. In the initial quiet stage, corresponding to the elastic stage of the CTB specimen, the acoustic emission ringing energy is sporadically distributed, the signal is weak and disordered, the numerical value is small, and only the compaction phenomenon of the initial crack appears locally. In the pre-peak active stage, the active stage of acoustic emission of CTB specimens mostly appears before the peak compressive strength of the specimens. Corresponding to the yield failure stage of CTB, the structural performance of CTB begins to release, and the energy accumulated in the initial quiet stage begins to release. In the post-peak quiet stage, after the CTB reaches peak strength, its internal cracks expand greatly, resulting in a significant decrease in the bearing capacity of the specimen. However, the post-peak strength of the CTB is large, and the fragment structure still has bearing capacity. In addition, by comparing the curves, it was found that with an increase in intermediate principal stress, the intensity of the acoustic emission signal increases, and that the cumulative energy of acoustic emission increases gradually at the same time.

3.1.3. Analysis of Failure Characteristics

Figure 8 shows the failure fracturing diagram under various conditions in the true triaxial test of the CTB. It can be seen that when the intermediate principal stress is small, the specimen is mainly shear failure. At this time, with the application of axial load and an increase in strength, the CTB appears to penetrate cracks. When the intermediate principal stress increases, the specimen appears as obvious tensile failure. From the microscopic point of view inside the CTB, it is caused by the transverse tensile stress exceeding the tensile limit of the CTB, and the main cracks appear in the form of tensile failure. With an increase in the intermediate principal stress, the sample transforms from brittle failure to plastic failure.

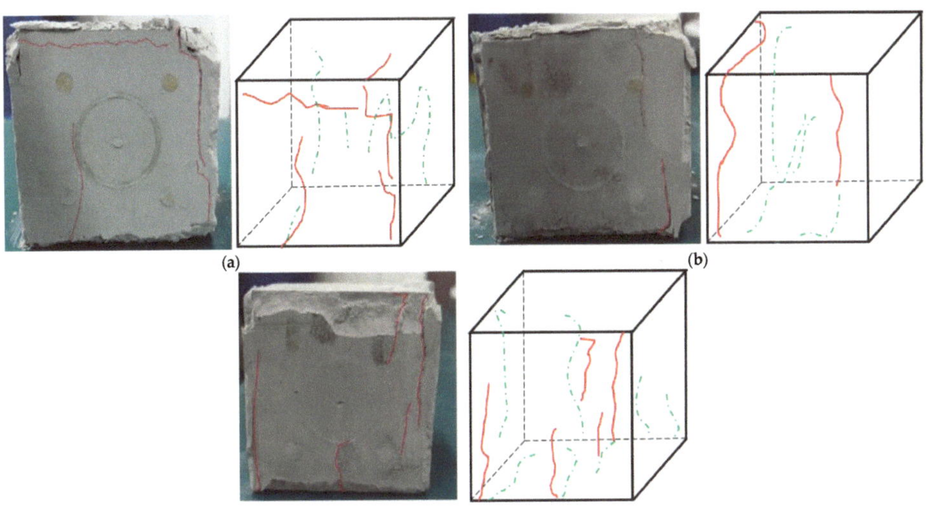

Figure 8. CTB true triaxial dynamic load disturbance test under various conditions damaging fracturing diagram: (**a**) σ_2 = 0.3 MPa; (**b**) σ_2 = 0.5 MPa; and (**c**) σ_2 = 0.7 MPa. (Red line: cracks on the front of the specimen, green line: cracks on other surfaces of the specimen).

3.2. Study on Long-Term Mechanical Properties of Backfill Under Unidirectional Disturbance

The results of the unidirectional unloading test of the CTB are shown in Table 1 [34]. The strength of the CTB in this state is 4.61 MPa~5.46 MPa, which is smaller than in the static compression results. The Poisson's ratio is largest at a lateral pressure of 0.5 MPa, indicating that tensile failure is most significant at this time.

Table 1. Unidirectional disturbance test results of CTB.

Unloading Pressure (MPa)	Length (mm)	Breadth (mm)	Height (mm)	Peak Strength (MPa)	Elastic Modulus (MPa)	Poisson's Ratio
0.3	99.96	100.02	99.90	4.61	802.33	0.2
0.5	99.95	99.95	99.99	5.22	722.1	0.41
0.7	99.95	99.98	99.98	5.46	336.97	0.05

3.2.1. Stress–Strain Curve Analysis

Figure 9 shows the stress–strain curves of CTB under unidirectional unloading under different intermediate principal stresses, which are similar to those under true triaxial compression. The deformation and failure of the CTB specimens can be divided into the following stages: the compaction stage of micro-pores and cracks, at this time, the original micro-pores and cracks inside the CTB gradually closed; the linear deformation stage, at this stage, the stress and strain of the material showed an approximate linear relationship; the yield stage, at which time the specimen began to undergo irreversible plastic deformation, the microstructure was gradually destroyed, and the deformation rate was accelerated; and the unloading and pressure-bearing stage, when the bearing capacity of the material gradually decreased and the deformation tended to be stable until it completely failed.

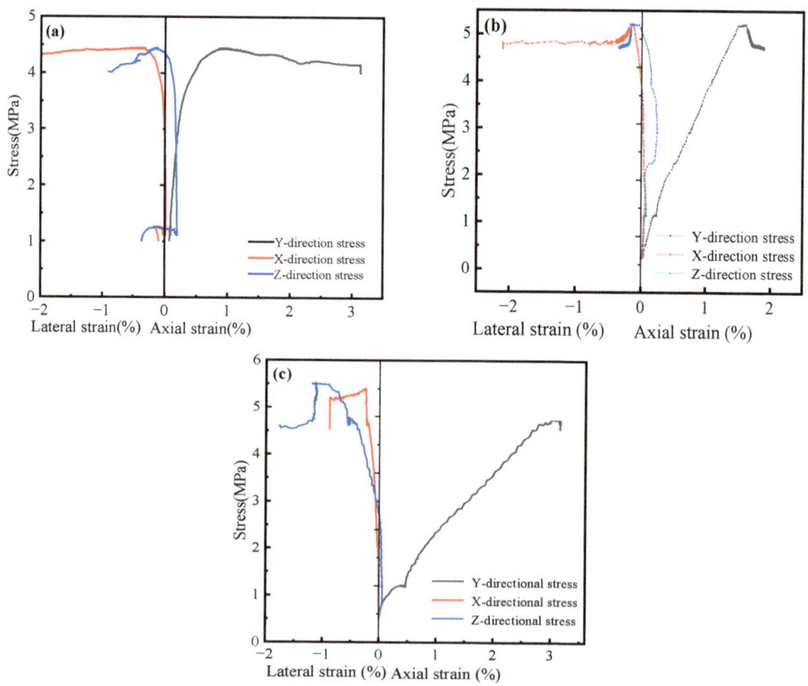

Figure 9. Complete stress–strain curves of unidirectional disturbance test of CTB: (**a**) $\sigma_2 = 0.3$ MPa; (**b**) $\sigma_2 = 0.5$ MPa; and (**c**) $\sigma_2 = 0.7$ MPa.

3.2.2. Analysis of Acoustic Emission Characteristics

Figure 10 [34] shows the acoustic emission energy curve generated via the unidirectional disturbance of CTB under different intermediate principal stresses. Under the conditions of unidirectional unloading, the acoustic emission activity law of the CTB specimen is similar to that under the conditions of triaxial compression. However, the intensity of the acoustic emission signal is significantly enhanced, and the peaks of each stage are more concentrated. This shows that under the conditions of unidirectional unloading, the failure of the CTB specimen is more thorough and rapid, and that its compressive performance is fully exerted and decreases rapidly with the rapid occurrence of failure. This is also in line with the conclusion found by Zhao et al. [20] that when the unloading reaches a certain level, the overall level of the acoustic emission b value under the triaxial unloading path is lower than that of the conventional triaxial path, and that the damage degree of the specimen is greater.

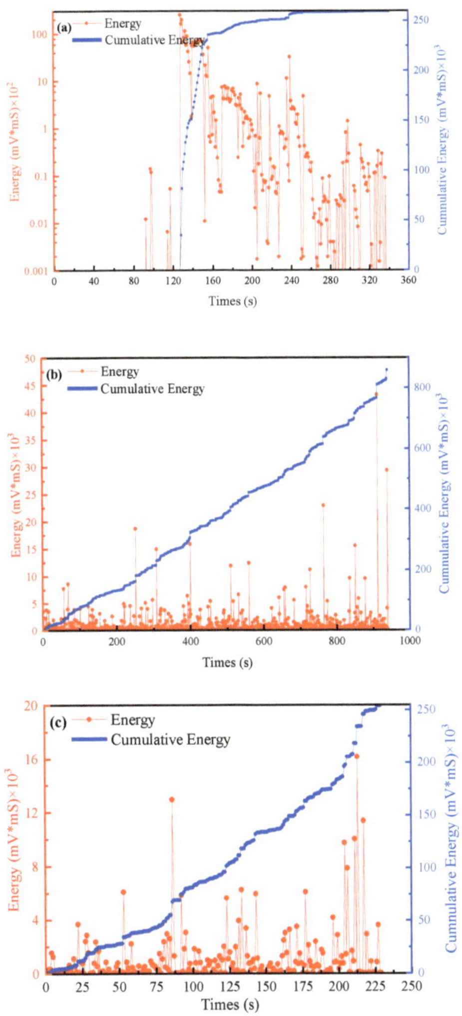

Figure 10. Unidirectional acoustic emission energy curves of CTB: (**a**) σ_2 = 0.3 MPa; (**b**) σ_2 = 0.5 MPa; and (**c**) σ_2 = 0.7 MPa.

3.2.3. Analysis of Failure Characteristics

Figure 11 shows the failure characteristics of the CTB under unloading under different intermediate principal stresses. By comparison, it was found that the change in intermediate principal stress can have a significant impact on the unloading failure. When the intermediate principal stress is small, the CTB begins to break as a whole, and many long tensile cracks appear on the surface. The main reasons for this are that the stress redistribution caused by unloading and the deformation trend of the same micro-unit are different, and there is large damage in the relatively weak position. When the intermediate principal stress is large, the crushing of the specimen is aggravated; with multiple small pieces peeling off, the size of the crack is obviously increased, and the bearing performance of the specimen is weakened, but due to the existence of confining pressure, it still has a certain bearing capacity. The confining pressure can make the broken small pieces form a new bearing structure and have a new mechanical response in the process of failure.

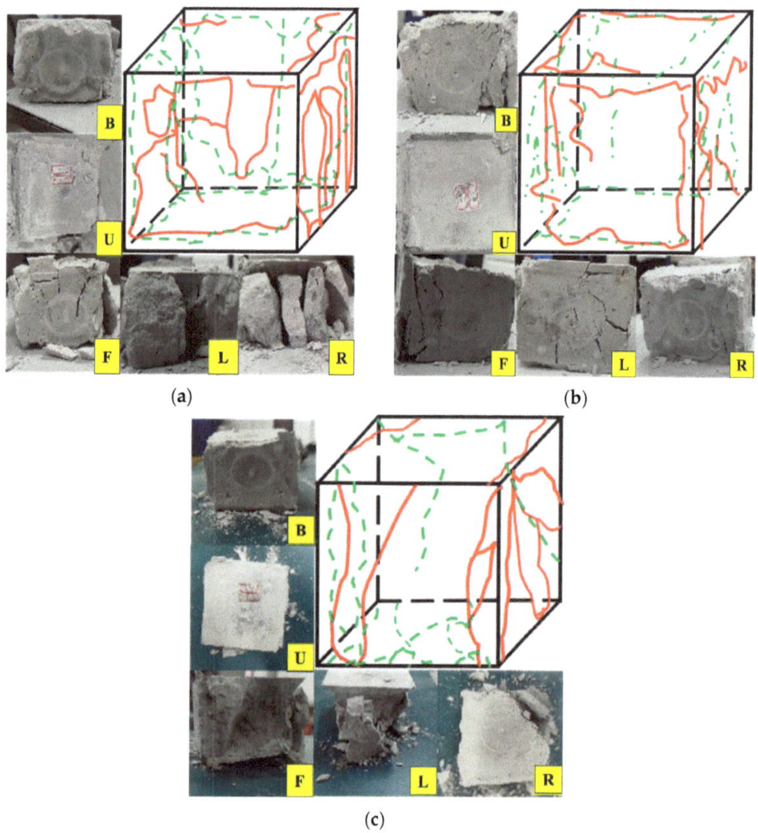

Figure 11. Unidirectional disturbance macroscopic failure characteristics of CTB: (**a**) σ_2 = 0.3 MPa; (**b**) σ_2 = 0.5 MPa; and (**c**) σ_2 = 0.7 MPa. (B, U, F, L and R represent the back, up, front, left and right sides respectively. Red line: cracks on the front, upper and right sides of the specimen; green line: cracks on other surfaces of the specimen).

3.3. *Study on Long-Term Mechanical Properties of Backfill Under Bidirectional Disturbance*

The results of the bidirectional unloading test of the CTB are shown in Table 2 [34]. The strength of the CTB in this state is 4.61 MPa~5.51 MPa, which is slightly higher than that in the unidirectional unloading state. The Poisson's ratio is largest when the lateral pressure is 0.7 MPa, indicating that the tensile failure is most significant.

Table 2. Bidirectional disturbance test results of CTB.

Unloading Pressure (MPa)	Length (mm)	Breadth (mm)	Height (mm)	Peak Strength (MPa)	Elastic Modulus (MPa)	Poisson's Ratio
0.3	99.94	100.01	99.99	4.61	231.49	0.02
0.5	100.02	99.91	100.00	5.17	567.50	0.08
0.7	100.00	100.05	100.00	5.51	454.65	0.48

3.3.1. Stress–Strain Curve Analysis

According to the test results, the stress–strain relationship curve of the CTB under bidirectional disturbance is shown in Figure 12. Similarly to under the true triaxial compression conditions, the deformation and failure of the CTB have experienced the micro-pore fracture compaction stage, the linear deformation stage, the yield stage, and the unloading pressure stage. The greater the intermediate principal stress, the higher the peak strength. The higher the confining pressure, the more the linear elastic deformation stage of the CTB specimen can be extended to a higher level.

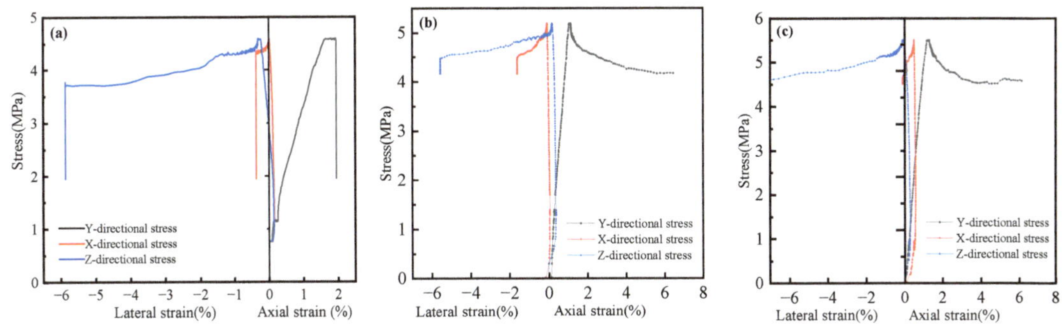

Figure 12. Complete stress–strain curves of bidirectional disturbance test of CTB: (**a**) σ_2 = 0.3 MPa; (**b**) σ_2 = 0.5 MPa; and (**c**) σ_2 = 0.7 MPa.

3.3.2. Analysis of Acoustic Emission Characteristics

Figure 13 shows the time–ringing energy–cumulative ringing energy relationship diagram of the CTB under the influence of bidirectional dynamic load disturbance under different intermediate principal stresses. According to the change process of acoustic emission ringing count over time, there are still four stages. With an increase in the confining pressure, due to the increase in the strength of the CTB, relatively more micro cracks are produced during the failure process, and the friction effect between the internal particles increases. Therefore, the ringing energy monitored via acoustic emission before failure increases with an increase in curing age and shows an increasing trend in each stage before failure. The acoustic emission activity of the CTB specimen increases significantly under the conditions of bidirectional unloading, which indicates that the CTB specimen has experienced rapid instability and failure under these conditions.

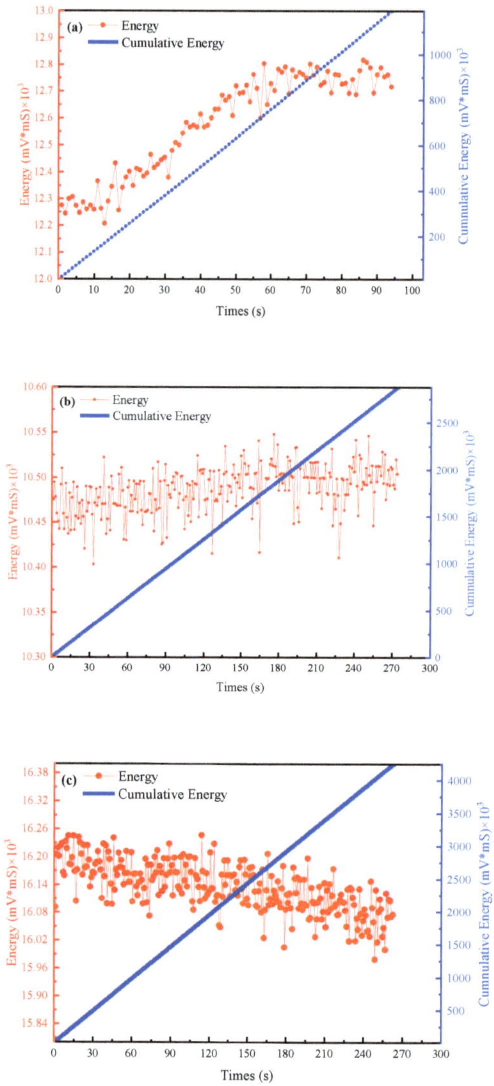

Figure 13. Bidirectional acoustic emission energy curves of CTB: (**a**) σ_2 = 0.3 MPa; (**b**) σ_2 = 0.5 MPa; and (**c**) σ_2 = 0.7 MPa.

3.3.3. Analysis of Failure Characteristics

Figure 14 shows the failure characteristics of the CTB under different confining pressures during bidirectional unloading. By comparison, it was found that the change in confining pressure under bidirectional unloading shows obvious azimuth, and that the angle of sample failure is more comprehensive. Under a confining pressure of 0.3 MPa, the unloading effect is not obvious, and the damage degree of the specimen is light. However, due to the stress concentration at the end, the end of the specimen is damaged. When the confining pressure is increased to 0.5 MPa, the specimen is severely broken during unloading, the size of the crack on the specimen is significantly increased, and the phenomenon of multiple small pieces accompanied by large pieces of spalling is presented. Under a confining pressure of 0.7 MPa, the failure of the specimen is that large pieces of debris and some small pieces of debris appear at the same time. At this time, the CTB loses its bearing capacity in structure and performance.

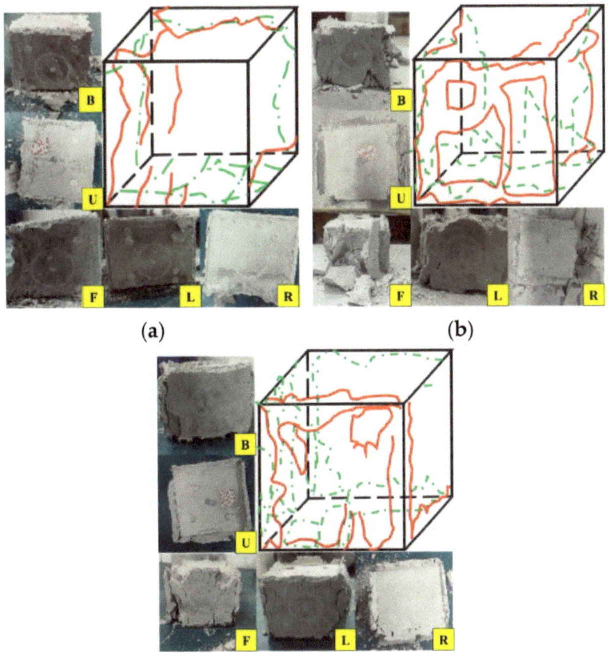

(a) (b)

(c)

Figure 14. Macroscopic failure characteristics of bidirectional disturbance of CTB: (**a**) σ_2 = 0.3 MPa; (**b**) σ_2 = 0.5 MPa; and (**c**) σ_2 = 0.7 MPa.(B, U, F, L and R represent the back, up, front, left and right sides respectively. Red line: cracks on the front, upper and right sides of the specimen; green line: cracks on other surfaces of the specimen).

4. Construction of Constitutive Model

The CTB is transported via slurry and undergoes a complex hydration reaction. There are a large number of micro cracks, holes, and other defects in the CTB. These defects are randomly distributed inside the CTB, which changes the mechanical properties of the CTB and reduces the deformation and strength parameters of the CTB. The damage and failure of the CTB under the action of external load manifest as the generation of micro cracks, the expansion to the intersection of each crack, and nucleation to form macro cracks. At the initial stage of loading, the generation of such cracks is also random, which is due to the randomness of the nature of the micro-element of the CTB. With an increase in load, the cracks continue to expand and cross each other, and finally localize and lead to damage. Therefore, it can be concluded that for the alternative material of CTB, its damage is essentially random.

Based on this understanding, the damage constitutive model of the CTB is established using the stochastic statistical theory. The following derivation is based on the following assumptions: It is assumed that the damage to CTB is isotropic. It is assumed that the failure of the micro-element of the CTB is controlled by its ultimate compressive strain, that is, the failure of the micro-element is considered when its strain reaches the ultimate strain. The stress–strain relationship of rock medium conforms to the generalized Hooke's law.

Assuming that the number of failed infinitesimals under a certain level of load is N_f, the statistical damage variable D_s is defined as the ratio of the number of failed infinitesimals to the total number of infinitesimals N.

$$D_s = \frac{N_f}{N} \tag{1}$$

In Equation (1), Ds is the damage variable of rock under external load; N_f is the number of rock infinitesimal failures under external load; and N is the total number of rock infinitesimals without external load.

Assuming that the strength F of each infinitesimal element satisfies a certain probability distribution, the maximum value that the stress level S in each infinitesimal element can reach also satisfies the probability distribution. When the stress level in an infinitesimal element is in a certain stress level interval [S, S + dS], the probability of its failure is shown in Equation (2).

$$P_f = P(S)\mathrm{d}S \tag{2}$$

In Equation (2), P is the density function of the probability distribution satisfied by the infinitesimal strength F. The infinitesimal number dN_d of failure in this interval in the whole rock mass is obtained by multiplying the above probability by the total infinitesimal number N:

$$\mathrm{d}N_{\mathrm{d}} = NP_f = NP(S)\mathrm{d}S \tag{3}$$

When loading to a certain stress level S, the infinitesimal whose strength is less than the stress level has also failed. The number of failed infinitesimals N_d in the rock is the sum of the number of failed infinitesimals in the previous intervals [35]:

$$N_{\mathrm{d}} = \int_0^S NP(x)\mathrm{d}x = NP(S) \tag{4}$$

In Equation (4), P is the distribution function of the probability distribution satisfied by the infinitesimal strength F.

The combined Equations (1) and (4) can be obtained as follows:

$$D_s = P(S) \tag{5}$$

Equation (4) is the statistical damage evolution equation derived from the statistical strength theory. For any distribution, with an increase in the stress level S, the distribution function value $P(S)$ changes from 0 to 1, which is consistent with the change rule of damage variable D_s. From this point of view, the rationality of Equation (5) can be explained.

The randomness of the infinitesimal properties of CTB can be described using some statistical distribution. Some studies have shown that Weibull distribution is more suitable for describing the statistical distribution of CTB. Taking the CTB as the research object, Sun et al. found that the failure of the backfill was caused by the damage to and failure of the meso-element by establishing the meso-damage evolution equation and constitutive equation in accordance with the Weibull statistical distribution law and combining it with the numerical calculation of the meso-scale [36]. It is assumed that the ultimate strain of the CTB infinitesimal conforms to the two-parameter Weibull distribution function $f(\varepsilon)$:

$$f(\varepsilon) = \eta\beta(\eta\varepsilon)^{\beta-1}e^{-(\eta\varepsilon)^{\beta}} \tag{6}$$

In Equation (6):

$f(\varepsilon)$—Distribution density function;

β—Homogeneity of ultimate strain of infinitesimal element;

η^{-1}—Overall average of the ultimate strain of all infinitesimals.

Since the rock damage process under unloading is not a simple superposition of pore damage and stress damage, it is necessary to derive the total damage variable D in the unloading process of sandstone under unloading according to the equivalence principle:

$$D = D_s + D_w - D_s D_w \tag{7}$$

The following equation for calculating the comprehensive damage variable of rock during unloading can be obtained:

$$D = 1 - \frac{1-n}{1-n_0}\exp\left[-\left(\frac{k}{F}\right)^m\right] \tag{8}$$

In this derivation, the Drucker–Prager criterion is introduced as a criterion to measure the strength of rock micro elements:

$$f(\sigma) = k = \alpha I_1 + \sqrt{J_2} \tag{9}$$

In Equation (9):

I_1 is the first invariant of the stress tensor;

J_2 is the second invariant of the stress deviator;

α is a parameter related to the material properties in the variable.

$k = \alpha I_1 + \sqrt{J_2}$;

$\alpha = \frac{\sin\varphi}{\sqrt{3}\sqrt{3+\sin^2\varphi}}$;

$I_1 = \sigma_1^* + \sigma_2^* + \sigma_3^*$;

$J_2 = \frac{1}{6}\left[\left(\sigma_1^* - \sigma_2^*\right)^2 + \left(\sigma_2^* - \sigma_3^*\right)^2 + \left(\sigma_3^* - \sigma_1^*\right)^2\right]$

Equation (9) can be written as follows:

$$D = 1 - \frac{1-n}{1-n_0}\exp\left\{-\left[\left(\frac{\alpha I_1 + \sqrt{J_2}}{F}\right)^m\right]\right\} \tag{10}$$

Considering that the rock still has a certain ability to transfer load after failure, it is considered that the infinitesimal can still transfer part of the load after failure, and has a certain bearing capacity [35]. Combined with the concepts of the strain equivalence principle and effective stress principle, proposed by Lemaitre, the relationship equation between nominal stress and effective stress with the introduction of the damage variable correction factor is obtained.

$$\sigma^* = \frac{\sigma}{1-D} \tag{11}$$

According to the generalized Hooke's law, the following equation can be obtained:

$$\varepsilon_i = \frac{1}{E}\left[\sigma_i^* - \mu\left(\sigma_j^* + \sigma_k^*\right)\right] \tag{12}$$

In Equation (12), i, j, k = 1,2,3; ε_i is the strain in the i direction; E is the elastic modulus of rock in a certain state; σ_i*, σ_j*, and σ_k* are directional stresses in the directions of i, j, and k; and μ is the Poisson's ratio of rock.

Based on the stress conditions of the triaxial test $\sigma_2 = \sigma_3$ and Equation (12), the deformation is obtained:

$$\sigma_1 = E\varepsilon_1(1-D) + 2\mu\sigma_3 \tag{13}$$

When substituting Equation (10) into Equation (13), the damage evolution equation of the CTB sample under the actions of loading and unloading is obtained as follows:

$$\sigma_1 = E\varepsilon_1 \frac{1-n}{1-n_0} \exp\left\{-\left[\left(\frac{\alpha I_1 + \sqrt{J_2}}{F}\right)^m\right]\right\} + 2\mu\sigma_3 \tag{14}$$

$$\begin{cases} I_1 = E\varepsilon_1 \frac{\sigma_1+2\sigma_3}{\sigma_1-2\mu\sigma_3} \\ \sqrt{J_2} = \frac{1}{\sqrt{3}} E\varepsilon_1 \frac{\sigma_1-\sigma_3}{\sigma_1-2\mu\sigma_3} \\ \alpha = \frac{\sin\varphi}{\sqrt{3}\sqrt{3+\sin^2\varphi}} \end{cases} \tag{15}$$

Equations (14) and (15) are the damage constitutive equations of the CTB under different porosities.

5. Conclusions

1. Under the conditions of static true triaxial compression and with an increase in the intermediate principal stress σ_2, the axial stress of the CTB increases continuously, indicating that an increase in the confining pressure can improve the bearing capacity of the CTB. When the intermediate principal stress is small, the specimen is mainly shear failure. When the intermediate principal stress increases, the specimen shows obvious tensile failure. At the same time, with the application of dynamic load disturbance, the failure of the CTB is more severe, and the specimen changes from brittle failure to plastic failure;
2. The deformation and failure strength of CTB are closely related to its loading and unloading methods. Under the actions of loading and unloading, it generally experiences the process of volume compression first, and then volume expansion. Unloading must cause the phenomenon of rock mass expansion. Different unloading methods lead to the different deformation and failure characteristics of the rock mass. Under a certain stress intensity, bidirectional unloading produces greater deformation than the unidirectional unloading of the rock mass, and the failure strength of the rock mass is higher;
3. The distribution law of the acoustic emission energy signal shows that for the dataset generated during the triaxial compression process of the CTB, it is dominant in the early compaction stage of loading, and that the local damage is shear failure. After the fracture begins to develop and the CTB enters the yield stage, the overall failure mode of the specimen is replaced by tensile failure, because the CTB is a low-porosity and no-pre-existing-crack structure material. The results show that the useful information from the acoustic emission source characteristics can effectively provide information on related damage in deformed rock.

Author Contributions: Conceptualization, Y.T.; formal analysis, Y.L.; resources, Y.K.; writing—original draft preparation, Y.D. and J.Y.; writing—review and editing, J.Y.; supervision, W.S.; project administration, Y.T. All authors have read and agreed to the published version of the manuscript.

Funding: This work was funded by the National Natural Science Foundation of China (52274110), the National Key Research and Development Program of China (2022YFC2905003), and the China Scholarship Council fund (202206465005).

Data Availability Statement: The data are contained within the article.

Acknowledgments: The authors thank the journal editors and anonymous reviewers for their constructive comments.

Conflicts of Interest: Yiding Li is an employee of Beijing Guoxin Anke Technology Co., Ltd., Beijing 100160, China. This paper reflects the views of the scientists and not the company.

References

1. Wen, X.; Fu, J.; Zheng, W. Research on New Solid Waste Heat Insulation Material for Deep Mining. *Minerals* **2023**, *13*, 838. [CrossRef]
2. Ranjith, P.G.; Zhao, J.; Ju, M.; De Silva, R.V.S.; Rathnaweera, T.D.; Bandara, A.K.M.S. Opportunities and Challenges in Deep Mining: A Brief Review. *Engineering* **2017**, *3*, 546–551. [CrossRef]
3. Cai, M.; Xue, D.; Ren, F. Current status and development strategy of metal mines. *Chin. J. Eng.* **2019**, *41*, 417–426. [CrossRef]
4. Xie, H.; Gao, F.; Ju, Y. Research and development of rock mechanics in deep ground engineering. *Chin. J. Rock Mech. Eng.* **2015**, *34*, 2161–2178.
5. Cai, M.; Li, P.; Tan, W.; Ren, F. Key Engineering Technologies to Achieve Green, Intelligent, and Sustainable Development of Deep Metal Mines in China. *Engineering* **2021**, *7*, 1513–1517. [CrossRef]
6. Zhao, Y.; Yang, T.; Liu, H.; Wang, S.; Zhang, P.; Jia, P.; Wang, X. A Path for Evaluating the Mechanical Response of Rock Masses Based on Deep Mining-Induced Microseismic Data: A Case Study. *Tunn. Undergr. Space Technol.* **2021**, *115*, 104025. [CrossRef]
7. Yu, J.; Liu, S.; Zhou, Z.; Jiao, H.; Wang, J.; Zhou, L. Influence of Filling Mining on Ground Stability in a Metal Mine. *Arab. J. Geosci.* **2024**, *17*, 18. [CrossRef]
8. Xue, G.; Yilmaz, E.; Wang, Y. Progress and Prospects of Mining with Backfill in Metal Mines in China. *Int. J. Miner. Metall. Mater.* **2023**, *30*, 1455–1473. [CrossRef]
9. Zhang, Q.; Zhang, P.; Chen, Q.; Li, H.; Song, Z.; Tao, Y. Study of the Critical Safe Height of Goaf in Underground Metal Mines. *Minerals* **2024**, *14*, 227. [CrossRef]
10. Wang, J.; Yang, T.; Qiao, D.; Xu, P.; Zhang, X. Strength Model of Surface Backfill in Upward Slicing and Filling Method and Its Application. *Min. Metall. Explor.* **2024**, *41*, 997–1011. [CrossRef]
11. Yu, X.; Tan, Y.; Song, W.; Wang, J. Strength and failure mechanism of the rock-encased by backfills structure. *China Univ. Min. Technol.* **2023**, *52*, 30–42. [CrossRef]
12. Qi, W.; Zhang, J.; Zhou, N.; Wu, Z.; Zhang, J. Mechanism by Which Backfill Body Reduces Amount of Energy Released in Deep Coal Mining. *Shock Vib.* **2019**, *2019*, 8253269. [CrossRef]
13. Song, X.; Fan, B.; Wang, S.; Zhang, H. Reloading mechanical characteristics of cemented tailings backfill after dynamic loading action. *J. China Coal Soc.* **2024**, *49*, 4785–4797. [CrossRef]
14. Wang, Z.; LI, W.; Feng, G.; Du, J.; Hao, R. Mechanical properties of sandstone and prediction model of shear deformation band under true triaxial unloading conditions. *J. China Coal Soc.* **2023**, *48*, 3700–3712. [CrossRef]
15. Feng, F.; Chen, S.; Wang, Q.; Rostami, J.; Khoreshok, A.A. Experimental study on failure characteristics of natural and saturated sandstone under true triaxial unloading and dynamic disturbance condition. *Chin. J. Rock Mech. Eng.* **2022**, *41*, 2240–2253.
16. Duan, M.; Jiang, C.; Guo, X.; Yang, K.; Zhang, X.; Ma, H. Experimental study on mechanical and damage characteristics of coal under cyclic true triaxial loading. *Chin. J. Rock Mech. Eng.* **2021**, *40*, 1110–1118. [CrossRef]
17. Sha, P.; Wu, F.; Chang, J. Unloading strength and failure pattern of marble under true triaxial test. *Chin. J. Rock Mech. Eng.* **2018**, *37*, 2084–2092. [CrossRef]
18. Li, B.; Gao, P.; Cao, Y.; Gong, W.; Zhang, S.; Zhang, J. Damage and Failure Characteristics of Surrounding Rock in Deep Circular Cavern under Cyclic Dynamic Load: A True Triaxial Experiment Investigation. *Minerals* **2022**, *12*, 134. [CrossRef]
19. Rong, H.; Li, G.; Zhao, G.; Liang, D.; Liang, J. True triaxial test study on mechanical properties of deep rock mass in different stress paths. *J. China Coal Soc.* **2020**, *45*, 3140–3149. [CrossRef]
20. Zhao, J.; Fan, Q.; Li, P.; Li, S.; Wang, G.; Xie, M. Acoustic emissiom B value characteristics and failure precursor of the dacite under different stress paths. *J. Eng. Geol.* **2019**, *27*, 487–496. [CrossRef]
21. Wu, W.; Gong, F.; Jiang, Q.; He, L. Strength Reduction of Initial High-Stress Rock Pillars under Different Triaxial Unloading Paths. *Rock Mech. Rock Eng.* **2023**, *56*, 3519–3537. [CrossRef]
22. Wang, S.; Wang, L.; Tian, J.; Fan, H.; Jiang, C.; Ding, K. An Experimental Study on the Effects of True Triaxial Loading and Unloading Stress Paths on the Mechanical Properties of Red Sandstone. *Minerals* **2022**, *12*, 204. [CrossRef]
23. Miao, S.; Liu, Z.; Zhao, X.; Huang, Z. Energy dissipation and damage characteristics of Beishan granite under cyclic loading and unloading. *Chin. J. Rock Mech. Eng.* **2021**, *40*, 928–938.
24. Meng, Q.; Zhang, M.; Zhang, Z.; Han, L.; Pu, H. Research on Non-Linear Characteristics of Rock Energy Evolution under Uniaxial Cyclic Loading and Unloading Conditions. *Environ. Earth Sci.* **2019**, *78*, 650. [CrossRef]
25. Lu, H.; Wang, Y.; Gan, D.; Wu, J.; Wu, X. Numerical Investigation of the Mechanical Behavior of the Backfill—Rock Composite Structure under Triaxial Compression. *Int. J. Miner. Metall. Mater.* **2023**, *30*, 802–812. [CrossRef]
26. Fu, J.; Zhang, B.; Tan, Y.; Wang, J.; Song, W. Study on Creep Characteristics and Damage Evolution of Surrounding Rock and Filling Body (SR-FB) Composite Specimens. *J. Mater. Res. Technol.* **2023**, *23*, 5382–5399. [CrossRef]

27. Yang, W.; Geng, Y.; Zhou, Z.; Li, L.; Ding, R.; Wu, Z.; Zhai, M. True Triaxial Hydraulic Fracturing Test and Numerical Simulation of Limestone. *J. Cent. South Univ.* **2020**, *27*, 3025–3039. [CrossRef]
28. Meng, Q.; Zhang, M.; Han, L.; Pu, H.; Nie, T. Effects of Acoustic Emission and Energy Evolution of Rock Specimens under the Uniaxial Cyclic Loading and Unloading Compression. *Rock Mech. Rock Eng.* **2016**, *49*, 3873–3886. [CrossRef]
29. Zhang, C.; Song, X.; Fu, Y.; Lei, D.; She, W.; Zhu, W. Acoustic Emission-Based Modeling of Fiber Tailings Cementation and Filling Body Dynamics and Damage Ontology. *Minerals* **2023**, *13*, 1499. [CrossRef]
30. Cheng, A.; Shu, P.; Zhang, Y.; Wang, P.; Wang, M. Acoustic emission characteristics and damage constitution of backfill-surrounding rock combination. *J. Min. Saf. Eng.* **2020**, *37*, 1238–1245. [CrossRef]
31. Gao, Y.; Guo, P.; Li, X.; Li, Y.; Xu, D.; Zou, Z.; Qin, J.; Shi, G.; Li, S. Investigation of triaxial compression failure and acoustic emission characteristics of different reservoir rocks. *J. Eng. Geol.* **2022**, *30*, 1169–1178. [CrossRef]
32. Zhao, X.G.; Wang, J.; Cai, M.; Cheng, C.; Ma, L.K.; Su, R.; Zhao, F.; Li, D.J. Influence of Unloading Rate on the Strainburst Characteristics of Beishan Granite under True-Triaxial Unloading Conditions. *Rock Mech. Rock Eng.* **2014**, *47*, 467–483. [CrossRef]
33. Chen, P.; Li, N.; Fang, L.; Cai, C.; Huang, J.; Li, Y.; Liu, X. Acoustic Emission Characteristics and Damage Evolution of Sandstone under True Triaxial Step-loading. *Met. Mine* **2023**, *3*, 106–113. [CrossRef]
34. Jing, X.; Song, Z.; Kou, Y.; Zhu, G. Mechanical Characteristics of Backfill Under True Triaxial Compression and Unloading Conditions. *Min. Res. Dev.* **2024**, *44*, 56–64.
35. Tan, Y.; Yu, X.; Song, W.; Wang, H.; Cao, S. Experimental study on combined pressure-bearing mechanism of filling body and surrounding rock. *J. Min. Saf. Eng.* **2018**, *35*, 1071–1076.
36. Sun, G.; Wei, S.; Liu, X.; Lu, H. Study on heterogeneous mico-mechanics damage in compress performance of cemented filling body. *Ind. Miner. Process.* **2015**, *44*, 41–44. [CrossRef]

minerals

MDPI

Article

Mechanical Properties, Failure Modes, and Damage Development of Stratified Cemented Tailings Backfill under Uniaxial Compression

Wenbin Xu *, Yalun Zhang, Wei Chen, Tong Sun and Yilin Sang

School of Energy and Mining Engineering, China University of Mining & Technology (Beijing), Beijing 100083, China; zhangyalun05@163.com (Y.Z.); chenw0626@163.com (W.C.); 2111102222@student.cumtb.edu.cn (T.S.); 15060096900@163.com (Y.S.)

* Correspondence: xuwenbin@cumtb.edu.cn

Abstract: Layered cemented filling leads to a layered composite structure of cemented tailings backfill (CTB) composed of high-strength top and bottom layers, as well as a low-strength middle layer. To solve the problem of the low mechanical properties of the middle layer caused by layered filling, this study proposes the concept of an enhance layer, that is, an enhance layer is added to the middle weak layer to improve its overall mechanical properties. To explore the characteristics of strength, failure modes, energy dissipation, and progressive damage of stratified cemented tailings backfill (SCTB) with varying layered structures, the uniaxial compressive tests of SCTB specimens with enhance layers c/t of 1:15, 1:10, and 1:6, as well as height proportions of 0.1, 0.2, and 0.3, are examined. The results show that the elastic modulus and uniaxial compressive strength (UCS) of SCTB samples increase with the height ratio and cement-to-tailings ratio of the enhance layer. The elastic modulus and strength of SCTB specimens is more sensitive to the height ratio of the enhance layer than the c/t ratio. Moreover, the SCTB specimens mainly manifested as tensile failure of the upper layer and lower layer, but they did not penetrate the entire specimen. The propagation of cracks is limited by the addition of the enhance layer. The SCTB specimens have stronger plastic deformation ability, and a large part of the all-strain energy is dissipated in the shape of plastic failure. In addition, a constitutive model for damage in SCTB samples has been developed. The SCTB samples with a reasonable structure can also achieve sufficient strength compared to directly increasing the c/t ratio of CTB specimens while reducing the cost of cemented tailings backfill preparation. This approach reduces the carbon footprint of the mining industry and improved the overall mechanical properties and stability of the stratified cemented tailings backfill. This study provides a new approach for high-stage subsequent stope backfilling. The findings will offer guidance for the design of a layered filling mining method.

Keywords: stratified cemented tailings backfill; uniaxial compressive tests; energy dissipation; failure modes; damage progression

check for updates

Citation: Xu, W.; Zhang, Y.; Chen, W.; Sun, T.; Sang, Y. Mechanical Properties, Failure Modes, and Damage Development of Stratified Cemented Tailings Backfill under Uniaxial Compression. *Minerals* **2024**, *14*, 917. https://doi.org/10.3390/min14090917

Academic Editors: Abbas Taheri, Yuye Tan, Xun Chen and Yuan Li

Received: 17 August 2024

Revised: 3 September 2024

Accepted: 3 September 2024

Published: 6 September 2024

1. Introduction

The mining industry boom has fueled an unprecedented economic growth for many countries around the world, but the extraction of minerals, especially metal resources, has brought many problems to human beings, such as solid wastes disposal and underground goafs management [1–3]. The filling mining method is favored for its efficacy in stabilizing ground pressure and managing mine waste [4,5]. The cemented tailings backfill is a kind of composite medium with a certain strength, which is prepared by tailings, water, and a binding agent in a certain proportion [6,7]. The strength of the CTB is affected by the amount of binding agent. Blindly increasing the amount of binding agent will indeed enhance the strength of the CTB, but it will significantly increase the material cost. Reducing the amount of cement will reduce the cost of filling, but it will bring instability factors. Therefore, it is very important to determine a reasonable strength of cemented tailings backfill [8–10].

At present, in order to simultaneously take into account the cost saving and the strength of the CTB, the layered filling mining method has been adopted by numerous mines, namely, the top layer and bottom layer are filled with a high cement–tailings ratio slurry, and a low cement–tailings ratio slurry is used in the middle layer [11–13], as shown in Figure 1. However, this structure entails certain problems in practical application. Due to the relatively weak strength of the CTB in the middle layer, it often breaks down and falls off under the mining disturbance of the neighboring stope, which affects the safety and stability of the stope [14–16].

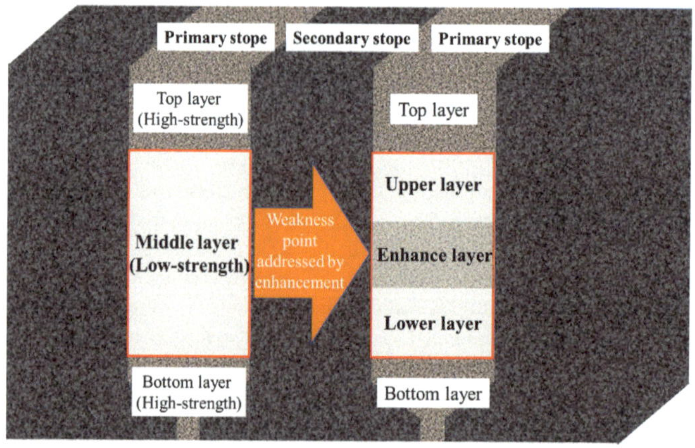

Figure 1. Background of layered filling mining method.

The strength and stability of SCTB are mainly affected by internal and external factors. Xu et al. [17,18] have demonstrated that the strength development of the CTB was inhibited by low temperature, and the strength escalated as the temperature rose. Zheng et al. [19] revealed that the UCS of the sulfur-containing CTB will decrease to a certain extent in the later stage of curing. Moreover, the larger the proportion of fine tailings, the lower the strength of the CTB. Nochaiya and Xu [20,21] investigated the silica fume on the UCS of cemented tailings backfill. The silica fume improves the mechanical property and stability of the cemented tailings backfill. Wang et al. [22] examined the influence of the cement-to-tailings (c/t) ratio and height of the intermediate layer on the mechanical properties of the SCTB. The results show that the strength of the CPB increases with decreases in the height of the intermediate layer and increases with the cement-to-tailings ratio. Zhang et al. [23] obtained that the peak strength and elastic modulus of the SCTB increase with the decrease of the number of layers. Cao et al. [11] reported that the UCS of the layered cemented tailings backfill decreases with the extension of the filling interval time in different layers. Wu et al. [24] experimentally studied the influence of interface angle on CPB–rock deformation characteristics and shear strength. The interface angle is crucial in determining the failure of the specimen.

In the above study, it has been confirmed that the failure mode, damage progression, and strength of the SCTB are mainly affected by the number of layers, cement–tailings ratio, and the proportion of intermediate layers [25,26]. However, no effective measures have been put forward to enhance the mechanical properties of the middle layer. The middle layer in the layered cemented tailings backfill is still the most vulnerable area [27,28].

Therefore, this study proposes the concept of a enhance layer, that is, adding a enhance layer to the middle weak layer to improve its mechanical properties, thereby increasing the mechanical property and stability of the entire SCTB, as shown in Figure 1. To uncover the mechanical characteristics and deformation behavior of the middle layer containing the enhance layer, the effect of the c/t ratio and height ratio of the enhance layer on it were studied using a uniaxial compression test. The height of the enhance layer (h/H) was set as 0.1, 0.2, and 0.3; the position of the enhance layer was set as 1/2 of the middle layer; the cement–tailings ratio of the enhance layer was set as 1:15, 1:10, and 1:6; and the cement–tailings ratio of the middle layer was set as 1/20. The purpose was to clarify the reinforcing effect of enhance layer on the stratified cemented tailings backfill.

2. Materials and Experimental Program

2.1. Materials

The silica tailings (ST), composed of 99.6% silica, were utilized in the fabrication of the stratified cemented tailings backfill (SCTB) specimens. The distribution of grain size for ST is illustrated in Figure 2. The content of fine particles was more than 30%, which was intended to avert the segregation of fresh CTB slurry. The cement used for fresh SCTB preparation was Ordinary Portland Cement (OPC, P.O. 42.5R). The primary chemical constituents of Ordinary Portland Cement are detailed in Table 1. As shown in Table 1, it is mainly composed of Calcium Oxide (64.78%), Silicon Oxide (20.34%), and Aluminium Oxide (5.02%). Its specific gravity is 3.0–3.2. The cement and ST were mixed with tap water.

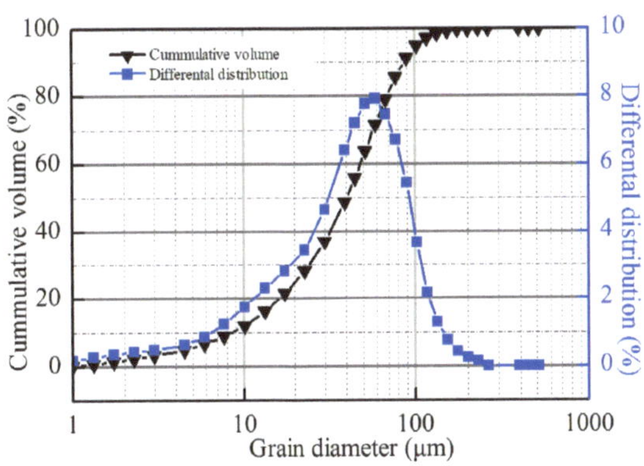

Figure 2. The distribution of grain size for ST.

Table 1. The chemical constituents of the cement.

Component	CaO	Al_2O_3	Fe_2O_3	SiO_2	MgO	K_2O	Na_2O	SO_3	TiO_2	LOI
Content (wt. %)	64.78	5.02	3.11	20.34	1.09	0.35	0.10	2.20	0.26	2.75

2.2. Mix Proportions and Specimen Preparation

The SCTB specimens included three layers: upper layer, enhance layer, and lower layer. The c/t ratio for both the lower and upper layers was designed to be 1/20. Meanwhile, the enhance layer was designed with three different height ratios and three distinct c/t ratios. The fresh CTB mixtures concentration of all layers was 70%. In all 11 groups of specimens were prepared, three samples were made for each case. The specific experimental proposal and parameters of SCTB samples are offered in Figure 3 and Table 2. The cement, tailings, and tap water were thoroughly blended using an agitator for over 5 min. After homogeneously mixing, the fresh slurry was poured into the cylindrical molds measuring 50 mm in diameter and 100 mm in height. During the preparation of each layer, the fresh mixtures should be stamped with a stirring rod to eradicate the air bubbles as far as possible. Then, the SCTB samples were cured in a chamber with relative humidity controlled at 95 ± 5% and temperature controlled at 20 ± 5 °C.

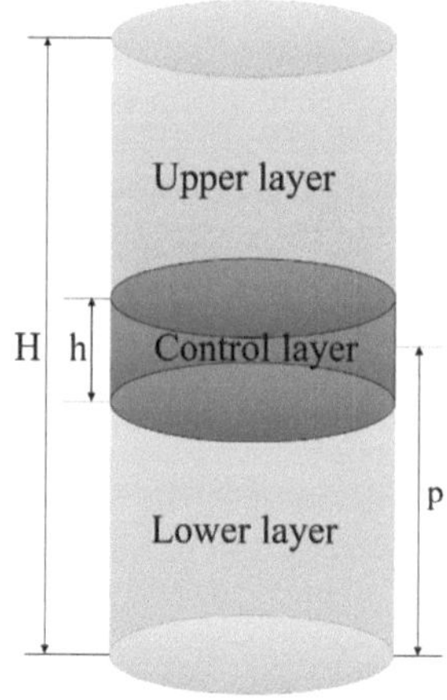

Figure 3. Schematic diagram of SCTB specimen.

Table 2. The mix proportions for SCTB samples.

Sample ID	Cement/Tailings (c/t) Ratio of Middle Layer	Height Ratio (h/H)	Cement/Tailings (c/t) Ratio of Enhance Layer	Curing Age (d)
CTB-20	1/20			28
CTB-15	1/15			28
SC-0.1-15	1/20	0.1	1/15	28
SC-0.1-10	1/20	0.1	1/10	28
SC-0.1-6	1/20	0.1	1/6	28
SC-0.2-15	1/20	0.2	1/15	28
SC-0.2-10	1/20	0.2	1/10	28
SC-0.2-6	1/20	0.2	1/6	28
SC-0.3-15	1/20	0.3	1/15	28
SC-0.3-10	1/20	0.3	1/10	28
SC-0.3-6	1/20	0.3	1/6	28

2.3. Test Apparatus and Procedure

Unconfined compressive strength tests were conducted on CTB specimens in accordance with ASTM C39 (ASTM C39/C39M-18, 2001) [29]. The uniaxial compressive strength of CTB sample was tested by a computer-controlled press (SLB-1) with a deformation rate of 0.5 mm/min and a loading capacity of 60 kN. The pressing apparatus had a full-scale precision of 1%. The stress–strain curves of the samples were recorded automatically by the computer during the test. A group of four SCTB specimens was utilized for the compressive strength tests, and the mean value of four specimens was acquired for following examination.

3. Result and Discussion

3.1. Mechanical Characteristics of SCTB Specimens

The UCS of SCTB specimens with different height ratios and c/t ratios of the enhance layer are shown in Figure 4. Based on Figure 4, the results shown that the UCS of the SCTB specimens all increased with the height ratio of the enhance layer. For example, for SCTB specimen with a c/t ratio of 1/15, the UCS of the SCTB sample with a height ratio of 0.1 was 321.3 kPa. When the height ratio was increased to 0.2 and 0.3, the UCS of the SCTB specimens increased to 375.5 kPa and 451.1 kPa, and the UCS incremental rate came out to 16.9% and 40.5%, respectively. Likewise, when the cement-to-tailings ratio of the SCTB specimens increased to 1/10, the UCS values were 354.5 kPa, 393.3 kPa, and 562.6 kPa at height ratios of 0.1, 0.2, and 0.3, which increased by 10.9% and 58.7%, respectively. Similar to the law of height ratio, the UCSs of the SCTB specimens all increased with the c/t ratio of the enhance layer. For the SCTB specimen with a height ratio of 0.1, the UCSs were 321.3 kPa, 354.5 kPa, and 432.3 kPa at cement-to-tailings ratios of 1/15, 1/10, and 1/6, which increased by 10.3% and 34.5%, respectively. In summary, the UCS values of the SCTB specimens increased as both the height ratio and the cement-to-tailings ratio were increased.

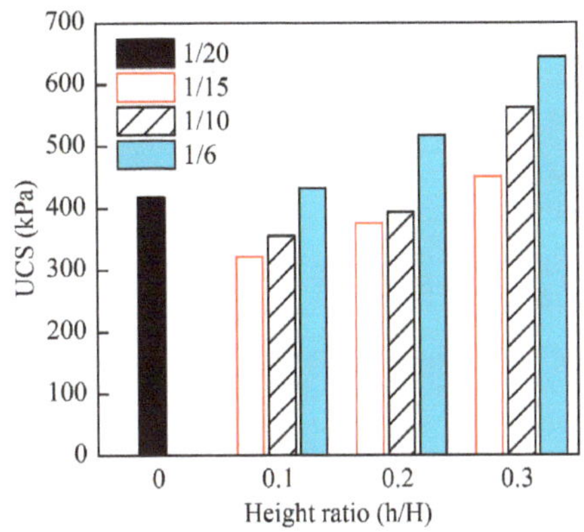

Figure 4. The UCSs of SCTB samples.

The above outcomes indicate a strong correlation between the UCS values of the SCTB specimens and the height ratio and cement-to-tailings ratio of the enhance layer. The failure of the SCTB samples resulted from the generation and propagation of cracks. Generally, failure of the SCTB samples initiated in the low-intensity regions and then gradually spread to the enhance layer, culminating in comprehensive failure of the specimens [22]. The ratio of the low-strength region decreased with the increase in the enhance layer height ratio, and the crack was more difficult to expand, leading to the improvement of the whole strength of the SCTB samples. The internal strength difference of the SCTB samples increased with the cement-to-tailings ratio of the enhance layer, which made the overall tensile failure of the specimen less likely to occur.

It is noteworthy that the UCS values of some SCTB specimens were lower than those of complete CTB specimens. The addition of enhance layer divided the CTB specimen into an upper layer, enhance layer, and lower layer, which damaged the overall structure of the CTB specimen. The stratification effect caused by the addition of the enhance layer will reduce the strength of CTB specimens. The reduction in CTB specimen strength caused by the stratification effect was higher than the enhancement effect caused by the addition of the enhance layer, which reduced the strength of the CTB specimens.

The UCS contour map of the SCTB specimens with different height ratios and cement-to-tailings ratios is shown in Figure 5. The strength contours intersect more with the X axis, indicating a significant range of strength values at different height ratio levels. By adjusting the height ratio, the strength can be more effectively optimized at a specific cement-to tailings ratio. This suggests that the impact of the height ratio of the enhance layer on the UCS of the SCTB specimens is greater than that of the c/t ratio. It can be concluded that the strength of SCTB specimens is more sensitive to the height ratio of the enhance layer.

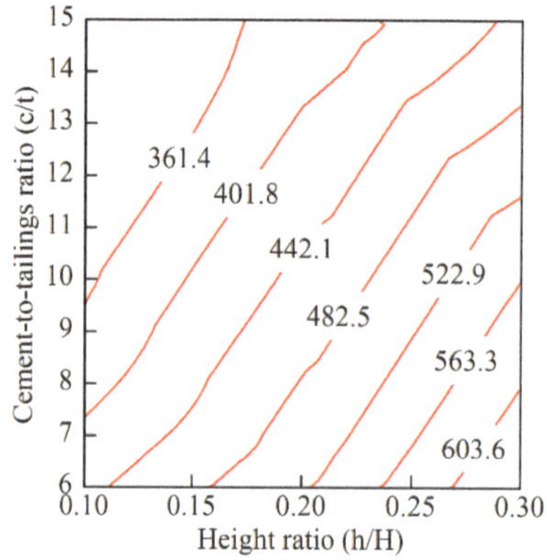

Figure 5. The UCS contour map of SCTB specimens.

Figure 6 displays the conventional UCS values of CTB specimens with a c/t ratio of 1/15 and SCTB specimens containing height ratios of 0.1, 0.2, and 0.3; as well as c/t ratios of 1/15, 1/10, and 1/6, respectively. The UCS values of the specimens of SC-0.2-6, SC-0.3-10, and SC-0.3-6 reached UCSs of 517.4 kPa, 562.6 kPa, and 643.8 kPa, while the UCS of CTB-15 reached 483.5 kPa, which is lower compared to the above SCTB samples. Based on the preceding results, it can be deduced that SCTB samples with a reasonable structure can also achieve sufficient strength compared to directly increasing the c/t ratio of the CTB specimens while also reducing the cost of CTB preparation.

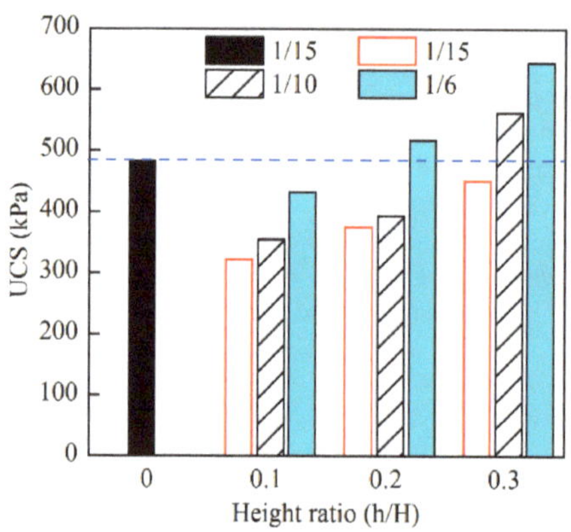

Figure 6. The UCSs of different SCTB samples.

Figure 7 shows the elastic modulus of SCTB specimens with various enhance layer c/t ratios and height ratios. The elastic modulus values of the SCTB samples all increased with the height ratio of the enhance layer given a constant c/t ratio. For example, for the SCTB specimen with a cement-to-tailings ratio of 1/15, the elastic modulus of the SCTB sample with a height ratio of 0.1 was 12.4 MPa. As the height ratio was elevated to 0.2 and 0.3, the elastic modulus of the SCTB specimens increased to 17 MPa and 23.7 MPa, and the elastic modulus incremental rate came out to 37.1% and 91.1%, respectively. Similar to the law of height ratio, the elastic modulus values of the SCTB specimens all increased with the c/t ratio of the enhance layer for a given height ratio. For the SCTB specimen with a height ratio of 0.1, the elastic modulus values were 12.4, 14.5, and 17.9 MPa at cement-to-tailings ratios of 1/15, 1:10, and 1:6, which increased by 16.9% and 44.4%, respectively. This phenomenon can be comprehended, as the stiffness of the lower layer and upper layer of the SCTB sample is relatively small, and the stiffness of the enhance layer is relatively large. The section of the high stiffness region escalates as the height ratio and cement-to-tailings ratio of the enhance layer augment, which results in an enhancement in the overall stiffness of the SCTB samples. In summary, the elastic modulus of SCTB specimens escalates as both the height ratio and the cement-to-tailings ratio increase.

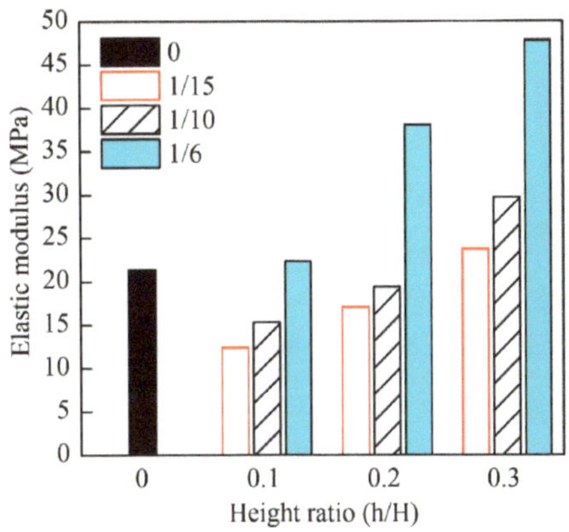

Figure 7. The elastic modulus values of SCTB samples.

However, the elastic modulus values of the SCTB samples were inferior to those of the CTB sample under a certain cement-to-tailings ratio and height ratio of the enhance layer. It is evident that when the height ratio was 0.1, the elastic modulus of the SCTB sample was inferior to that of the CTB specimen regardless of the cement-to-tailings ratio. As the height ratio increased to 0.2, the elastic modulus of the SCTB specimen surpassed that of the CTB sample only as the cement-to-tailings ratio was 1:6. When the height ratio increased to 0.3, the elastic modulus of the SCTB sample was inferior to that of the CTB sample only when the cement-to-tailings ratio was 1:15. Therefore, the elastic modulus values of the SCTB specimens with a height ratio of 0.1 or a cement-to-tailings ratio of 1:15 were both inferior to those of the CTB specimens.

Figure 8 exhibits the contour map depicting the elastic modulus values of SCTB specimens across various h/H ratios and c/t ratios. The elastic modulus is often used to characterize the deformation resistance of a SCTB specimen under varied conditions. An elevated elastic modulus indicates a reduced propensity for the specimens to undergo deformation, which significantly augments the structural stability. It can be seen from Figure 8 that the strength contours intersect more with the X axis. By adjusting the height ratio, the elastic modulus can be more effectively optimized at a specific cement-to tailings ratio. This manifested that the influence of the enhance layer height ratio on the elastic modulus values of the SCTB specimens was more significant than the influence of the cement-to-tailings ratio on the elastic modulus values of the specimens. It can be concluded that the elastic modulus of a stratified cemented tailings backfill specimen is more sensitive to the height ratio of the enhance layer.

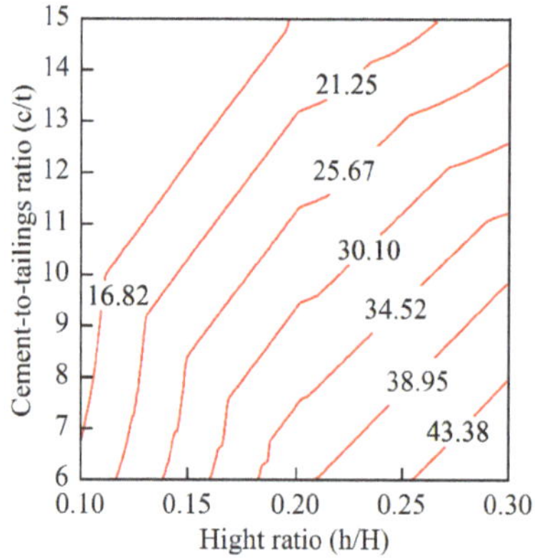

Figure 8. The contour map depicting the elastic modulus values of SCTB samples.

3.2. Energy Evolution Characteristics

The failure and instability of SCTB specimens were examined through the perspective of energy progression. It is assumed that in a closed system, the SCTB unit deforms under the action of external load, which is the total input energy U caused by the work of external force that can be calculated by the following [30–32]:

$$U = U^e + U^d \tag{1}$$

where U represents the total energy of the SCTB specimen, and U^d and U^e are the dissipated energy and elastic energy during the deformation of SCTB specimen, respectively.

The U and U^e can be calculated by the following:

$$U = \int \sigma_1 d\varepsilon_1 \tag{2}$$

$$U^e = \frac{\sigma_1^2}{2E} \tag{3}$$

where σ_1 represents the axial stress, ε_1 represents the axial strain, and E represents the elastic modulus.

Figure 9 shows the energy evolution and stress–strain curves of SCTB specimens with different cement-to-tailings ratios and height ratios of the enhance layer. In the elastic deformation phase, no new cracks were incubated. The energy absorbed by the work was primarily stored as elastic energy. Meanwhile, the dissipated energy curve almost stayed the same or even decreased. In the plastic deformation stage, the specimen produced irreversible plastic deformation. At this phase, the internal crack of the sample began to expand; meanwhile, the elastic strain energy increased slowly. Correspondingly, the growth rate of the dissipated energy began to increase more quickly, and the proportion of energy dissipated began to increase progressively. After the peak stress, the stored elastic strain energy in the SCTB specimen started to be released, leading to a substantial increase in the dissipated energy of the specimen rupture damage, and the specimen gradually destabilized and destroyed.

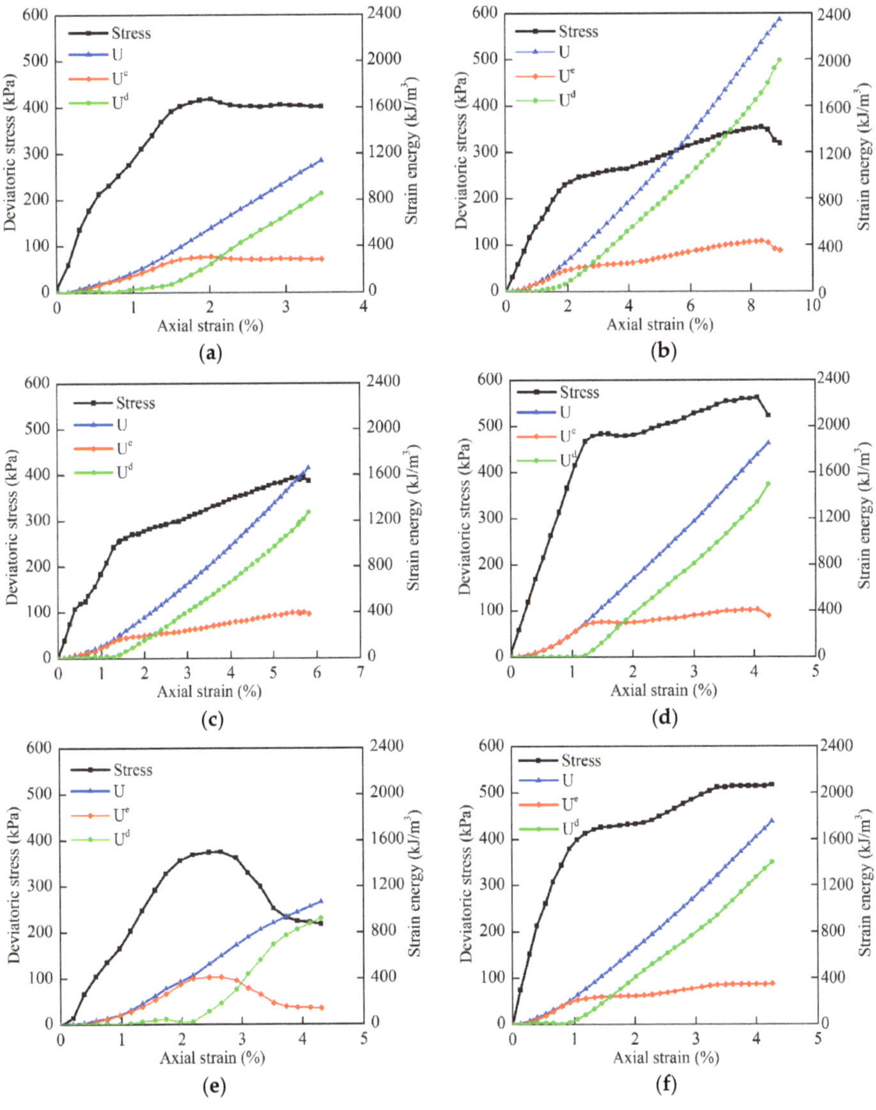

Figure 9. Stress–strain and energy evolution of SCTB specimens. (**a**) CTB-20; (**b**) SC-0.1-10; (**c**) SC-0.2-10; (**d**) SC-0.3-10; (**e**) SC-0.2-15; (**f**) SC-0.2-6.

The escalation in dissipated energy fosters crack formation and growth inside the SCTB samples. The dissipative energy of SCTB specimens and CTB specimens exhibited different evolution modes. The dissipative energy of the CTB samples began to increase significantly as the specimens approached yield failure. The SCTB specimen with a cement/tailings ratio of 1/15 also showed the same pattern. However, the dissipative energy of other SCTB samples started to increase rapidly during the plastic deformation stage. During the compression process, the propagation of crack first occurred in the CTB with a low c/t ratio of the SCTB samples. However, the addition of the enhance layer restricted the propagation of cracks, and the overall SCTB specimens did not undergo damage, allowing the specimens to continue to be compressed. Therefore, the plastic deformation and peak strain of the SCTB specimens were found to be larger than those of CTB specimens. Since the upper and lower layers of the SCTB sample had been partially destroyed in the plastic deformation phase, the dissipated energy began to increase rapidly in this stage.

The energy distributions of SCTB specimens with different height ratios and c/t ratios of the enhance layer are shown in Figure 10. Through the analysis of energy distribution, the dissipation energy and elastic strain energy of the specimens were found to be closely related to the enhance layer. The ability of SCTB specimens with a cement-to-tailings ratio of 1/15 in the enhance layer and complete CTB specimens to accumulate elastic strain energy was stronger than that of the other SCTB specimens, and the energy accumulation rates reached 60.2 and 55.6, respectively. A significant portion of the total strain energy was transformed into elastic strain energy, which was then stored within the sample. The failure and instability of the samples will be aggravated as the elastic energy is released. The energy dissipation rates of the SCTB samples were between 74.5%~79.8%, which indicates that the SCTB specimens have stronger plastic deformation ability, and a significant portion of the overall strain energy is dissipated through plastic failure.

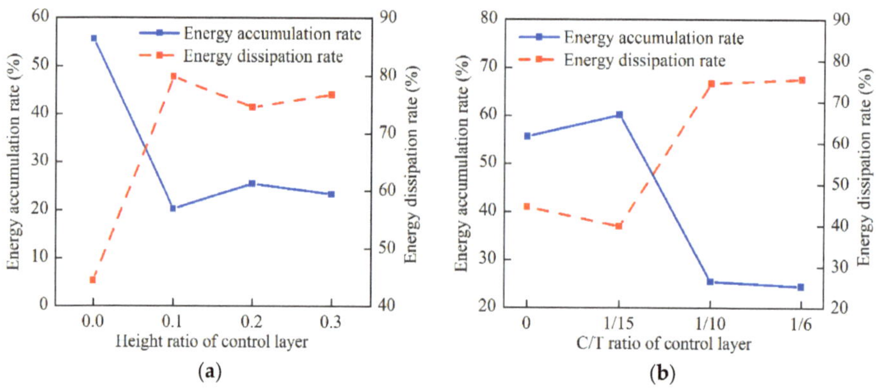

Figure 10. The energy distribution of the samples as the elastic energy touches the peaks: (**a**) cement-to-tailings ratio = 1:10; (**b**) height ratio = 0.2.

3.3. Failure Modes

Figure 11 shows the various failure modes of SCTB specimens with different height ratios and cement-to-tailings ratios of the enhance layer. The CTB samples mainly manifested as a shear failure through the whole specimen. Keeping the height ratio of the enhance layer at 0.2, when the cement/tailings ratio was 1/15, such failure primarily appeared as tensile failure of the upper layer, enhance layer, and lower layer. As the cement-to-tailings ratio increased to 1/10 and 1/6, it primarily appeared as tensile failure of the upper layer. As observed in Figure 11b, keeping the c/t ratio of the enhance layer at 1/10, regardless of the height ratio of the enhance layer, it primarily appeared as tensile failure of the upper layer. The difference is that as the height ratio of the enhanced layer increased, the number of cracks gradually decreased.

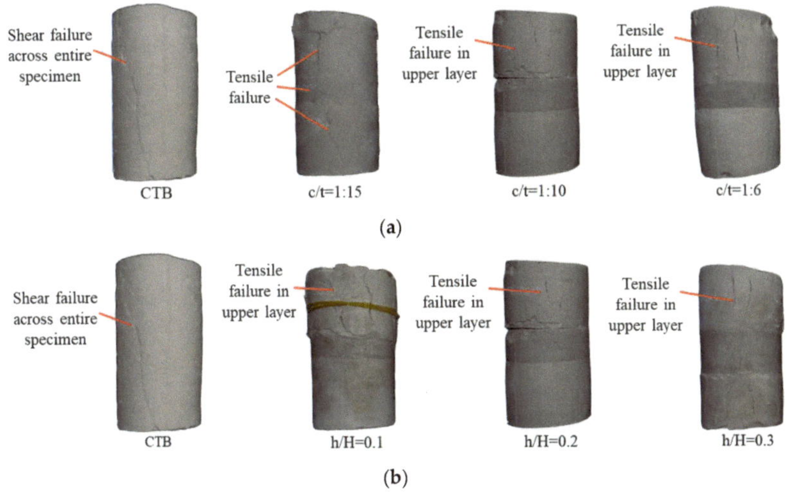

Figure 11. Failure modes of SCTB samples: (**a**) height ratio = 0.2 (**b**); cement-to-tailings ratio = 1:10.

Generally, the initial signs of failure in SCTB samples appears in the low-intensity regions, and then the extension progresses to the enhanced layer, ultimately resulting in failure of the specimens as a whole. Due to the different volume expansion coefficients of each layer of the SCTB specimen, there is a bonding effect between the layers. The layer with a large coefficient will drive the layer with a small expansion coefficient to expand outward during the stress process. At the same time, the layer with a large expansion coefficient (upper or lower layer) will be limited by the layer with a small expansion coefficient (enhance layer) [33]. The propagation of cracks is limited by the enhance layer, thereby improving the mechanical properties of the SCTB specimens.

3.4. Damage Constitutive Equation

Following the Lemaitre strain equivalence principle [34,35], the damage constitutive model for CTB can be defined as

$$\sigma = (1 - D_d)E\varepsilon \tag{4}$$

where D_d and E respectively indicate the loaded damage variable and the elastic modulus of the CTB before loading.

The progression of damage in a CTB can be analyzed by introducing failure criterion. Assuming that the Weibull distribution governs the strength of each micro-element [36,37], the distribution density function is

$$P_l = \frac{m}{\varepsilon_0}\left(\frac{\varepsilon}{\varepsilon_0}\right)^{m-1} e^{-\left(\frac{\varepsilon}{\varepsilon_0}\right)^m} \tag{5}$$

where ε represents the strain of the CTB, P_l indicates the failure probability of the CTB samples, ε_0 represents the holistic influence of the micro-element strength in the Weibull distribution, and m represents the non-uniformity degree of the Weibull distribution.

When the strain of the CTB is ε, the load damage variable D_d is the proportion of the damaged area to the non-destructive material area, which can be expressed as

$$D_d = \frac{S}{S_{\mathrm{m}}} = \int_0^{\varepsilon} P_l(x)d(x) = 1 - e^{-\left(\frac{\varepsilon}{\varepsilon_0}\right)^m} \tag{6}$$

It can be found from the results of the above research that the mechanical properties of SCTB samples are directly influenced by the height ratio and the cement-tailings ratio of the enhance layer. Therefore, Equation (14) is no longer suitable for characterizing the constitutive relationship of a SCTB specimen. The SCTB specimen examined consisted of high-strength structures in the middle layer and low-strength structures at both ends. Its structural characteristics resemble those of layered cemented tailings backfill and coal–rock assemblage [22,38,39]. Therefore, the constitutive relationship of the SCTB specimen can be calculated by imitating the constitutive relationship of coal and rock combinations or the SCTB. Therefore, the constitutive relationship of the SCTB specimen can be depicted as a series of intermediate elastomer and damage body at both ends, as shown in Figure 12. The constitutive equation for the elastomer is described by Equation (7).

$$\sigma = E\varepsilon \tag{7}$$

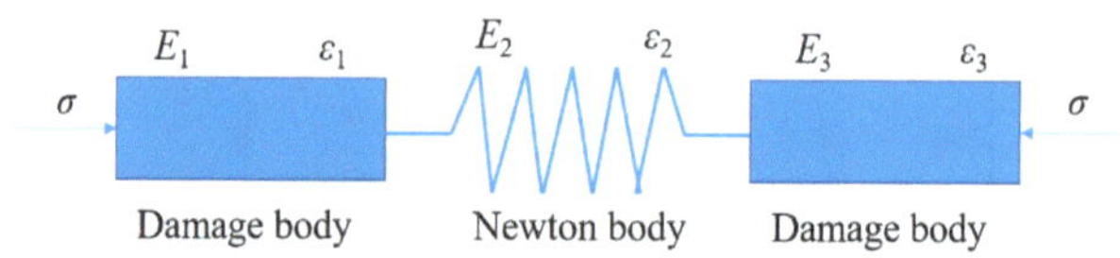

Figure 12. Damage constitutive models of SCTB specimens.

The SCTB specimen satisfies the following relationship:

$$\begin{cases} \sigma = \sigma_1 = \sigma_2 = \sigma_3 \\ \varepsilon = \varepsilon_1 + \varepsilon_2 + \varepsilon_3 \end{cases} \tag{8}$$

Combination of Equations (4), (7) and (8) yield the following:

$$\sigma = \frac{k(1 - D_d)E_1E_2}{(1 - D_d)E_1 + 2E_2}\varepsilon \tag{9}$$

where E_1 and E_2 are the elastic modulus values of the damage body and elastomer, respectively.

The obtained damage constitutive equation was used to fit the stress–strain curves of SCTB samples with various height ratios and cement-to-tailings ratios of the enhance layer, and the results are shown in Figure 13. When the cement-to-tailings ratio of the enhance layer was constant, as the height ratio of the enhance layer increased from 0.1 to 0.3, the R-square values of the fitting curve came out to 0.9406, 0.9872, and 0.975, respectively. The R-square values of all fitting results are higher than 0.94, and the average value is 0.9676. In addition, when the height ratio of enhance layer was immobile, as the c/t ratio of enhance layer rose from 1/15 to 1/6, the R-square values for the fitting curve came out to 0.9673, 0.9872, and 0.9804. The R-square values of all fitting results exceed 0.96, with an average value of 0.9783. According to the fitting results above, it can be concluded that the damage constitutive equation established in this study effectively characterizes the stress–strain development of SCTB specimens.

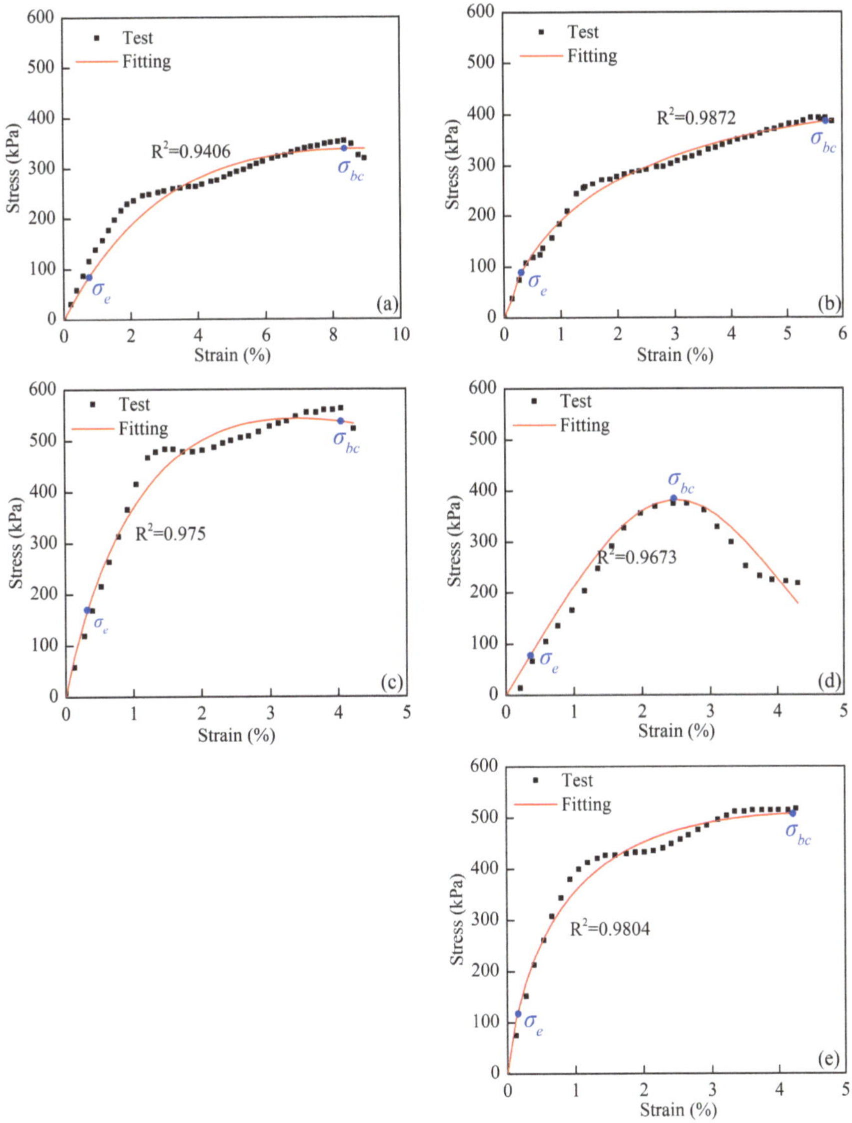

Figure 13. The fitting results of stress-strain: (**a**) SC-0.1-10; (**b**) SC-0.2-10; (**c**) SC-0.3-10; (**d**) SC-0.2-15; (**e**) SC-0.2-6.

4. Discussion

As shown in Figure 14, the total height of the SCTB specimen model was represented by H, the height of the upper layer was H_u, the height of the enhance layer was indicated by H_c, and the height of the lower layer was denoted by H_l. The elastic modulus of the lower layer, the enhance layer, the upper layer, and SCTB samples are expressed as E_l, E_c, E_u, and E_e, respectively. The axial deformation of the SCTB model under the action of axial stress σ was defined as ΔH. The axial strain of the entire SCTB model and its components can be defined as

$$\begin{aligned} \varepsilon_u &= \sigma_u / E_u \\ \varepsilon_c &= \sigma_c / E_c \\ \varepsilon_l &= \sigma_l / E_l \\ \varepsilon_e &= \sigma_e / E_e \end{aligned} \tag{10}$$

where the axial strain of the upper layer, the enhance layer, the lower layer, and SCTB samples are expressed as ε_u, ε_c, ε_l, and ε_e, respectively; the axial stress of the upper layer, the enhance layer, the lower layer, and SCTB samples are expressed as σ_u, σ_c, σ_l, and σ_e.

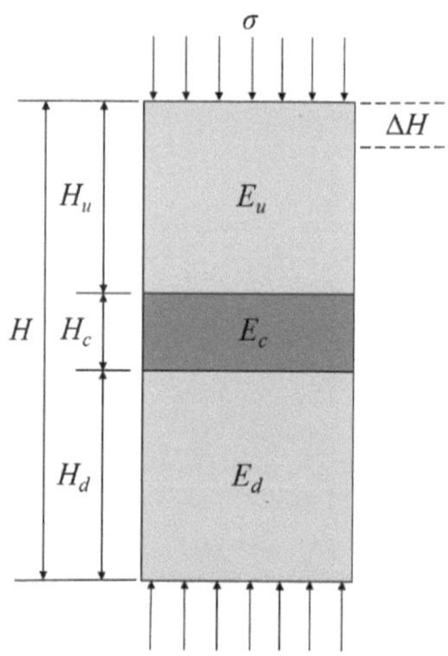

Figure 14. Schematic representation of deformation analysis for the SCTB model.

The axial deformation of the SCTB model can be defined as

$$\Delta H = \varepsilon_e H = \varepsilon_u H_u + \varepsilon_c H_c + \varepsilon_l H_l \tag{11}$$

By combining Equations (10) and (11), the axial deformation of the SCTB can be expressed as

$$\frac{\sigma}{E_e} H = \frac{\sigma_u}{E_u} H_u + \frac{\sigma_c}{E_c} H_c + \frac{\sigma_l}{E_l} H_l \tag{12}$$

During the loading process, the axial stresses of the three parts of the SCTB model were equal, and the elastic model of the lower layer was the same as that of the upper layer as follows:

$$\begin{aligned} \sigma &= \sigma_u = \sigma_c = \sigma_l \\ E_u &= E_l \end{aligned} \tag{13}$$

By combining of Equations (12) and (13), the elastic modulus of the SCTB model can be expressed as

$$E_e = \frac{E_u E_c H}{(H - H_c) E_c + H_c E_u} \tag{14}$$

Equation (14) can also be expressed as

$$E_e = \frac{E_u H}{H + \left(\frac{E_u}{E_c} - 1\right) H_c} \tag{15}$$

The equivalent elastic modulus of the combined SCTB can be expressed by Equation (15). Based on Equation (15), it is apparent that the elastic modulus of the SCTB sample was affected by the height and the elastic modulus of the enhance layer, and it escalated as both the height and elastic modulus of the enhance layer augmented.

The SCTB composite specimen was composed of CTB with different mechanical properties. Thus, the mechanical properties were influenced by the interface confinement effect and the characteristics of the CTB, as shown in Figure 15. It is assumed that the various components are tightly bonded at the interface without relative slippage, so the thickness of the interface was ignored. In addition, the CTB of each part in the SCTB composite model is treated as a homogeneous isotropic medium. The stress conditions at various positions within the SCTB model were examined. The lateral strains of the upper layer and enhance layer in the elastic stage were defined as follows:

$$\begin{cases} \varepsilon_{2u} = \varepsilon_{3u} = \mu_u \frac{\sigma_{1u}}{E_u} \\ \varepsilon_{2c} = \varepsilon_{3c} = \mu_c \frac{\sigma_{1c}}{E_c} \end{cases} \quad (16)$$

where σ_{1u} and σ_{1c} are the axial stress of the upper layer and enhance layer; μ_c and μ_u are the Poisson's ratio of the enhance layer and upper layer, respectively.

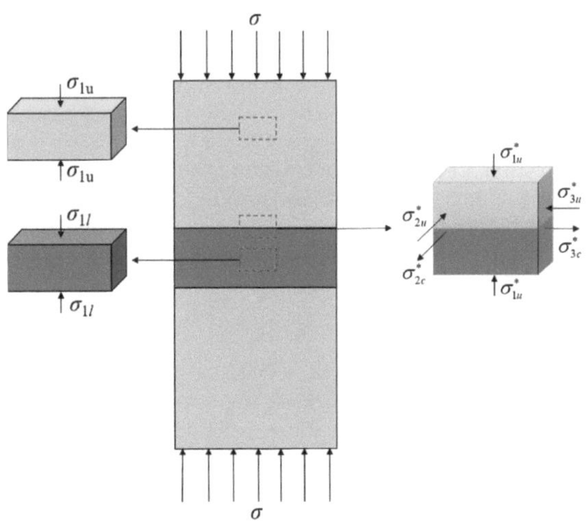

Figure 15. Stress distribution characteristics in the SCTB combination model at various position.

The elastic modulus and Poisson's ratio of the enhance layer and upper layer in the SCTB model meet the requirements of $\mu_c < \mu_u$ and $E_u < E_c$, respectively. Then, the lateral strain of the enhance layer and the upper layer under the same axial stress meet the requirements of $\varepsilon_{2c} = \varepsilon_{3c} < \varepsilon_{2u} = \varepsilon_{3u}$. Thus, the lateral cohesive force would arise near the interface for the SCTB model. It can be seen from Equation (16) that under the same stress conditions, the upper layer exhibits greater lateral deformation compared to the enhanced layer. Furthermore, so as to achieve the uniform deformation of the upper layer and the enhance layer, the upper layer near the interface will be exposed to compressive stress, and the enhance layer will be exposed to tensile stress.

Based on the conditions of the continuous deformation and static equilibrium, the following relationship was established:

$$\begin{array}{l} \sigma_{1u}^* = \sigma_{1c}^* = \sigma_1^* \\ \sigma_{2u}^* = \sigma_{2c}^* = \sigma_{3u}^* = \sigma_{3c}^* \\ \varepsilon_{2u}^* = \varepsilon_{2c}^* = \varepsilon_{3u}^* = \varepsilon_{3c}^* \end{array} \quad (17)$$

where σ_1^* is the axial stress near the interface; σ_{1c}^* and σ_{1u}^* indicate the axial stress of the enhance layer and upper layer near the interface, respectively; σ_{2u}^* and σ_{3u}^* are the stress of the upper layer near the interface in two lateral principal directions, respectively; σ^*_{2c} and σ^*_{3c} are the stress of the enhance layer near the interface in two lateral principal directions, respectively; ε^*_{2u} and ε^*_{3u} indicate the lateral strain of the upper layer near the interface in two lateral principal directions, respectively; ε^*_{2c} and ε^*_{3c} indicate the lateral strain of the enhance layer near the interface in two lateral principal directions, respectively.

The lateral strain of the upper layer and the enhance layer near the interface can be respectively represented as

$$\begin{aligned} \varepsilon^*_{2u} &= \frac{-\sigma^*_{2u} + \mu_u(\sigma^*_{1u} + \sigma^*_{3u})}{E_u} \\ \varepsilon^*_{2c} &= \frac{\sigma^*_{2c} + \mu_c(\sigma^*_{1u} - \sigma^*_{3c})}{E_c} \end{aligned} \tag{18}$$

By combining of Equations (17) and (18), the following relationship is obtained:

$$\begin{aligned} \sigma^*_{2u} &= \sigma^*_{2c} = \sigma^*_{3u} = \sigma^*_{3c} = k_{uc}\sigma^*_1 \\ k_{uc} &= \frac{E_c\mu_u - E_u\mu_c}{E_c(1-\mu_u) + E_u(1-\mu_c)} \end{aligned} \tag{19}$$

where k_{uc} denotes the coefficient representing the cohesive force at the interface between the upper and enhance layers in the SCTB model.

From Equation (19), it can be seen that the magnitude of the cohesive force between the upper and enhance layer is related to their Poisson's ratio and elastic modulus values. According to the above calculation, it can be seen that the stress state in the upper layer near the interface transitions from being under uniaxial compression to experiencing triaxial compression; moreover, the mechanical properties of the upper layer are enhanced. Therefore, the addition of the enhance layer can expand the mechanical properties of the CTB.

5. Conclusions

It is very crucial to comprehend the mechanical properties of SCTB specimens containing the enhance layer, which are of great significance for the reasonable design of the CTB. In this study, the UCS test was proceeded for SCTB specimens with different height ratios and c/t ratios of the enhance layer. The main results were obtained as follows:

(1) The stratification effect caused by the addition of the enhance layer will reduce the mechanical properties of CTB samples. However, the mechanical properties of the enhance layer improve the overall strength of the SCTB specimen. The elastic modulus and uniaxial compressive strength of SCTB samples increase with the height ratio and cement-to-tailings ratio of the enhance layer. Moreover, the UCS and elastic modulus of SCTB specimens are more sensitive to the height ratio of the enhance layer than the cement-to-tailings ratio.

(2) The failure of SCTB samples first appears in the low-intensity regions and then gradually extends to the enhance layer, leading to the overall failure of the specimens. The SCTB specimen mainly manifested as tensile failure of the upper layer and lower layer, but this did not penetrate the entire specimen. The addition of the enhance layer limits the crack propagation, thereby improving the mechanical properties of the SCTB samples.

(3) The dissipative energy of the CTB samples began to increase significantly as the specimen approaches yield failure. However, the dissipative energy of other SCTB samples started to increase rapidly during the plastic deformation stage. The SCTB specimens had stronger plastic deformation ability, and a large part of the total strain energy was dissipated in the form of plastic failure.

(4) SCTB samples with a reasonable structure can also achieve sufficient strength compared to directly increasing the c/t ratio of CTB samples while reducing the cost of

CTB preparation. Therefore, it is necessary to actively explore the application of the enhance layer in mine filling.

Author Contributions: Conceptualization, W.X.; formal analysis, Y.Z.; data curation, Y.Z., W.C., T.S. and Y.S.; writing—original draft preparation, Y.Z.; writing—review and editing, W.X.; supervision, W.X. All authors have read and agreed to the published version of the manuscript.

Funding: This research was funded by the Fundamental Research Funds for the Central Universities (Grant number: 2023JCCXNY01) and the Key Technologies Research & Development Program (Grant number: 2022AB31022).

Data Availability Statement: Data are contained within the article.

Conflicts of Interest: The authors declare no conflict of interest.

References

1. Mohammad, H.G.; Christopher, A.B. Sustainable Reuse of Mine Tailings and Waste Rock as Water-Balance Covers. *Minerals* **2017**, *7*, 128. [CrossRef]
2. Yang, L.; Xu, W.; Yilmaz, E.; Wang, Q.; Qiu, J. A combined experimental and numerical study on the triaxial and dynamic compression behavior of cemented tailings backfill. *Eng. Struct.* **2020**, *219*, 110957. [CrossRef]
3. Wang, Y.; Wang, H.; Zhou, X.; Yi, X.; Xiao, Y.; Wei, X. In Situ X-ray CT Investigations of Meso-Damage Evolution of Cemented Waste Rock-Tailings Backfill (CWRTB) during Triaxial Deformation. *Minerals* **2019**, *9*, 52. [CrossRef]
4. Mu, W.; Li, L.; Guo, Z.; Du, Z.; Wang, S. Novel Segmented Roadside Plugging-Filling Mining Method and Overlying Rock Mechanical Mechanism Analyses. *Energies* **2019**, *12*, 2073. [CrossRef]
5. Chmykhalova, S.V. Toward a reclamation effort estimation procedure for open pit mines. *Eurasian Min.* **2023**, *1*, 83–85. [CrossRef]
6. Fang, K.; Cui, L. Experimental investigation of fiber content and length on curing time-dependent mode-I fracture behavior and properties of cemented paste backfill and implication to engineering design. *Fatigue Fract. Eng. Mater. Struct.* **2022**, *45*, 3302–3318. [CrossRef]
7. Nan, X.; Liu, X.; Wu, B.; Zhang, H.; Song, K.; Wang, X. Coupled CFD-DEM simulation and experimental study of particle distribution and accumulation during tailings seepage process. *J. Clean. Prod.* **2023**, *427*, 139229. [CrossRef]
8. Zhao, Y.; Soltani, A.; Taheri, A.; Karakus, M.; Deng, A. Application of Slag–Cement and Fly Ash for Strength Development in Cemented Paste Backfills. *Minerals* **2019**, *9*, 22. [CrossRef]
9. Liu, L.; Fang, Z.; Qi, C.; Zhang, B.; Guo, L.; Song, K.I.-I.L. Experimental investigation on the relationship between pore characteristics and unconfined compressive strength of cemented paste backfill. *Constr. Build. Mater.* **2018**, *179*, 254–264. [CrossRef]
10. Zhang, J.; Deng, H.W.; Taheri, A.; Deng, J.; Ke, B. Effects of Superplasticizer on the Hydration, Consistency, and Strength Development of Cemented Paste Backfill. *Minerals* **2018**, *8*, 381. [CrossRef]
11. Cao, S.; Song, W.; Yilmaz, E. Influence of structural factors on uniaxial compressive strength of cemented tailings backfill. *Constr. Build. Mater.* **2018**, *174*, 190–201. [CrossRef]
12. Chen, Q.; Zhang, Q.; Qi, C.; Fourie, A.; Xiao, C. Recycling phosphogypsum and construction demolition waste for cemented paste backfill and its environmental impact. *Constr. Build. Mater.* **2018**, *186*, 418–429. [CrossRef]
13. Deng, D.Q.; Liu, L.; Yao, Z.L.; Song, K.I.-I.L.; Lao, D.Z. A practice of ultra-fine tailings disposal as filling material in a gold mine. *J. Environ. Manag.* **2017**, *196*, 100–109. [CrossRef] [PubMed]
14. Hamberg, R.; Maurice, C.; Alakangas, L. Lowering the water saturation level in cemented paste backfill mixtures-Effect on the release of arsenic. *Miner. Eng.* **2017**, *112*, 84–91. [CrossRef]
15. Qin, X.; Wang, P.; Liu, L.; Wang, M.; Xin, J. Sensitivity Analysis of Microstructure Parameters and Mechanical Strength during Consolidation of Cemented Paste Backfill. *Math. Probl. Eng.* **2018**, *112*, 84–91. [CrossRef]
16. Alireza, G.; Mamadou, F. Strength evolution and deformation behaviour of cemented paste backfill at early ages: Effect of curing stress, filling strategy and drainage. *Int. J. Min. Sci. Technol.* **2016**, *26*, 809–817.
17. Xu, W.; Li, Q.; Zhang, Y. Influence of temperature on compressive strength, microstructure properties and failure pattern of fiber-reinforced cemented tailings backfill. *Constr. Build. Mater.* **2019**, *222*, 776–785. [CrossRef]
18. Xu, W.B.; Zhang, Y.L.; Liu, B. Influence of silica fume and low curing temperature on mechanical property of cemented paste backfill. *Constr. Build. Mater.* **2020**, *254*, 119305. [CrossRef]
19. Zheng, J.; Guo, L.; Sun, X.; Li, W.; Jia, Q. Study on the strength development of cemented backfill body from lead-zinc mine tailings with sulphide. *Adv. Mater. Sci. Eng.* **2018**, *2018*, 7278014. [CrossRef]
20. Nochaiya, T.; Jeenram, T. Microstructure, compressive strength, and permeability of Portland-condensed silica fume cement. *Monatshefte Chem.-Chem. Mon.* **2017**, *148*, 1363–1370. [CrossRef]
21. Xu, W.; Zhang, Y.; Zuo, X.; Hong, M. Time-dependent rheological and mechanical properties of silica fume modified cemented tailings backfill in low temperature environment. *Constr. Build. Mater.* **2020**, *114*, 103804. [CrossRef]

22. Wang, J.; Fu, J.; Song, W.; Zhang, Y.; Wang, Y. Mechanical behavior, acoustic emission properties and damage evolution of cemented paste backfill considering structural feature. *Constr. Build. Mater.* **2020**, *261*, 119958. [CrossRef]
23. Zhang, C.; Fu, J.; Song, W.; Kang, M.; Li, T.; Wang, N. Analysis on mechanical behavior and failure characteristics of layered cemented paste backfill(LCPB) under triaxial compression. *Constr. Build. Mater.* **2022**, *324*, 126631. [CrossRef]
24. Wu, W.; Xu, W.; Zuo, J. Effect of inclined interface angle on shear strength and deformation response of cemented paste backfill-rock under triaxial compression. *Constr. Build. Mater.* **2021**, *279*, 122478.
25. Qin, H.; Cao, S.; Yilmaz, E. Mechanical, energy evolution, damage and microstructural behavior of cemented tailings-rock fill considering rock content and size effects. *Constr. Build. Mater.* **2024**, *411*, 134449. [CrossRef]
26. Koohestani, B.; Darban, A.K.; Mokhtari, P. A comparison between the influence of superplasticizer and organosilanes on different properties of cemented paste backfill. *Constr. Build. Mater.* **2018**, *173*, 180–188. [CrossRef]
27. Liu, Z.; Ming, L.A.; Xiao, S.; Guo, H. Damage failure of cemented backfill and its reasonable match with rock mass. *Trans. Nonferrous Met. Soc. China* **2015**, *25*, 954–959. [CrossRef]
28. Gao, R.; Zhou, K.; Yang, C. Damage mechanism of composite cemented backfill based on complex defects influence. *Mater. Werkst.* **2017**, *48*, 893–904. [CrossRef]
29. *ASTM C39/C39M-18*; Standard Method for Compressive Strength of Cylindrical Concrete Specimens. ASTM: West Conshohochen, PA, USA, 2001.
30. Zhao, K.; Zhou, Y.; Yin, S.; Yan, Y.; Wu, J.; Zhao, H.; Wang, S. Effect of initial defects on the microstructure, mechanics, and energy dissipation characteristics of cemented paste backfill. *Mater. Today Commun.* **2023**, *35*, 105785. [CrossRef]
31. Gu, Q.; Ma, Q.; Tan, Y.; Jia, Z.; Zhao, Z.; Huang, D. Acoustic emission characteristics and damage model of cement mortar under uniaxial compression. *Constr. Build. Mater.* **2019**, *213*, 377–385. [CrossRef]
32. Zhao, K.; Yu, X.; Zhu, S.; Zhou, Y.; Wang, Q.; Wang, J. Acoustic emission investigation of cemented paste backfill prepared with tantalum-niobium tailings. *Constr. Build. Mater.* **2020**, *237*, 117523. [CrossRef]
33. Xiu, Z.; Wang, S.; Ji, Y.; Wang, F.; Ren, F. Experimental study on the triaxial mechanical behaviors of the Cemented Paste Backfill: Effect of curing time, drainage conditions and curing temperature. *J. Environ. Manag.* **2022**, *301*, 113828. [CrossRef] [PubMed]
34. Zhang, H.; Yang, G. Research on damage model of rock under coupling action of freeze-thaw and load. *Chin. J. Rock Mech. Eng.* **2010**, *29*, 471–486.
35. Zhang, H.; Lei, L.; Yang, G. Characteristic and representative model of rock damage process under constant confining stress. *China Univ. Min. Technol.* **2015**, *44*, 59–63.
36. Liu, M.; Li, Y.; Xu, F.; Xu, Z.; Cheng, L. Dynamic compressive mechanical properties and a new constitutive model of 2D-C/SiC composites. *Mater. Sci. Eng. A* **2008**, *489*, 120–126.
37. Liu, X.; Ning, J.; Tan, Y.; Gu, Q. Damage constitutive model based on energy dissipation for intact rock subjected to cyclic loading. *Int. J. Rock Mech. Min. Sci.* **2016**, *85*, 27–32. [CrossRef]
38. Liu, H.; Liu, S.; Xia, Z.; Liu, J.; Guo, H. Study on mechanical properties and damage features of Rock-Coal-Rock combination models with defects and fillings. *Geomech. Eng.* **2021**, *27*, 239–251.
39. Dai, L.; Zhang, Z.; Sun, H.; Gao, H. Research on Mechanical Properties and Energy Evolution Law of Coal-Rock Assemblage with Different Gas Pressures. *Sustainability* **2022**, *14*, 9904. [CrossRef]

minerals

MDPI

Article

Two-Step Shear Flocculation for High-Efficiency Dewatering of Ultra-Fine Tailings

Ying Yang [1,2], Xiaohui Liu [1,*], Liqiang Zhang [1] and Miaomiao Guo [1]

1 School of Safety Engineering, North China Institute of Science and Technology, Langfang 065201, China; 18810607452@163.com (Y.Y.); zhang@ncist.edu.cn (L.Z.); miaomiaoguo@ncist.edu.cn (M.G.)
2 JCHX Mining Management Co., Ltd., Beijing 101500, China
* Correspondence: liuxiaohui28@ncist.edu.cn

Abstract: The high-efficiency dewatering of ultra-fine tailings is one of the most prominent challenges in tailings thickening. The two-step shear flocculation process represents a promising practical method to achieve the dewatering of ultra-fine tailings. In this paper, a small self-made experimental device was used to simulate the two-step shear flocculation process of ultra-fine tailings, strengthening the effect of shear failure and shear coagulation, and we explored the mass fraction of ultra-fine tailings, floc structure size, floc strength, and regeneration performance of tailings slurry according to the variation in shear action in different stages. In addition, the synergistic mechanism of shear failure and shear coagulation in the two-step high-efficiency dewatering process of ultra-fine tailings was proposed. The results show that the dewatering of ultra-fine tailings was significantly improved by two-step flocculation, and the mass fraction of tailings can reach more than 71%. In the primary floc failure stage, the value of G_1T_1 should be higher than 100,000, and in the secondary floc regeneration stage, the value of G_2T_2 should be in the range of 7000~11,000. This paper provides a reference for the regulation of the shear mode and action range in the two-step flocculation process of ultra-fine tailings.

Keywords: two-step flocculation; ultra-fine tailings; tailings thickening; shear failure; shear coagulation

check for updates

Academic Editors: Sunil Kumar Tripathy, Yuye Tan, Xun Chen and Yuan Li

Received: 18 December 2024
Revised: 5 February 2025
Accepted: 11 February 2025
Published: 14 February 2025

Citation: Yang, Y.; Liu, X.; Zhang, L.; Guo, M. Two-Step Shear Flocculation for High-Efficiency Dewatering of Ultra-Fine Tailings. *Minerals* **2025**, *15*, 176. https://doi.org/10.3390/min15020176

1. Introduction

With the ever-growing demand for mineral products and the continuous innovations in relation to beneficiation technology, the production of ultra-fine tailings is gradually increasing [1–3]. Ultra-fine tailings dewatering has become one of the most prominent challenges in the tailings thickening process due to difficulties such as a poor flocculation effect, low sedimentation velocity, substandard overflow and underflow, etc., restricting the development of green tailings disposal technology [4,5]. The two-step flocculation process for ultra-fine tailings has emerged as one of the most promising practical approaches for the efficient dewatering and safe disposal of tailings [6–8].

In order to clarify the mechanism of the two-step flocculation process to improve the dewatering of ultra-fine tailings, scholars have analyzed the effects of different factors [9–11], such as polyacrylamide charge density and molecular weight [12], vacuum filtration parameters [13], dosages of flocculants [14,15], surface chemical and electrochemical factors [16,17], magnetic flocculation [18], the catalysts used [19], salinity [20], the microbes present [21], polymer structure type, and other factors [22–25]. For instance, Lemanowicz et al. (2011) used two kinds of flocculants treated ultrasonically to study the influence of the flocculant addition sequence and mixing mode on the dewatering of ultra-fine tailings.

It was found that the floc formed by the double flocculation system was larger than that under the action of the optimal dosage of any flocculant alone [26]. Lu et al. (2016) used two flocculants with opposite charges to dehydrate ultra-fine oil sand tailings, which improved the settling speed of flocculating tailings and the clarity of supernatant. They also analyzed the action mechanism of double flocculants, but did not consider the influence of shear action [27].

As one of the most common external mechanical actions, shear has an important impact on the flocculation effect of tailings [28–30]. Appropriate shear action can improve the flocculation effect of tailings by promoting particle collision and changing the floc structure. According to the famous collision rate formula of flocculating particles, the shear velocity gradient and shear action time are the most important influencing parameters [31,32]. Yang Y et al. (2020) studied the variation in the yield stress, the floc strength factor, and the fractal dimensions of iron ore fine tailings with different shear rates and shear times in a two-step flocculation process and explained the internal reasons for the stronger shear resistance of the secondary floc network [33]. Petra Bubakova et al. (2013) studied the influence of shear velocity gradient on floc size and structure via the image analysis method and proposed a relationship between fractal dimensions and shear velocity gradient [34]. Guerin (2019) used high and low shear rates to promote the breakage and aggregation of flocs, respectively. In addition, he also observed full reproducibility after one or more cycles of experiments [35]. However, the shearing mechanism in the two-step flocculation process for ultra-fine tailings is still not clear, which may have an adverse impact on the development of ultra-fine tailings dewatering technology.

Therefore, in this paper, a small self-made experimental device is used to simulate the two-step flocculation process of ultra-fine tailings by means of a "stirring blade and pipe flocculation" to strengthen the effect of shear failure and shear coagulation promotion. The variation laws of the ultra-fine tailings' mass fraction, floc structure and size, and floc strength factor and regeneration factor with shear action in different stages are explored. The synergistic mechanism of shear failure and coagulation promotion in the two-step flocculation process for ultra-fine tailings is proposed, which will provide a reference for the regulation of the shear mode and action range in the high-efficiency dewatering process of ultra-fine tailings.

2. Materials and Methods

2.1. Materials

In this study, the tailings were obtained from an iron mine. The particle size distribution was measured using the Malvern Mastersizer 2000 Laser Diffraction Particle Size Analyzer (Figure 1). The volume fraction of tailings with particle sizes less than 20 μm, 37 μm, and 74 μm was 55.66%, 76.43%, and 94.58%, respectively. Therefore, the experimental tailings were ultra-fine tailings. The non-uniformity coefficient of these tailings was 17.78 and the curvature coefficient was 1.41. The particle size distribution curve could meet the two conditions that "the non-uniformity coefficient is greater than or equal to 5 and the curvature coefficient is 1~3", indicating that the grading of tailings was good. The results of X-ray diffraction are listed in Table 1.

In order to avoid introducing other impurities that may affect the hydrolysis process of flocculant molecules, deionized water was used in all the tests that required water. The two kinds of commercial flocculants used in this study were from SNF in Vaughan, ON, Canada and their basic physical and chemical properties are reported in Table 2.

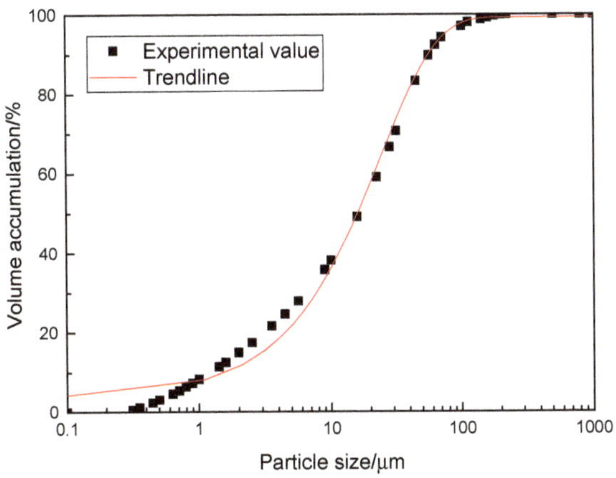

Figure 1. Particle size distribution of tailings.

Table 1. Mineral components of the tailings [22].

Mineral	Ideal Formula	Iron Tailings (wt%)
Anatase	TiO_2	0.2
Barite	$BaSO_4$	0.2
Goethite	α-$Fe^{3+}O(OH)$	0.2
Hematite	α-Fe_2O_3	0.1
Iron—alpha	α-Fe	0.1
Maghemite	γ-Fe_2O_3	0.1
Plagioclase	$NaAlSi_3O_8$—$CaAl_2Si_2O_8$	0.8
Quartz	SiO_2	97.1
Rutile	TiO_2	1.2

Table 2. Basic physical and chemical properties of flocculants [22].

Flocculant	Polymer	Charge	Molecular Weight	Charge Density	Relative Density	Solubility in Water
923VHM	Polyacrylamide	Anionic	16–18 million	Medium	0.6~0.9	Yes
4800 SSH	Polyacrylamide	Cationic	5–7 million	Very high	0.6~0.9	Yes

2.2. Experimental Device

According to the relevant literature and previous explorations, it has been found that the shear failure effect provided by a mixing blade is stronger, and the shear coagulation effect realized by a flocculation pipeline is better. Therefore, in order to enhance the shear effect, the shear failure effect was provided with mixing blades, and the shear coagulation effect was provided by a flocculation pipeline. As shown in Figure 2, a small self-made experimental device was used to carry out a two-step flocculation experiment. This device mainly included the following three parts:

1. The primary flocculation part (the first experimental part) was mainly composed of a large beaker, a variable-speed mixer, a single-layer impeller with a diameter of 75 mm, and a small peristaltic pump, which was responsible for completing the primary flocculation of tailings slurry and providing shear failure for the underflow of primary flocculating slurry.
2. The secondary flocculation part (the second experimental part) was mainly composed of a PVC plastic pipe with a total length of 6 m and a diameter of 6.35 mm,

five secondary flocculant addition points, a small peristaltic pump, etc. It was responsible for completing the secondary flocculation of tailings slurry and providing shear coagulation for secondary flocculation.

3. The deposition dewatering part (the third experimental part) was mainly composed of a sieve and a tray, which was responsible for completing the deposition and dewatering of two-step flocculating slurry.

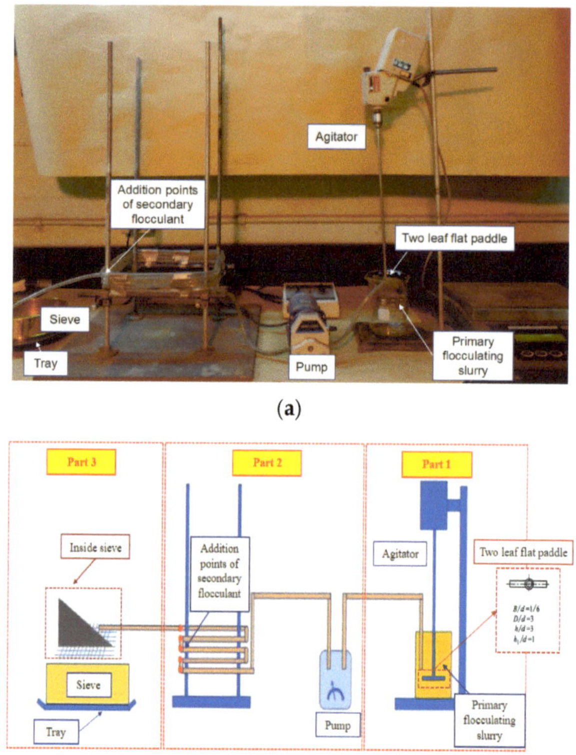

(**a**)

(**b**)

Figure 2. The two-step flocculation experimental device. (**a**) Photo of experimental device; (**b**) schematic diagram of experimental device.

2.3. Experimental Method and Measurement

The stock solutions of two kinds of flocculants were freshly prepared at a concentration of 0.1% solids (w/v %) in deionized water. A slurry of 13.5% solids (w/v %) was prepared in a large beaker using 135 g of tailings and 865 mL of deionized water. The agitation speed was set to about 600 rpm for 30 s and then decreased and maintained at 280 rpm for 10 min to homogenize the slurry, at which point 2.7 mL of the primary flocculant, a 923 VHM stock solution (the polymer's optimum dosage was 20 g/t according to the pre-experiment), was added to the slurries and then agitation was stopped immediately to avoid breaking flocs. We allowed the slurry to stand for 5 min and then removed the supernatant. As shown in Table 3, we changed the rotating speed and action time of the agitator to apply shear damage to the underflow of the flocculating slurry. Then, a small peristaltic pump was used to send the primary flocculating underflow, after shear failure, into the pipeline for secondary flocculation.

Table 3. The value of shear action G_1T_1 in the primary floc failure stage.

No.	1	2	3	4	5	6	7	8
Rotation speed/rpm	30	30	30	30	30	30	30	30
G_{11}/s^{-1}	29.12	29.12	29.12	29.12	29.12	29.12	29.12	29.12
Time/s	30	60	90	120	150	180	210	240
G_1T_1	14,611	15,485	16,358	17,232	18,106	18,979	19,853	20,727
No.	**9**	**10**	**11**	**12**	**13**	**14**	**15**	
Rotation speed/rpm	30	30	30	50	75	100	125	
G_{11}/s^{-1}	29.12	29.12	29.12	62.66	115.12	177.23	247.69	
Time/s	270	300	180	180	180	180	180	
G_1T_1	21,600	22,474	18,979	25,016	34,458	45,639	58,322	

Three different flow velocities and five different flocculant addition points were designed for pipeline flocculation. As shown in Table 4, when the slurry flowed from the beaker into the specified addition point in the pipeline, we changed the pipeline flow velocity to shear the slurry for secondary flocculation. Meanwhile, 2.7 mL of the secondary flocculant, a 4800SSH stock solution, was gradually injected into the pipeline at a uniform speed through a syringe until all the slurry in the beaker had flowed through the specified addition point in the pipeline. Finally, a small peristaltic pump was used to discharge the secondary flocculating slurry onto the ultra-fine screen to realize deposition, evaporation, and dewatering. Because the underflow concentration of the primary flocculating slurry was significantly higher than the mass concentration of the low-concentration tailings slurry, the two-step shear flocculation process required a longer sedimentation and dehydration time. Therefore, after 48 h, we took samples and measured the concentration of slurry.

Table 4. The value of shear action G_2T_2 in the secondary floc regeneration stage.

No.	1	2	3	4	5	6	7	8
Flow velocity/m/s	0.3	0.3	0.3	0.3	0.3	0.4	0.4	0.4
Pipe length/m	1	2	3	4	5	1	2	3
G_2/s^{-1}	626.21	647.74	659.40	665.15	668.58	934.86	969.00	987.46
G_2T_2	2087	4318	6594	8869	11,143	2337	4845	7406
No.	**9**	**10**	**11**	**12**	**13**	**14**	**15**	
Flow velocity/m/s	0.4	0.4	0.5	0.5	0.5	0.5	0.5	
Pipe length/m	4	5	1	2	3	4	5	
G_2/s^{-1}	996.57	1001.99	1277.98	1326.72	1353.05	1366.02	1373.74	
G_2T_2	9966	12,525	2556	5307	8118	10,928	13,737	

During the above experiment, the macro and micro variables to be sampled and measured were as follows. The primary flocculating slurry, the shear-damaged primary flocculating slurry, and the secondary flocculating slurry were sampled. The floc size of different slurries was measured using a Malvin laser particle sizer. In the measurement process, two measures were taken to ensure the accuracy of the measurement results. Firstly, the water dilution method was used instead of ultrasound for sample dispersion to avoid damage to the flocs caused by the ultrasonic waves from the laser particle size analyzer. Secondly, 3 samples were used in each experiment to measure the floc size and calculate the average value. The microstructure of flocs of different slurries was observed by means of a scanning electron microscope.

In the two-step flocculation process, the primary flocculation and secondary flocculation affect each other and jointly determine the dewatering efficiency of ultra-fine tailings.

Therefore, in the primary floc failure stage, it was necessary to ensure that the shear coagulation effect in the secondary floc regeneration stage did not change—that is, the pipeline flow rate was set at 0.5 m/s, and the secondary flocculant addition point was kept 1 m away from the outlet. The suitable agitator speed and mixing time values were designed to change the G_1T_1 value of shear failure and explore the variation laws of the mass fraction and micro characteristics of ultra-fine tailings with shear action in the two-step flocculation process for ultra-fine tailings (Table 3).

Similarly, in the secondary floc regeneration stage, it was necessary to ensure that the shear failure effect in the primary floc failure stage did not change—the rotating speed of the agitator was set to 50 rpm and the mixing time was set to 180 s. The suitable pipeline flow velocity value and secondary flocculant addition point were designed to change the G_2T_2 value of the shear coagulation effect and explore the variation laws of the mass fraction and micro characteristics of ultra-fine tailings with the shear effect in the two-step flocculation process for ultra-fine tailings (Table 4).

In many flocculation studies, the GT value, a product of the shear velocity gradient G and shear time T, is usually regarded as the key parameter determining the influence of shear action on flocculation effect. The specific calculation method of the GT value of shear action in the above two stages of this experiment is as follows.

1. The shearing action of the mixer.

The velocity gradient (G_m) of slurry under the shear action of a stirring blade is

$$G_m = \sqrt{\frac{P}{\mu W}} \tag{1}$$

where G_m is the velocity gradient of slurry under the shear action of the stirring blade, s^{-1}; P is the power consumed by slurry flow under the shear action of the mixing blade, W; μ is the dynamic viscosity of the slurry, being about 1.01×10^{-3} Pa·s; and W is the volume of slurry in the mixing vessel, m^3.

The power consumed by slurry flow under the shear action of the stirring blade is

$$P = N_p \rho N^3 {D_m}^5 \tag{2}$$

where N_p is the power number and depends on the impeller type and the impeller Reynolds number, and N_p = 1.90; ρ is the density of primary flocculating underflow, being about 1519.86 kg/m^3; N is the impeller velocity, r/s; and D_m is the impeller diameter, m.

2. The shearing action of the pipeline flow.

Under the shear action of pipeline flocculation, the shear time (T) of the slurry is

$$T = \frac{l}{v} \tag{3}$$

where T is the time that the slurry is in the pipeline flocculation equipment, s; l is the length of the flocculation pipeline, m; and v is the average flow velocity at a cross-section of the flocculation pipeline, m/s.

The velocity gradient of slurry under the shear action of pipeline flocculation is

$$G_p = \sqrt{\frac{\gamma h}{\mu T}} \tag{4}$$

where G_p is the velocity gradient of the slurry in the flocculation pipeline, s^{-1}; γ is the specific weight of the primary flocculation underflow, N/m^3; h is the head loss, m; μ is the dynamic viscosity of the slurry, being about 1.01×10^{-3} Pa·s; and T is the time that the slurry is in the flocculation equipment, s.

Due to the presence of bent sections in the flocculation pipeline, the head loss includes frictional head loss and local head loss. For the calculation of frictional head loss along the full pipe flow, Darcy's formula is used.

$$h_f = \lambda \frac{l}{D} \frac{v^2}{2g} \tag{5}$$

where D is the diameter of the flocculation pipeline, m, and λ is the frictional resistance coefficient.

Regarding the calculation of the frictional resistance coefficient, the following formula is used. When the flow velocity is less than 1.2 m/s,

$$\lambda = 0.001824 \frac{g}{D^{0.3}} \left(1 + \frac{0.867}{v}\right)^{0.3} \tag{6}$$

The calculation formula for the local head loss is as follows:

$$h_j = \zeta \frac{v^2}{2g} \tag{7}$$

where h_j is the local head loss, m, and ζ is the local resistance coefficient. The local resistance coefficient of a right-angled elbow is 0.9.

Therefore, the calculation formula for head loss is as follows:

$$h = h_f + h_j \tag{8}$$

3. The value of shear action G_1T_1 in the primary floc failure stage.

The shear action in the primary floc failure stage is divided into two parts. The first part is the shear action applied by the mixing blades in the beaker, $G_{11}T_{11}$, while the second part is the shear action during the flow of the slurry before the addition of the secondary flocculant solution in the flocculation pipeline, $G_{12}T_{12}$.

$$G_1T_1 = G_{11}T_{11} + G_{12}T_{12} \tag{9}$$

Therefore, the calculation of G_{11} is based on Equations (1) and (2), the calculation of G_{12} is based on Equations (4)–(8), and the calculation of T_{12} is based on Formula (3).

4. The value of shear action G_2T_2 in the secondary floc regeneration stage

The shear action in the secondary floc regeneration stage only occurred in the pipeline. Therefore, the calculation of G_2 is based on Equations (4)–(8), and the calculation of T_2 is based on Equation (3).

3. Results

In view of the different adding positions, adding methods, and effects of the two flocculants in the two-step flocculation process, the variation laws of the dewatering of ultra-fine tailings and the internal flocs with shear in the primary floc failure stage and the secondary floc regeneration stage were investigated, respectively. Through analyzing the effects of shear action in different stages on the floc structure size, strength factor, and regeneration factor of ultra-fine tailings slurry, the mechanism of the two-step flocculation and efficient dewatering of ultra-fine tailings is explained.

3.1. Effect of Shear on the Concentration of Two-Step Flocculating Slurry

Two-step flocculation relies on the interaction of two flocculant molecules to increase the binding sites on the floc surface, which is conducive to the formation of a dense floc network structure. This is an effective way for ultra-fine tailings to dehydrate efficiently by further releasing the internal water in the flocs structure and promoting the formation of a dense floc network structure.

In the primary floc failure stage, the shear action mainly releases the water in the floc network structure and improves the mass fraction of tailings. The variation law of the mass fraction of ultra-fine tailings with the shear action was explored to provide a reference for obtaining a suitable shear failure range and realizing the efficient dewatering of ultra-fine tailings.

It can be seen from Figure 3 that the mass fraction of ultra-fine tailings obtained by means of two-step flocculation was 69%~72%, and the dewatering efficiency of ultra-fine tailings was significantly improved compared with that produced using single flocculation in the previous experiment (the mass fraction was 65%~67%). In the primary floc failure stage, with the increase in shear action (G_1T_1), the mass concentration of ultra-fine tailings slurry increased and then became stable. When G_1T_1 was in the range of 0~100,000, the mass fraction of slurry increased from 69% to about 71%. When G_1T_1 was in the range of 100,000~150,000, the growth rate of the slurry mass fraction slowed down gradually. When G_1T_1 was higher than 150,000, the mass fraction of the slurry was stable close to the highest value.

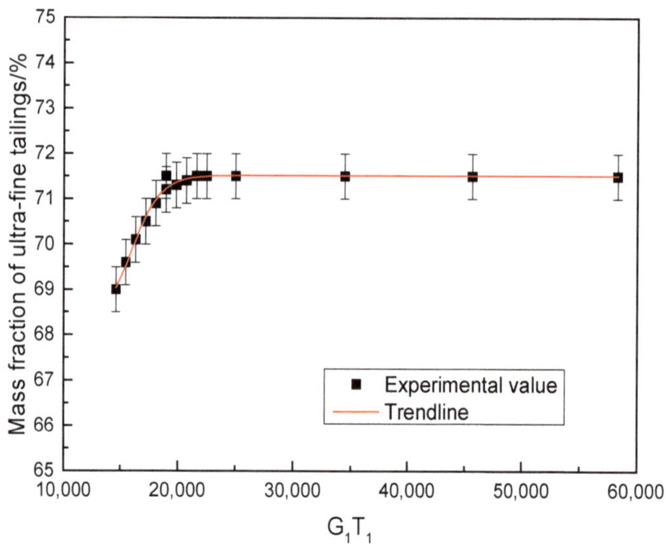

Figure 3. Variation in the mass fraction of ultra-fine tailings with shear action (G_1T_1) in the primary floc failure stage.

In the secondary floc regeneration stage, shear mainly promotes the combination of flocculant molecules and tailings particles to form a dense floc network structure and improve the mass fraction of tailings. The variation law of the mass fraction of ultra-fine tailings with the shear action was explored to provide a reference for obtaining a suitable shear coagulation range and realizing the efficient dewatering of ultra-fine tailings.

Figure 4 shows the variation in the mass fraction of ultra-fine tailings with shear action (G_2T_2) in the secondary floc regeneration stage. As introduced in Section 2.3, the shear action in the primary floc failure stage of each experiment was basically the same and the primary floc structure was fully broken. Therefore, it can be considered that the difference in the mass fraction of the slurry was more due to the influence of shear coagulation in the secondary floc regeneration stage.

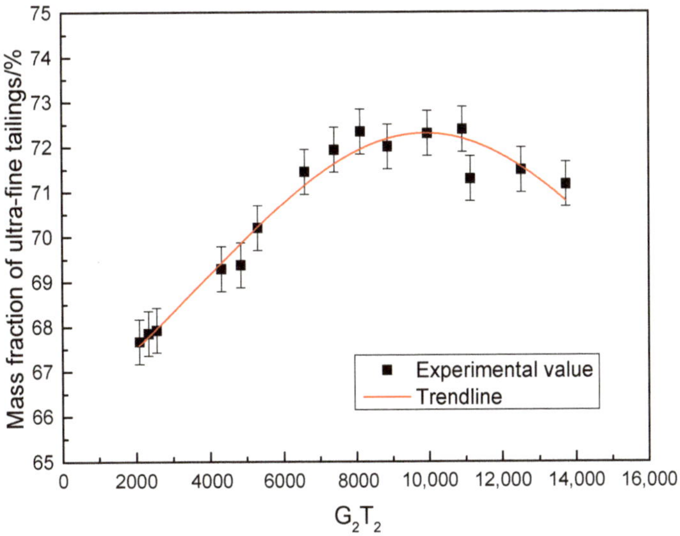

Figure 4. Variation in the mass fraction of ultra-fine tailings with shear action (G_2T_2) in the secondary floc regeneration stage.

In the secondary floc regeneration stage, with the increase in shear action (G_2T_2), the mass fraction of ultra-fine tailings slurry increased, then became stable, and finally decreased slightly. When G_2T_2 was in the range of 2000~11,000, the mass concentration increased gradually, from 68% to about 72%. When G_2T_2 was in the range of 7000~11,000, the mass concentration was stable close to the highest value. When G_2T_2 was higher than 11,000, the mass concentration began to decrease.

It is worth noting that there were pronounced dips in the curve at the points with G_2T_2 values of 9000 and 11,000. At the point with a G_2T_2 value of 9000, the decrease in the mass concentration was about 0.5%. This is because the G_2T_2 value of the previous point was 8118, corresponding to a flow velocity of 0.5 m/s, representing the fastest group in the experiment, resulting in better coagulation effect. The point with a G_2T_2 value of 9000 corresponded to a flow velocity of 0.3 m/s, representing the slowest group in the experiment, reducing its coagulation effect and obtaining a lower mass concentration of slurry. At the point with a G_2T_2 value of 11,000, the dip in the curve was due to the excessive G_2T_2 value and strong shear force, which had a destructive effect on the formed floc network structure and reduced the mass concentration of the slurry.

The variation law of the mass fraction of ultra-fine tailings with shear action in the two-step flocculation process was analyzed. The two-step flocculation process significantly improved the dewatering efficiency of ultra-fine tailings, and the mass fraction of slurry can reach more than 71%. In order to achieve the efficient and ideal dewatering described above, the shear action G_1T_1 in the primary floc failure stage should be higher than 100,000, and the shear action G_2T_2 in the secondary floc regeneration stage should be in the range of 7000~11,000.

3.2. Effect of Shear on Floc Structure and Size

Flocculant molecules adsorb ultra-fine tailings particles, increase floc size, and form the floc network structure, which is the fundamental reason why flocculating ultra-fine tailings accelerates sedimentation and helps to realize efficient dewatering. In the flocculation process, shear can not only promote the collision between flocculant molecules and tailings particles but also destroy the floc network structure. Therefore, the change law of floc structure and size under shear in different stages was analyzed to further determine the internal mechanisms driving two-step flocculation and the efficient dewatering of ultra-fine tailings.

In the primary floc failure stage, the variation law of the average particle size of flocs with the shear action was explored and the failure form of the floc structure was analyzed to obtain the suitable shear failure range in two-step flocculation.

As shown in Figure 5, in the primary floc failure stage, when the shear action G_1T_1 was less than 100,000, with the increase in shear action, the average particle size of broken flocs decreased rapidly. When G_1T_1 increased from 50,000 to 100,000, the average floc particle size reduced from 69 μm to 54 μm. When G_1T_1 was greater than 100,000 and less than 150,000, the average particle size of broken flocs gradually decreased. When G_1T_1 was higher than 150,000, the average particle size of broken flocs was basically stable at 50 μm.

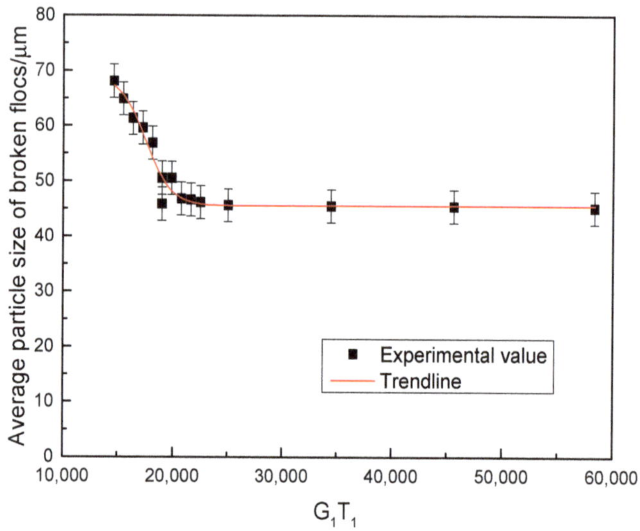

Figure 5. Variation in average particle size of broken flocs with shear failure (G_1T_1).

This may be because the original loose and fragile primary floc network structure contained many weak connections in addition to the connections between the polymer flocculant and particles through bridging. These weak connections were very sensitive to external disturbances. The low shear action mainly destroyed the weak connections in the floc structure, which promoted the large-scale structural fracture failure of the primary floc structure. High shear action could not only destroy the weak connections in the primary floc structure, resulting in structural fracture failure, but also destroy the connections formed by bridging between particles and the flocculant, resulting in surface stripping failure. This produced a large number of small-size broken flocs, quickly increasing the number of flocs, thus reducing the average particle size of flocs. With the continuous improvement in shear action, the above two failure forms were basically completed, and the size of broken flocs basically remained stable.

Figure 6 shows the SEM of flocs under different shear failure values in the primary floc failure stage. When G_1T_1 = 30,000, the floc structure was obviously damaged, but the size of the broken flocs was still large. When G_1T_1 = 120,000, the floc network structure was completely broken into small-size flocs, and there were basically no large-size flocs. This is consistent with the change law of floc average particle size.

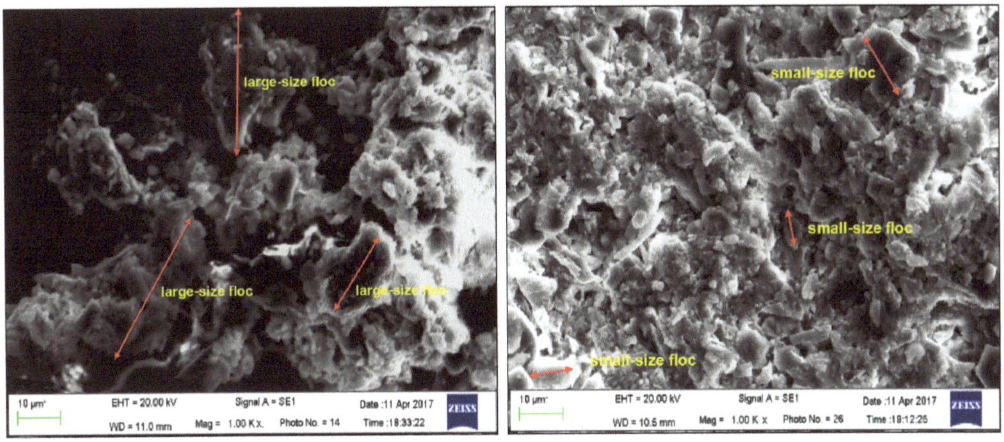

(a) (b)

Figure 6. SEM of flocs under different shear failure values in the primary floc failure stage. (**a**) Low shear (G_1T_1 = 30,000); (**b**) high shear (G_1T_1 = 120,000).

In the secondary floc regeneration stage, the variation law of the average particle size of flocs with shear action was explored and the influence of different shear action ranges on the secondary floc regeneration was analyzed to obtain the suitable shear coagulation range in two-step flocculation.

In the secondary floc regeneration stage (Figure 7), with the increase in G_2T_2, the average particle size of secondary flocs increased and then decreased. When the G_2T_2 was 2000~7000, the average particle size of secondary flocs increased slowly from 84 μm to 88 μm. When the G_2T_2 was 7000~9000, the average particle size of secondary flocs significantly increased from 88 μm to 99 μm. When G_2T_2 was higher than 9000, the average particle size of secondary flocs decreased rapidly, from 99 μm to 87 μm. In addition, when G_2T_2 was 7000~11,000, the average particle size of secondary flocs was greater than 90 μm, which indicates the high degree of secondary floc regeneration.

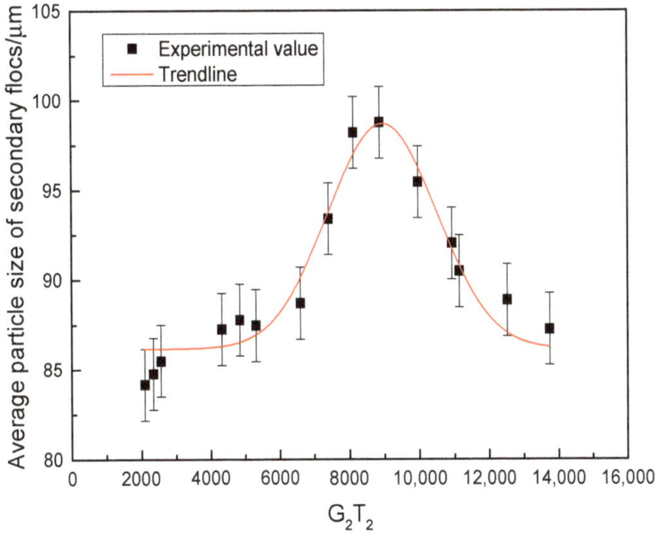

Figure 7. Variation in average particle size of secondary flocs with shear coagulation (G_2T_2).

In the secondary floc regeneration stage, shear can promote the combination of broken flocs and secondary flocculant molecules, increase floc size, and form a secondary floc network structure. However, excessive shear will damage the secondary flocculation. The reason why the average particle size of secondary flocs changed with G_2T_2 may be that after the addition of the secondary flocculant, the broken flocs and secondary flocculant were flocculated mainly through the adsorption of heterogeneous charges on their binding sites under the action of Brownian motion. The relatively weak shear action can promote the combination of the broken flocs and the secondary flocculant, but the action time and intensity were not sufficient, and the increase in the average particle size of the secondary flocs was small. When G_2T_2 increased to 7000~11,000, shear action within an appropriate range fully promoted the collision between broken flocs and secondary flocculants, and the collision rate and binding efficiency were significantly improved. Combined with the bridging effect of flocculant molecules, a dense secondary floc network structure was formed, and the average particle size of flocs increased rapidly. However, when G_2T_2 continued to increase to higher than 11,000, the excessive shear action destroyed the floc structure and the average particle size of secondary floc decreased rapidly.

Figure 8 shows the SEM of flocs under different shear coagulation levels in the secondary floc regeneration stage. When G_2T_2 = 3000, the secondary floc structure was gradually formed, but was relatively loose, with a large number of free broken flocs or tailings particles. When G_2T_2 = 9000, the secondary floc network structure was dense, and there were basically no free broken flocs or tailings particles. This is consistent with the change law of floc average particle size.

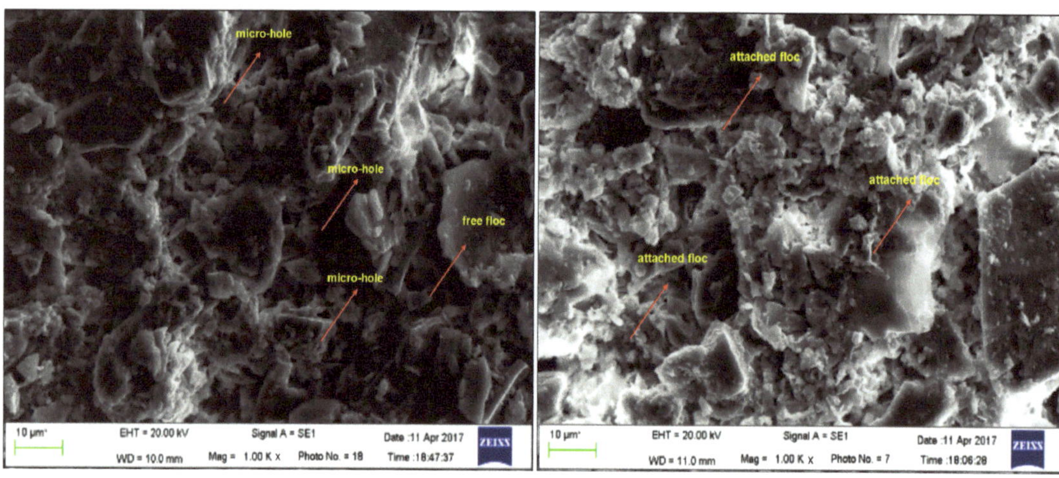

(a) (b)

Figure 8. SEM of flocs under different shear coagulation levels in the secondary floc regeneration stage. (**a**) Low shear (G_2T_2 = 3000); (**b**) high shear (G_2T_2 = 9000).

The variation law of the floc structure and size of ultra-fine tailings with shear action in the two-step flocculation process was analyzed. In the primary floc failure stage, the shear failure action caused structural fracture and surface stripping of the floc network structure and released water in the floc network structure, and thus the shear action G_1T_1 should be higher than 100,000. In the secondary floc regeneration stage, the shear coagulation action promoted the combination of the flocculant and flocs to form a dense floc network structure, and thus the shear action G_2T_2 should be in the range of 7000~11,000.

3.3. Effect of Shear on Floc Strength Factor and Regeneration Factor

In the primary floc failure stage, in order to quantitatively analyze the variation law of floc structure strength with shear action, the concept of the floc strength factor was introduced. The floc strength factor mainly reflects the strength of the floc structure (Equation (10)). Generally, for the same kind of floc, under the same shear action, the greater the floc strength factor, the stronger the floc structure, the denser the floc structure, the lower the internal moisture content, and the greater the interaction force between particles in the floc.

The floc strength factor can be calculated as follows:

$$F_1 = \frac{d_2}{d_1} \tag{10}$$

where d_1 is the average particle size of the floc before shear failure, μm, and d_2 is the average particle size of the floc after shear failure, μm. Both were obtained in this this study through experimental measurements.

As can be seen from Figure 9, in the primary floc failure stage, when G_1T_1 was less than 100,000, the floc strength factor decreased rapidly. When G_1T_1 was 30,000, the floc strength factor was about 0.82 of the original floc structure strength, indicating that the shear action can only overcome 0.18 of the total strength of the original floc structure, and the lower shear action can only destroy the weak connections in the original floc structure. When G_1T_1 was greater than 100,000 and less than 150,000, the floc strength factor decreased slowly. When G_1T_1 increased to 150,000, the floc strength factor was about 0.50 of the original floc structure strength. The shear action can thus overcome 50% of the total strength of the original floc structure, meaning that it can not only destroy the weak connections in the floc structure, but also overcome a small portion of the strong connections. With the continuous increase in shear action, the two failure forms of the floc structure were basically completed, and the floc strength factor did not change significantly.

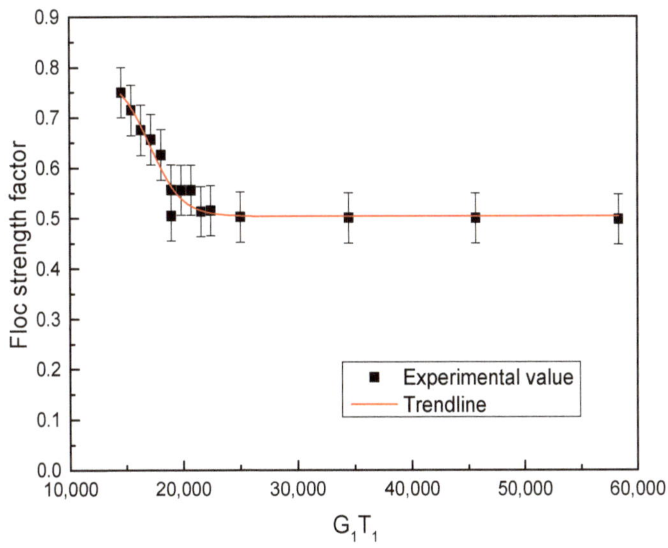

Figure 9. Variation in floc strength factor with shear failure (G_1T_1).

In the secondary floc regeneration stage, in order to quantitatively analyze the variation law of floc regeneration characteristics with shear action, the concept of the floc regeneration factor was introduced. The floc regeneration factor mainly reflects the recovery of floc size in the process of floc structure failure and re-flocculation, and can reflect the secondary flocculation effect (Equation (11)). Generally, when the sizes of the original floc and the broken floc are the same, the larger the floc regeneration factor, the higher the floc recovery performance and the better the secondary floc regeneration.

The floc regeneration factor can be calculated as follows:

$$F_2 = \frac{d_3 - d_2}{d_1 - d_2} \quad (11)$$

where d_1 is the average particle size of the floc before shear failure, μm; d_2 is the average particle size of the floc after shear failure, μm; and d_3 is the average particle size of the floc after secondary flocculation, μm.

As can be seen from Figure 10, in the secondary floc regeneration stage, the floc regeneration factor increased and then decreased with the increase in G_2T_2. When G_2T_2 was 2000~7000, the increase in floc regeneration factor was small, from 0.65 to 0.85. When G_2T_2 was 7000~9000, the floc regeneration factor increased significantly, up to about 1.4. When G_2T_2 was higher than 9000, the floc regeneration factor decreased rapidly. Overall, when G_2T_2 was 7000~11,000, the floc regeneration factor was higher than 1.0, which shows that the broken flocs displayed good regeneration and could realize the complete recovery of their size before shear failure.

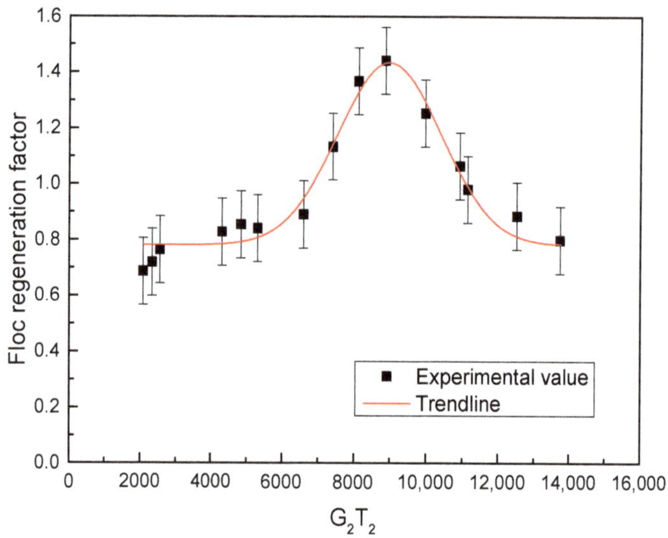

Figure 10. Variation in floc regeneration factor with shear coagulation (G_2T_2).

The variation in the floc strength factor and regeneration factor with shear action in the two-step flocculation process was analyzed. In the primary floc failure stage, with the increase in shear action, the floc structure strength that shear overcame gradually increased, and thus the shear action G_1T_1 should be higher than 100,000. In the secondary floc regeneration stage, shear can effectively promote floc regeneration, and G_2T_2 should thus be in the range of 7000~11,000.

3.4. Synergistic Mechanism of Shear Failure and Shear Coagulation in Two-Step Flocculation

Two-step flocculation is an effective way to achieve the high-efficiency dewatering of ultra-fine tailings. It comprises the joint effect of shear failure in the primary floc failure stage and shear coagulation in the secondary floc regeneration stage. In the primary floc failure stage, shear failure within an appropriate range promotes different failure forms of the primary floc structure, resulting in the ideal unstable state of the flocculating slurry, which provides the best conditions for the regeneration of the secondary flocs. In the secondary floc regeneration stage, shear coagulation within an appropriate range can promote the combination of flocculant molecules and tailings particles, form a dense floc network structure, achieve the best secondary floc regeneration effect, and improve the dewatering capacity of ultra-fine tailings.

Therefore, considering the effects of shear failure and shear coagulation on the slurry in different stages of the two-step flocculation process, a hypothesis regarding the synergism of shear failure and shear coagulation in the two-step flocculation process of ultra-fine tailings was proposed. This was expounded from the perspective of floc structure change and the flocculation principle to clarify the internal mechanism of the efficient dewatering of ultra-fine tailings.

From the perspective of floc structure change, in the primary floc failure stage, shear failure within an appropriate range causes structural fracture failure and surface stripping failure of the floc network structure, the number of broken flocs is high, and the binding sites in broken flocs are fully exposed. Large flocs produced by structural fracture failure and small flocs produced by surface stripping failure make the size distribution of broken flocs uniform. At the same time, shear failure disturbs the particle diffusion layer in the broken floc, changes the shape of the particle diffusion layer, reduces the repulsion between particles, and causes the flocculating slurry to be in an unstable state, providing the best conditions for secondary floc regeneration.

In the secondary floc regeneration stage, shear coagulation promotion in an appropriate range improves the probability of collisions between positively charged cationic flocculant molecules and negatively charged broken flocs. Both of these adsorb through electric neutralization, which reduces the charge on the particle surface, weakens the repulsion between particles, and increases the average particle size of regenerated flocs. In addition, shear coagulation can further reduce the repulsive force between particles through the disturbance effect on the particle diffusion layer. Based on the bridging effect of flocculant molecules, the broken flocs finally form a compact secondary floc network structure. This promotes the further separation of ultra-fine tailings particles from water and then realizes rapid dewatering in the tailings deposition stage.

In conclusion, shear can cause the flocculating slurry to be in an unstable state by changing the floc structure and size, thus improving the flocculation and dewatering efficiency of ultra-fine tailings.

The root cause of the shear failure and shear coagulation in two-step flocculation is double flocculation. Under the conditions of an imperfect primary flocculation effect, secondary flocculation is a favorable supplement for improving the dewatering capacity of tailings. Therefore, the hypothesis regarding shear failure and shear coagulation's synergism was further explained from the perspective of the flocculation principle.

From the perspective of the flocculation principle (Figure 11), tailings particles and primary flocculant molecules form primary flocs through adsorption and bridging. In the primary floc failure stage, the shear action within an appropriate range releases the binding sites between the particles and the primary flocculant molecules. The state of the primary flocculant molecules is more extended, the tail of the molecular chain extending into the solution becomes longer, the bridging probability increases, meaning that the particles and the primary flocculant molecules are in an unstable state, which provides the best conditions for flocculation. At this time, if no secondary flocculant molecules are added, the particles and primary flocculant molecules tend to adsorb again.

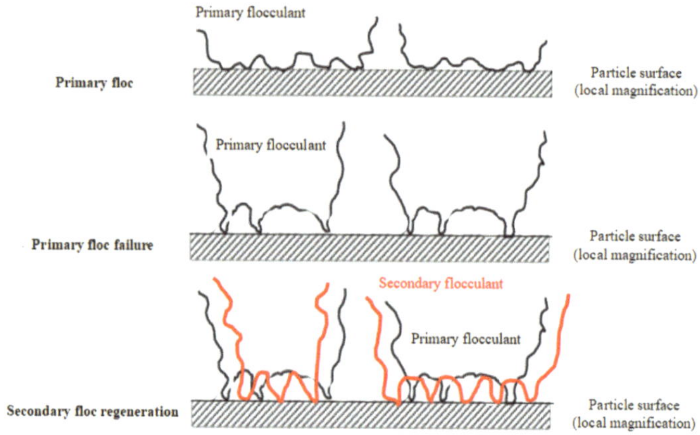

Figure 11. Schematic diagram of the flocculation principle in the hypothesis regarding shear failure and shear coagulation's synergism.

In the secondary floc regeneration stage, the addition of the secondary flocculant disturbed the unstable state of particles and primary flocculant molecules. Under shear action within an appropriate range, the secondary flocculant molecules adsorb with tailings particles and primary flocculant molecules at the spare binding sites, respectively. The primary and secondary flocculant molecules form a stable extension state on the surface of tailings particles, and the tails of their molecular chains extend to the solution at the same time, which increases the bridging probability of particles, so as to form flocs with a larger size and denser structure. The dense floc network structure is the internal reason for the dewatering of tailings, which is favorable for realizing the high-efficiency dewatering of ultra-fine tailings.

4. Conclusions

The two-step flocculation process is an effective method for the efficient dewatering of ultra-fine tailings. In this paper, the effects of shear action in different stages on the mass fraction of ultra-fine tailings, the floc structure and size, and the floc strength factor and regeneration factor of tailings slurry were investigated through a small self-made two-step flocculation experimental device. The hypothesis regarding the synergism of shear failure and shear coagulation in the two-step flocculation process for ultra-fine tailings was put forward. The main conclusions are as follows.

1. The dewatering efficiency of ultra-fine tailings is significantly improved by means of two-step flocculation, and the mass fraction of tailings can reach more than 71%. In order to achieve efficient and ideal dewatering as described above, the shear action G_1T_1 in the primary floc failure stage should be higher than 100,000, while the shear action G_2T_2 in the secondary floc regeneration stage should be in the range of 7000~11,000.
2. The shear action in the primary floc failure stage can cause structural fracture failure and surface peeling failure of the floc structure, and thus release the water in the floc structure. The shear action in the secondary floc regeneration stage can promote the re-combination of flocculant molecules and tailings particles to form a dense floc network structure.
3. With the increase in shear action in the primary floc failure stage, the floc strength factor rapidly decreased and then became stable. With the increase in shear action in the secondary floc regeneration stage, the floc regeneration factor increased and then

decreased. The broken flocs could realize the complete recovery of their size before shear failure.

4. In the primary floc failure stage, shear action disrupted the distribution of counter-ions in the diffusion layer of particles in the flocs, changed the shape of the particle diffusion layer, and weakened the repulsion between particles in different broken flocs. In the secondary floc regeneration stage, shear action increased the collision probability between flocculant molecules and particles and promoted the occurrence of flocculation.
5. Considering the effects of shear failure and shear coagulation on the slurry in different stages of the two-step flocculation process, a hypothesis regarding the synergism of shear failure and shear coagulation in the two-step flocculation process of ultra-fine tailings was proposed. It was expounded from the perspective of floc structure change and the flocculation principle to clarify the internal mechanism of the efficient dewatering of ultra-fine tailings.

Author Contributions: Conceptualization, Y.Y. and X.L.; methodology, Y.Y.; validation, Y.Y., L.Z. and M.G.; formal analysis, Y.Y.; investigation, Y.Y. and X.L.; resources, X.L.; data curation, Y.Y., L.Z. and M.G.; writing—original draft preparation, Y.Y.; writing—review and editing, Y.Y. and X.L.; supervision, X.L.; project administration, Y.Y. and X.L.; funding acquisition, X.L. All authors have read and agreed to the published version of the manuscript.

Funding: This research was funded by the National Natural Science Foundation of China, grant number 52074121.

Data Availability Statement: All data, models, and code generated or used during this study appear in the submitted article. No additional data are available.

Acknowledgments: The authors would like to thank the National Natural Science Foundation of China, JCHX Mining Management Co., Ltd., and the North China Institute of Science and Technology. The authors also thank the editor and the reviewers for the useful comments and suggestions on ways to improve the manuscript.

Conflicts of Interest: The author Ying Yang was employed by the company JCHX Mining Management Co., Ltd. The remaining authors declare that the research was conducted in the absence of any commercial or financial relationships that could be construed as potential conflicts of interest.

References

1. Jiao, H.; Chen, W.; Wu, A.; Yu, Y.; Ruan, Z.; Honaker, R.; Chen, X.; Yu, J. Flocculated unclassified tailings settling efficiency improvement by particle collision optimization in the feedwell. *Int. J. Miner. Metall. Mater.* **2022**, *29*, 2126–2135. [CrossRef]
2. Wu, A.; Ruan, Z.; Wang, J. Rheological behavior of paste in metal mines. *Int. J. Miner. Metall. Mater.* **2022**, *29*, 717–726. [CrossRef]
3. Ishigami, T.; Karasudani, T.; Onitake, S.; Shirzadi, M.; Fukasawa, T.; Fukui, K.; Mino, Y. Effect of liquid volume fraction and shear rate on rheological properties and microstructure formation in ternary particle/oil/water dispersion systems under shear flow: Two-dimensional direct numerical simulation. *Soft Matter* **2022**, *18*, 4338–4350. [CrossRef] [PubMed]
4. Zhang, C.; Liu, R.; Chen, M.; Li, J.; Wang, X.; Liu, Y.; Zhu, Z.; Wang, M.; Fan, F. Influence of the flocculation effect on the rheological properties of cement slurry. *Powder Technol.* **2022**, *398*, 117118. [CrossRef]
5. Wang, Z.; Wang, Y.; Wu, L.; Wu, A.; Ruan, Z.; Zhang, M.; Zhao, R. Effective reuse of red mud as supplementary material in cemented paste backfill: Durability and environmental impact. *Constr. Build. Mater.* **2022**, *328*, 127002. [CrossRef]
6. Chen, Q.; Tao, Y.; Zhang, Q.; Qi, C. The rheological, mechanical and heavy metal leaching properties of cemented paste backfill under the influence of anionic polyacrylamide. *Chemosphere* **2022**, *286*, 131630. [CrossRef] [PubMed]
7. Yang, X.; Zhang, Y.; Lin, C. Compressive and Flexural Properties of Ultra-Fine Coal Gangue-Based Geopolymer Gels and Microscopic Mechanism Analysis. *Gels* **2022**, *8*, 145. [CrossRef]
8. Saoudi, M. Plant-Based Flocculants as Sustainable Conditioners for Enhanced Sewage Sludge Dewatering. *Water* **2024**, *16*, 2949. [CrossRef]

9. Chen, Q.; Ran, F.; Wei, Q.; Zheng, X.; Zhao, M.; Liu, S.; Wang, L.; Fan, C. A Review on Dewatering of Dredged Sediment in Water Bodies by Flocculation Processes. *Water Air Soil Pollut.* **2024**, *235*, 67. [CrossRef]
10. Xu, R.; Zou, W.; Rao, B.; Zhao, W.; Wang, T.; Zhang, Z. In situ kinetics and flocs conformation studies of kaolinite flocculated by Chi-g-CPAM. *Int. J. Miner. Metall. Mater.* **2023**, *30*, 813–823. [CrossRef]
11. Demoz, A. Compressive Stress Dewaterability Limit in Fluid Fine Tailings. *Minerals* **2023**, *13*, 1514. [CrossRef]
12. Aldajani, M.; Alipoormazandarani, N.; Fatehi, P. Two-Step Modification Pathway for Inducing Lignin-Derived Dispersants and Flocculants. *Waste Biomass Valoriz.* **2022**, *13*, 1077–1088. [CrossRef]
13. Lu, C.; Xu, Z.; Zhang, Y.; Dong, B. Rapid and clean dewatering of dredged sediment via coupling of a two-step flocculation and geotextile filtration. *Chem. Eng. J.* **2024**, *480*, 147604. [CrossRef]
14. Dixon, D.V.; Soares, J.B.P. Flocculating oil sands tailings with dual anionic/cationic polymers: Dose and charge density effects. *Can. J. Chem. Eng.* **2024**, *102*, 88–101. [CrossRef]
15. Wen, L.; Yao, D. The Effect of Flocculants and Water Content on the Separation of Water from Dredged Sediment. *Water* **2023**, *15*, 2462. [CrossRef]
16. Xu, R.; Zou, W.; Wang, T.; Huang, J.; Zhang, Z.; Xu, C. Adsorption and interaction mechanisms of Chi-g-P(AM-DMDAAC) assisted settling of kaolinite in a two-step flocculation process. *Sci. Total Environ.* **2022**, *816*, 151576. [CrossRef]
17. Zeng, H.; Shi, C.; Huang, J.; Li, L.; Liu, G.; Zhong, H. Recent experimental advances on hydrophobic interactions at solid/water and fluid/water interfaces. *Biointerphases* **2015**, *11*, 018903. [CrossRef]
18. Sun, Y. Research Progress of Magnetic Flocculation in Water Treatment. *Magnetochemistry* **2024**, *10*, 56. [CrossRef]
19. Ali, A.; Meena, B.V.; Shah, N.A.; Kaushik, T.; Dolkar, T.; Ghoroi, C.; Dutta, A. Deploying a molecular copper catalyst for efficient degradation of commercial and industrial dyes under practical conditions. *Environ. Sci. Water Res. Technol.* **2023**, *9*, 2219–2225. [CrossRef]
20. Liu, Q.; Montoya, B. Effects of Microbially Induced Carbonate Precipitation on Diffuse Double Layer and Particle Fabric of Oil Sands Fine Tailings. *Can. J. Civ. Eng.* **2022**, *49*, 1310–1316. [CrossRef]
21. Wang, S.; Zhang, L.; Yan, B.; Xu, H.; Liu, Q.; Zeng, H. Molecular and Surface Interactions between Polymer Flocculant Chitosan- g -polyacrylamide and Kaolinite Particles: Impact of Salinity. *J. Phys. Chem. C* **2015**, *119*, 7327–7339. [CrossRef]
22. Yang, Y.; Wu, A.; Klein, B.; Wang, H. Effect of primary flocculant type on a two-step flocculation process on iron ore fine tailings under alkaline environment. *Miner. Eng.* **2019**, *132*, 14–21. [CrossRef]
23. Wang, Y.; Yang, Z.; Zhu, F.; Ma, Y.; Ren, J. Study on the Dewatering Effect of In-Situ Dredged Mud Under the Combined Action of Flocculant-Absorbent. In *Advances in Geoengineering Along the Belt and Road*; Zhu, H.-H., Garg, A., Zhussupbekov, A., Su, L.-J., Eds.; Springer: Singapore, 2022. [CrossRef]
24. Nguyen, C.V.; Nguyen, A.V.; Doi, A.; Dinh, E.; Nguyen, T.V.; Ejtemaei, M.; Osborne, D. Advanced solid-liquid separation for dewatering fine coal tailings by combining chemical reagents and solid bowl centrifugation. *Sep. Purif. Technol.* **2021**, *259*, 118172. [CrossRef]
25. Vedoy, D.R.; Soares, J.B. Water-soluble polymers for oil sands tailing treatment: A Review. *Can. J. Chem. Eng.* **2015**, *93*, 888–904. [CrossRef]
26. Lemanowicz, M.; Jach, Z.; Kilian, E.; Gierczycki, A. Ultra-fine coal flocculation using dual-polymer systems of ultrasonically conditioned and unmodified flocculant. *Chem. Eng. J.* **2011**, *168*, 159–169. [CrossRef]
27. Lu, Q.; Yan, B.; Xie, L.; Huang, J.; Liu, Y.; Zeng, H. A two-step flocculation process on oil sands tailings treatment using oppositely charged polymer flocculants. *Sci. Total Environ.* **2016**, *565*, 369–375. [CrossRef]
28. Guo, Z.; Liu, J.; Lei, Z.; Hu, P. Enhancement of flocculant-aided filtration performance of coal tailings under alternating low and high shear rates. *Powder Technol.* **2022**, *399*, 117253. [CrossRef]
29. Wang, Y.; Wu, A.-X.; Ruan, Z.-E.; Wang, Z.-H.; Wei, Z.-S.; Yang, G.-F.; Wang, Y.-M. Reconstructed rheometer for direct monitoring of dewatering performance and torque in tailings thickening process. *Int. J. Miner. Metall. Mater.* **2020**, *27*, 1430–1437. [CrossRef]
30. Ofori, P.; Nguyen, A.V.; Firth, B.; McNally, C.; Ozdemir, O. Shear-induced floc structure changes for enhanced dewatering of coal preparation plant tailings. *Chem. Eng. J.* **2011**, *172*, 914–923. [CrossRef]
31. Yu, W.-Z.; Gregory, J.; Campos, L.; Li, G. The role of mixing conditions on floc growth, breakage and re-growth. *Chem. Eng. J.* **2011**, *171*, 425–430. [CrossRef]
32. Appiah-Brempong, M.; Essandoh, H.M.K.; Asiedu, N.Y.; Dadzie, S.K.; Momade, F.Y. Performance Optimization of Chemical and Green Coagulants in Tannery Wastewater Treatment: A Response Surface Methodology Approach. *J. Optim.* **2023**, *2023*, 19. [CrossRef]
33. Yang, Y.; Wang, H.; Klein, B.; Wu, A. Shear-Dependent Yield Stress of Iron Ore Fine Tailings in Two-Step Flocculation Process. *Adv. Mater. Sci. Eng.* **2020**, *2*, 1–12. [CrossRef]

34. Bubakova, P.; Pivokonsky, M.; Filip, P. Effect of shear rate on aggregate size and structure in the process of aggregation and at steady state. *Powder Technol.* **2013**, *235*, 540–549. [CrossRef]
35. Guérin, L.; Frances, C.; Liné, A.; Coufort-Saudejaud, C. Fractal dimensions and morphological characteristics of aggregates formed in different physico-chemical and mechanical flocculation environments. *Colloids Surf. A Physicochem. Eng. Asp.* **2019**, *560*, 213–222. [CrossRef]

minerals

Article

Microwave Thermal Treatment on Enhanced Cemented Tailings Backfill: An Experimental Study

Xiaolong Cui [1], Keping Zhou [1] and Zheng Pan [1,2,*]

[1] School of Resources and Safety Engineering, Central South University, Changsha 410083, China; 225511034@csu.edu.cn (X.C.)
[2] The Robert M. Buchan Department of Mining, Queen's University, Kingston, ON K7L 3N6, Canada
* Correspondence: pan.zheng@queensu.ca

Abstract: Cemented tailings backfill (CTB), composed of tailings, binder, and water, is widely used for filling underground goaves in mining operations. Unmanaged tailings can occupy extensive land and pose significant environmental risks. Microwave technology offers a promising approach to enhance the utilization of tailings, reducing dependency on natural resources. However, limited research on microwave heating parameters has impeded its broader adoption. This study uses the orthogonal experimental method to study the influence of various factors on the strength of the CTB and to determine the impact capacity of each factor. Additionally, this study conducted a visual analysis of the microwave heating time (MHT), microwave delay time (MDT), cement-tailings ratio, slurry concentration and microwave power (MP) to verify the experimental results. The results show that microwave heating can enhance or diminish the mechanical properties of CTB samples at different curing ages, depending on the specific microwave parameter settings. Research indicates that microwave technology can be effectively applied to mine backfill materials to improve their early strength and the modulus of elasticity.

Keywords: cemented tailings backfill (CTB); microwave thermal treatment; uniaxial compressive strength (UCS); orthogonal experimental design (OED)

Academic Editors: Alexander Mikhailovich Kalinkin, Yuye Tan, Xun Chen and Yuan Li

Received: 12 December 2024
Revised: 22 January 2025
Accepted: 22 January 2025
Published: 24 January 2025

Citation: Cui, X.; Zhou, K.; Pan, Z. Microwave Thermal Treatment on Enhanced Cemented Tailings Backfill: An Experimental Study. *Minerals* **2025**, *15*, 115. https://doi.org/10.3390/min15020115

1. Introduction

The mining industry plays an invaluable role in promoting the national economy. However, while providing raw materials, it inevitably generates a large amount of solid waste, with tailings accounting for 29% of industrial solid waste [1,2]. Enhancing the early strength and elasticity modulus of CTB not only significantly improves the efficiency and safety of mine backfilling but also promotes the resource utilization of tailings, reducing environmental impact.

Existing studies have investigated ways to enhance the early strength of the CTB or cemented paste backfill (CPB). These studies found that factors such as water-to-cement ratio adjustment, curing temperature and humidity control, active binders, and the addition of fibrous materials significantly affect the uniaxial compressive strength (UCS). For example, Hou et al. [3] used steel slag and straw fibers to prepare backfill materials to improve the mechanical properties of the filling body. The adhesion between straw fibers and the mortar matrix is the key factor in enhancing the integration of the backfill, while steel slag improves the mechanical properties by enhancing the distribution of void structures within the backfill. Their results showed that the yield stress increases with the mass concentration, steel slag content, and straw fiber content. He et al. [4] used lithium slag (LS) and fly ash (FA) as binders for the CTB. Their results indicated that LS and FA significantly affect the strength of the binder. The optimal ratio of the binder is OPC (Portland cement)-LS-FA at 2:1:1, which yields the highest UCS. This is because fly ash has good pozzolanic activity, which reacts with cement to form denser hydration products, thereby enhancing the early strength. Chen et al. [5] found that under high-temperature curing conditions, the early strength of cement mortar with sulfur-containing tailings is higher, but the late strength gradually decreases with increasing curing temperature. This is due to the early strength development of the mortar mainly depending on the rate of hydration reactions, while the late strength changes are primarily influenced by the changes in the internal pore structure of the CTB. Existing studies indicate that adding different types and proportions of additives can improve the early strength of the CTB to a certain extent [6–8]. Although some progress has been made in the aforementioned fields, challenges remain, such as the diversity and complexity of material selection and the varying characteristics of tailings from different mines. These methods usually require longer curing times and have limited effects.

In recent years, microwave heating technology has been widely applied in materials science and engineering due to its efficiency, speed, and energy-saving characteristics. Therefore, introducing microwave heating technology into the treatment process of the CTB has significant research implications and application prospects [9–12]. Currently, research on the application of microwave heat treatment in the CTB is still in its early stages, and this just encourages the authors to commence the present study. Existing studies have mainly focused on the effects of microwave heat treatment on the performance of concrete and other construction materials, as well as aggregate recovery. For example, Wei et al. [13] studied the microwave heating and aggregate recovery of dry and saturated concrete, investigating the impact of moisture content on the heating results. Their findings provide a new method for enhancing aggregate recovery efficiency in microwave-assisted concrete by increasing moisture content. Tang et al. [11] studied the formation process of porosity, physical and mechanical properties, pore characteristics, and hydration products to clarify the mechanism of microwave pre-curing on carbide slag-fly ash autoclaved aerated concrete (CS-FA AAC). The findings demonstrate that the innovative microwave pre-curing technology significantly accelerates slurry foaming and thickening, thereby reducing the pre-curing time and enhancing the stability of CS-FA AAC green bodies. Wang et al. [14] used microwave heating technology to cure cement-fly ash (CFA) slurry. Their results showed that microwave curing significantly increased the early strength of CFA paste and maintained its compressive strength for up to 28 days. The pore structure of the hardened paste after 28 days of microwave curing was superior to that of thermal curing. Zhou et al. [15] combined microwave curing with secondary water curing for C30 concrete. Their results indicated that while microwave curing can enhance the early hydration degree of cement, it may reduce the later hydration degree. Secondary water curing can enhance the later hydration degrees, and microwave curing is suitable for both ordinary and reinforced concrete. Ding et al. [9] investigated the effects of microwave heating time and power on the mechanical properties of CTB and explored the improvement of CTB's mechanical properties. However, the impact of microwave delay time on CTB has been overlooked, even though delay time is also a critical parameter. Compared to traditional external heating methods, microwave heating offers advantages such as rapid temperature increase, uniform heating, and high energy efficiency. Therefore, studying the effects of microwave heat treatment on the performance of CTB can not only shorten curing time but also enhance early strength, providing a new method for mine tailings' treatment [16–18].

However, CTB differs significantly from concrete in aggregate properties, mix proportions, and operational conditions, making the findings from the aforementioned studies unsuitable for direct application to CTB materials. Currently, research on utilizing microwave technology in the curing process of CTB remains in its early stages, motivating the author to pursue this study. This study aims to investigate the effects of various microwave parameters on the changes in the UCS and elastic modulus of CTB and explore how microwave heat treatment can enhance the early strength of CTB, and refine appropriate microwave heating parameters. The structure of this study is as follows: Firstly, the orthogonal experimental design (OED) is presented in detail, including the materials used, the preparation of CTB specimens, and the testing procedures. This study employs an OED method to identify five factors influencing the mechanical performance of CTB, namely microwave power (MP), microwave heating time (MHT), microwave delay time (MDT), slurry concentration, and cement-tailings ratio and investigates the effects of these factors on the mechanical properties of CTB. Second, the results are analyzed, and the impact of microwave heating on CTB's mechanical performance is discussed. Finally, conclusions are drawn, and suggestions for future research are proposed. The advantage of orthogonal experimental design (OED) lies in its ability to analyze multiple factors and their interactions simultaneously [19]. By integrating OED with microwave heating processes, this study not only offers a novel approach to enhancing the early strength of cemented tailings backfill using microwaves but also broadens the application scope of orthogonal design in mining engineering. The OED is widely recognized and applied due to its efficiency and systematic approach [20]. The OED allows for the acquisition of substantial information with fewer experimental trials through scientifically designed schemes, thus enhancing experimental efficiency and effectiveness [21–23].

In summary, this study explores the innovative approach of accelerating slurry hydration through microwave heating, offering significant practical and theoretical value in addressing mine backfilling and tailings management issues. The innovation of this study lies in its first systematic exploration of the application of microwave heat treatment technology in the CTB, revealing the mechanisms by which microwave heating affects the internal structure and performance of the CTB. Through systematic experimental research and theoretical analysis, this study aims to provide a new approach and method for mine tailings' treatment, further promoting the development of the CTB technology, enhancing the efficiency and effectiveness of mine tailings' treatment, and advancing the green and sustainable development of mining engineering.

2. Experimental Program

2.1. Materials

The CTB samples were prepared by mixing tailings, cement, and water. The tailings were sourced from the Bainiuchang mine of Mengzi Mining and Metallurgy Co., Ltd. (Mengzi, Yunnan, China), with P.C. 42.5 Portland cement used as the binder and tap water as the mixing water. The chemical element content in the tailings and cement was analyzed using an X-ray fluorescence spectrometer. Table 1 shows the chemical composition of tailings and cement used. The particle size distribution of tailings significantly influences mine backfilling, as it is related not only to the dewatering process but also to the cementation performance and binder consumption of the cemented backfill. Key parameters include the median particle diameter and the coefficient of particle uniformity. The particle size distribution of tailings was measured using a Malvern MS3000 laser diffraction particle size analyzer. The particle size distribution of the tailings is shown in Figure 1, and it can be seen from Figure 1 that the tailings belong to the medium- and coarse-grained tailings.

The chemical composition of mine tailings reflects their mineralogical makeup and potential applications. The concentration and forms of each element significantly influence the physicochemical properties, environmental impact, and resource recovery potential of the tailings [24]. The tailings used in the experiment contain high concentrations of Si

(18.81%) and O (41.3%) elements, with a significant amount of Ca (11.48%) as well, which indicates the presence of minerals such as quartz and feldspar in the tailings, determining their physical stability. High-silicon-content tailings are suitable for the construction industry, such as for use as concrete aggregate or in glass manufacturing. High calcium content suggests that the tailings may possess excellent cementing properties, making them suitable for cement or construction material production. It may also enhance the alkalinity of tailings, aiding in the neutralization of acidic environments [25].

Table 1. Chemical composition of the materials.

Materials	SiO_2	Al_2O_3	K_2O	CaO	Fe_2O_3	SO_3	MgO	Others
Tailings (%)	82.00	6.96	2.66	2.03	1.30	0.26	0.25	4.54
Binder (%)	19.31	4.93	2.66	63.15	3.12	2.32	3.26	1.25

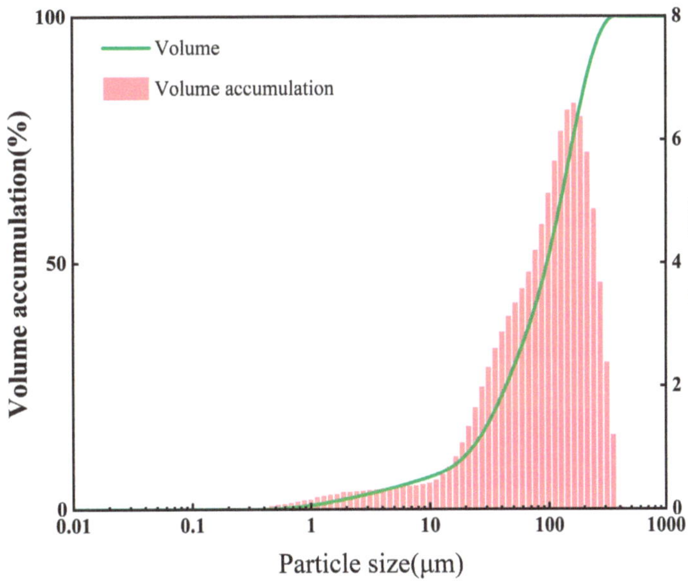

Figure 1. Particle size distribution of tailings.

2.2. Preparation of CTB Specimens

A mining concrete pump (HBMG30/21-110S) was used to mix the components of the CTB according to the experimental design ratios, ensuring a uniform composite material. The prepared CTB mixture was then cast into molds (7.07 cm × 7.07 cm × 7.07 cm) to form cubic specimens. The CTB samples were divided into two groups: one served as the control group, while the other was subjected to microwave heating after the slurry was poured into the molds, following the experimental design. The microwave heating device used in this experiment, an Intelligent Microwave Rock Breaker (WLK-A9), features adjustable output power with a maximum power of 3000 W. The heating conditions and microwave power settings are specified in Table 2. The specimens were then cured under controlled temperature conditions of 20 ± 1 °C and a humidity level of 95%. The curing equipment used is regularly calibrated, with calibration results documented to ensure its precision and accuracy. The samples were cured for 3 and 7 days using a standard constant temperature and humidity curing chamber (SHBY-60B). The preparation of the

CTB samples includes four parts: slurry mixing, mold casting, microwave heating, and sample curing. The complete process is shown in Figure 2.

Table 2. Orthogonal experimental factor level table.

NO.	Factors				
	Surry Concentration/%	Cement-Tailings Ratio	MP/W	MHT/min	MDT/h
1	70	1:4	60	4	0
2	72	1:6	180	6	2
3	74	1:8	300	8	4
4	76	1:12	420	10	6

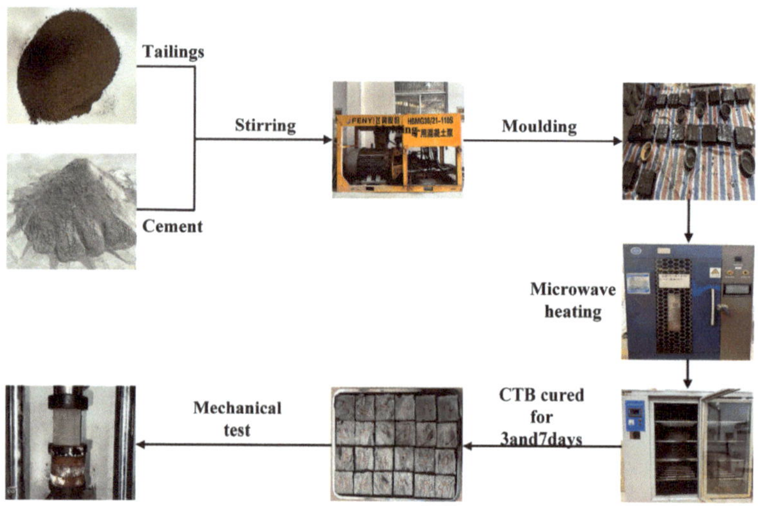

Figure 2. Experimental specimen preparation process.

2.3. Orthogonal Experimental Design

This study identified five factors influencing the UCS of the CTB: microwave power (MP), microwave heating time (MHT), microwave delay time (MDT), slurry concentration, and cement-tailings ratio. To investigate the influence patterns of these factors on the UCS of the CTB samples, each factor was designed with four levels. The levels of each factor are shown in Table 2.

Using the UCS of the CTB at 3 and 7 days of curing as evaluation indicators, a 5-factor, 4-level orthogonal experiment was conducted. A total of 16 experimental groups were conducted, along with a blank control group. The experimental design is shown in Table 3. Three samples were used for each set of tests. Using three samples per group helps ensure the reliability and stability of the data through repeated experiments. Individual experimental results may be influenced by random errors, equipment errors, and operational mistakes. Using multiple samples helps reduce the impact of random errors and improves the reliability of the results. Using three samples ensures that the experimental results are representative and minimizes the influence of individual sample variations on the conclusions. By calculating the average of multiple samples, the overall performance of the material under the given experimental conditions can be more accurately represented.

This experiment uses the L16 OED, a common type of orthogonal table, consisting of 16 experiments (i.e., 16 experimental units), aimed at studying the combinations of multiple factors and their levels. The L16 orthogonal table allows for a simultaneous analysis of the effects of multiple factors on the experimental results and explores the interaction effects between some factors. Typically, the L16 design is used to study the interaction between factors. Therefore, the L16 orthogonal experimental design is an efficient and resource-saving method, suitable for studying multiple factors and their interactions [26]. By properly arranging experimental combinations, the L16 design enables the acquisition of useful information about the effects of multiple factors with limited experiments, providing the scientific basis for parameter optimization and improving product or process performance. The experimental design is shown in Table 3. By performing range and variance analyses on the experimental data, the influence patterns of various factors on the UCS were determined.

Table 3. OED of factors affecting the mechanical properties of CTB.

NO.	Factors				
	MP/W	MHT/min	MDT/h	Cement-Tailings Ratio	Slurry Concentration/%
1	60	4	0	4	70
2	60	6	2	6	72
3	60	8	4	8	74
4	60	10	6	12	76
5	180	4	2	8	76
6	180	6	0	12	74
7	180	8	6	4	72
8	180	10	4	6	70
9	300	4	4	12	72
10	300	6	6	8	70
11	300	8	0	6	76
12	300	10	2	4	74
13	420	4	6	6	74
14	420	6	4	4	76
15	420	8	2	12	70
16	420	10	0	8	72

2.4. Mechanical Properties Test

Due to the irreplaceable advantages of low cost and simple operation, the UCS test is one of the most commonly used indicators to observe the mechanical properties of CTB [27–31]. During the test, a loading rate of 100 N/s is used, which is commonly adopted in mechanical property tests of various materials. This rate ensures that the material is not exposed to excessively rapid loading that could lead to brittle failure while still allowing the test to be completed within a reasonable timeframe. By selecting a loading rate of 100 N/s, the test effectively simulates the stress behavior of CTB in real-world engineering applications until the samples fail. After the test, the corresponding stress-strain curve is saved. From this curve, calculate the maximum stress value (i.e., maximum compressive strength) and modulus of elasticity of the CTB.

3. Results and Discussion

3.1. Orthogonal Experimental Results

After preparing the CTB samples, microwave heating and curing were performed according to the experimental design. After 3 and 7 days of curing, uniaxial compression tests were conducted on the CTB samples to obtain the UCS average of the three parallel samples, which was taken as the result under various influencing factors. The test results are shown in Table 4.

Table 4. UCS test results.

NO.	Factors					UCS/MPa (3d)	UCS/MPa (7d)
	MP/W	MHT/min	MDT/h	Cement-Tailings Ratio	Slurry Concentration/%		
1	60	4	0	1:4	70	1.29	1.98
2	60	6	2	1:6	72	0.9	1.55
3	60	8	4	1:8	74	0.53	0.9
4	60	10	6	1:12	76	0.36	0.44
5	180	4	2	1:8	76	0.69	1.16
6	180	6	0	1:12	74	0.24	0.37
7	180	8	6	1:4	72	1.62	2.77
8	180	10	4	1:6	70	0.65	1.17
9	300	4	4	1:12	72	0.25	0.41
10	300	6	6	1:8	70	0.41	0.69
11	300	8	0	1:6	76	1.16	1.57
12	300	10	2	1:4	74	1.6	2.82
13	420	4	6	1:6	74	0.93	1.42
14	420	6	4	1:4	76	2.07	3.42
15	420	8	2	1:12	70	0.23	0.41
16	420	10	0	1:8	72	0.46	0.68

The early stage of cement hydration refers to the process where cement reacts with water to form hydration products (such as calcium silicate hydrate, calcium hydroxide, etc.) and begins to form the cement matrix. The early hydration stage typically occurs between the initial and final setting times, generally within the first 24–72 h after mixing slurry. The early characteristics of cement hydration have a significant impact on the final performance of CTB, particularly on strength development, durability, and stability. Examining the early characteristics of cement hydration, especially under the influence of microwave heating, is crucial for understanding the performance of CTB. The specific significance is as follows: Influence the formation of early strength, predict long-term performance, optimize material proportions, improve construction quality and safety, and evaluate the effects of microwave heating.

The selection of 3 and 7 days as curing times is typically aimed at focusing on the early hydration stage, as these two time points are significant in the hydration process of CTB. The hydration reaction of cement is most significant in the early stage (typically within the first few days). During the first few hours to days, after cement and water are mixed, cement particles react with water to form hydration products (such as calcium silicate hydrate and calcium aluminate hydrate), which provide the foundation for the hardening of cement-based materials. Studying the early hydration stage, particularly with curing times of 3 and 7 days, helps us gain a deeper understanding of the early strength development and structural changes in tailings cemented backfill or other cement-based materials.

Day 3: This time point is typically used to capture the early stages of cement hydration. By the third day of curing, most of the hydration reactions have occurred. Although hydration is not fully complete, a relatively stable strength value can be obtained. For CTB, the 3-day curing time reflects the early strength development and helps evaluate the impact of microwave heating on the CTB's initial performance. Day 7: This time point is a common standard curing period, during which many cement products typically reach a relatively stable strength value. After 7 days of curing, the cement hydration reaction continues, but the rate of strength gain begins to slow down. Choosing 7 days as the testing period provides a more comprehensive assessment of the CTB's strength development and reflects the long-term effects of microwave heating during ongoing hydration.

The critical importance of early strength: In many applications, especially in the study of CTB, the rapid growth of early strength is crucial. Good early strength ensures the stability and load-bearing capacity of CTB during the construction process, especially when rapid curing and high early load-bearing capacity are required for the fill.

3.2. Analysis of Patterns

Analyzing the results of the OED can identify influence patterns of various factors on the UCS and rank the extent of their impact. Therefore, range analysis and variance analysis were performed on the results of this orthogonal experiment.

3.2.1. Range Analysis

Range analysis is simple to perform and provides intuitive results, but it cannot estimate experimental errors. By calculating the range of the experimental data, the range analysis results of this orthogonal experiment were obtained [32,33]. The analysis results are shown in Table 5. As seen from the range analysis in Tables 5 and 6, the influencing factors on the UCS, in descending order, are the cement-tailings ratio of the CTB, slurry concentration, MP, MHT, and MDT. Comparatively, internal factors of the CTB have a greater effect on the UCS than the external factors, and the influence of the cement-tailings ratio is greater than that of the slurry concentration.

Table 5. Range analysis in UCS of the CTB at 3 days.

Factors	MP	MHT	MDT	Cement-Tailings Ratio	Slurry Concentration
k_1	0.770	0.790	0.787	1.645	0.645
k_2	0.800	0.905	0.855	0.910	0.807
k_3	0.855	0.885	0.875	0.522	0.825
k_4	0.922	0.768	0.830	0.270	1.070
R	0.152	0.137	0.088	0.425	0.425

Table 6. Range analysis in UCS of the CTB at 7 days.

Factors	MP	MHT	MDT	Cement-Tailings Ratio	Slurry Concentration
k_1	1.218	1.242	1.150	2.748	1.063
k_2	1.367	1.507	1.485	1.427	1.353
k_3	1.373	1.413	1.475	0.858	1.377
k_4	1.482	1.277	1.330	0.407	1.647
R	0.264	0.265	0.335	2.341	0.584

In Table 6 here, k_n (n = 1, 2, 3, 4) represents the mean value of each level of the influencing factors, and R represents the range of each influencing factor. The larger the range, the greater the impact on the UCS of CTB.

As seen from the range analysis in Tables 5 and 6, the influencing factors on the UCS, in descending order, are the cement-tailings ratio of the CTB, slurry concentration, MP, MHT, and MDT. Comparatively, internal factors of the CTB have a greater effect on the UCS than the external factors, and the influence of the cement-tailings ratio is greater than that of the slurry concentration. The impact of microwave parameters on the CTB is typically manifested as local or short-term effects. For example, MP: the power of microwave heating affects the heating rate. Excessive power may lead to overheating, while insufficient power may hinder the progress of cement hydration. However, the impact of MP is generally limited to the heating process and does not directly or persistently affect the long-term strength of the material. MDT: MDT affects the cement hydration process. Too short a time may prevent sufficient hydration, while too long a time may lead to over-hydration. Overall, the impact of microwave heating is primarily concentrated in the early hydration stage, while the strength and performance of cement materials are more influenced by the long-term hydration process and cement mix. Therefore, the impact of microwave parameters is typically localized and limited; their variations may not influence the final strength of the material in the long term and broadly, as is the case with the sand-to-cement ratio and slurry concentration. The K-value of microwave parameters is generally small.

3.2.2. Variance Analysis

Compared to range analysis, variance analysis has the advantage of estimating experimental errors, analyzing the errors within and between groups, and avoiding the interference of experimental errors on the data, thereby ensuring the reliability and accuracy of the data [34]. By performing a variance analysis on the results of this experiment, the variance analysis results for the influencing factors on the UCS were obtained. The variance analysis results are shown in Tables 7 and 8, where * indicates that it has the greatest impact.

Table 7. Variance analysis in UCS of the CTB at 3 days.

Factors	Sum of Squared Deviations	Degrees of Freedom	F Ratio	F Critical Value	Significance
MP	0.054	3	1.000	9.280	
MHT	0.056	3	1.037	9.280	
MDT	0.017	3	0.315	9.280	
Cement-tailings ratio	4.314	3	79.889	9.280	*
Slurry concentration	0.369	3	6.833	9.280	
Error	0.05	3			

Table 8. Variance analysis in UCS of the CTB at 7 days.

Factors	Sum of Squared Deviations	Degrees of Freedom	F Ratio	F Critical Value	Significance
MP	0.142	3	1.000	9.280	
MHT	0.180	3	1.268	9.280	
MDT	0.295	3	2.077	9.280	
Cement-tailings ratio	12.358	3	87.028	9.280	*
Slurry concentration	0.686	3	4.831	9.280	
Error	0.14	3			

In analysis of variance (ANOVA), the F-ratio and the F-critical value are important statistics used to determine whether differences between groups are significant [23]. They have distinct meanings and roles. Meaning of F-ratio: the F-ratio in ANOVA is a ratio used to compare two different variances, typically the variance between groups and the variance within groups. Significance of F-ratio: a larger F-ratio indicates more pronounced differences between groups, suggesting that the factor has a stronger influence on the dependent variable. Conversely, a smaller F-ratio suggests smaller differences between groups, potentially indicating no significant factor effect [22]. The F-critical value is obtained from the F-distribution table and depends on the significance level (α) and degrees of freedom. The F-critical value is used to compare with the calculated F-ratio to determine whether the experimental results are significant. If the calculated F-ratio > F-critical value, it indicates significant differences between groups. If the calculated F-ratio < F-critical value, it indicates no significant differences between groups. This implies that the observed difference in group means may merely be due to random variation.

Tables 7 and 8 show that among the influencing factors, the cement-tailings ratio and slurry concentration of CTB have a significant effect on UCS, whereas MHT, MP, and MDT do not have a significant effect.

3.3. Effect of Microwave Heating on the UCS of CTB

As shown in Figure 3, during the three-day curing period, the UCS with a cement-tailings ratio of 1:4 and a slurry concentration of 70% increased by 300.13%. For a concentration of 72%, it increased by 326.32%. The average increase was 276.17%. For a cement-tailings ratio of 1:6, the UCS with a slurry concentration of 70% increased by 306.25%; for 72%, by 373.68. The average increase was 328.47%. For a cement-tailings ratio of 1:8, the UCS with a slurry concentration of 76% increased by 263.16%. The average increase was 196.11%. For a cement-tailings ratio of 1:12, the UCS with a slurry concentration of 70% increased by 187.5%. The average increase was 135.84%.

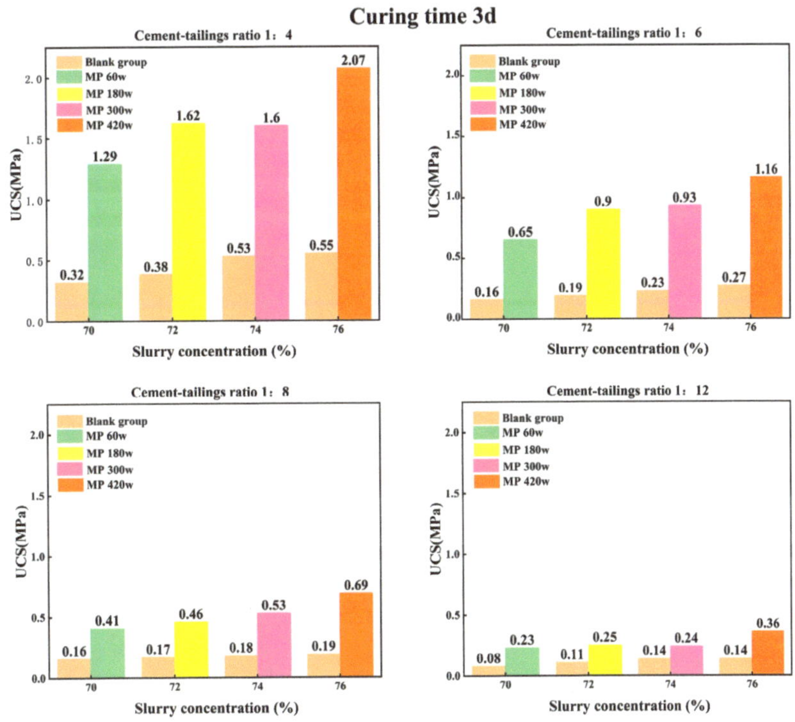

Figure 3. UCS comparison of CTB with a three-day curing period.

Although the orthogonal experimental analysis did not show significant effects of MHT, MP, and MDT on the UCS, the experimental results indicate that the UCS treated with microwave heating significantly improved under the same cement-tailings ratio, slurry concentration, and curing time. The higher the cement content, the greater the UCS of the CTB improvement. This is because microwave heating can significantly accelerate the cement hydration process. During hydration, cement produces hydration products, such as ettringite and calcium silicate hydrate, which form a solid framework within the CTB, thus increasing its compressive strength. Microwave heating rapidly raises the temperature of the cement paste, promoting early cement hydration and enabling the rapid generation of hydration products, thereby enhancing the early strength of the CTB [9,12,16]. Additionally, microwave heating can reduce the amount of free water in the cement paste, thus lowering the porosity of the CTB and further enhancing its compressive strength. Furthermore, microwave heating can effectively improve the microstructure of the CTB. In traditional curing processes, the uneven distribution of moisture within the CTB can lead to the formation of microcracks, which weaken its overall strength. Microwave heating, with its penetrating and selective heating characteristics, ensures more uniform moisture evaporation within the CTB, reducing the formation of microcracks and thereby improving the overall strength of the CTB. Wu et al. [12] found that the CTB treated with microwave heating exhibited a denser microstructure, significantly lower porosity, and fewer microcracks, resulting in a substantial increase in UCS. Additionally, microwave heating can effectively enhance the durability of the CTB. During the microwave heating process, the rapid temperature rise within the CTB accelerates the migration and diffusion of harmful substances, such as chloride ions and sulfates. Under the influence of microwave heating, these harmful substances can quickly migrate from the interior to the surface of the CTB and are expelled during subsequent cooling and moisture evaporation, thereby reducing their impact on the durability of the CTB.

As shown in Figure 4, during the seven-day curing period, the UCS with a cement-tailings ratio of 1:4 and a slurry concentration of 70% increased by 191.18%. The average increase was 123.55%. For a cement-tailings ratio of 1:6, the UCS with a slurry concentration of 70% increased by 129.41%. The average increase was 87.40%. For a cement-tailings ratio of 1:8, the UCS with a slurry concentration of 76% increased by 107.14%. The average increase was 74.08%. For a cement-tailings ratio of 1:12, the UCS with a slurry concentration of 70% increased by 105.00%. The average increase was 72.1%.

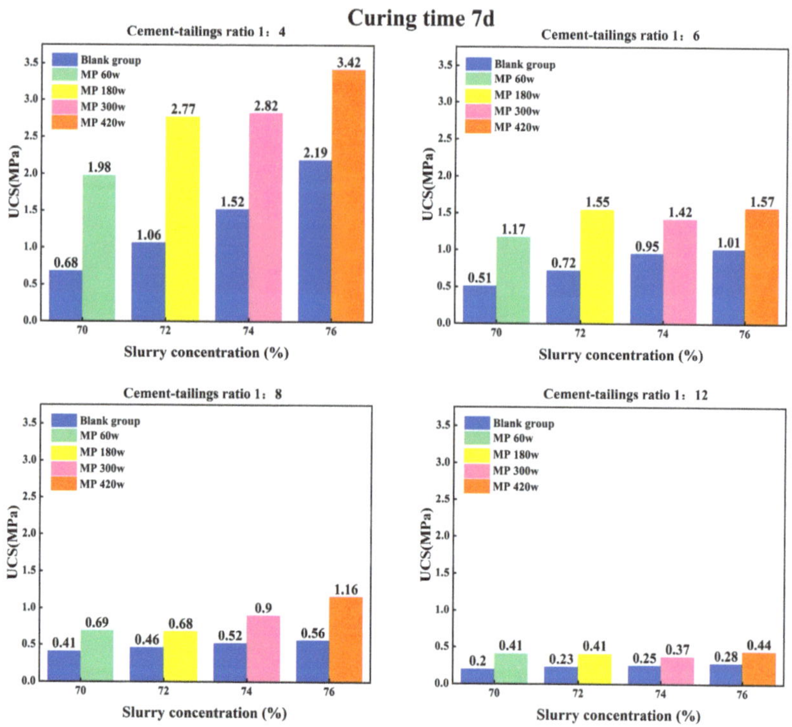

Figure 4. UCS comparison of CTB with a seven-day curing period.

From the above experimental results, it can be observed that the early strength of the CTB significantly improved after microwave heating. However, as the curing time extends, the rate of strength growth gradually slows down or even decreases. This phenomenon is mainly due to the following reasons. First, although microwave heating can quickly enhance the early strength of the CTB, the rapid temperature rise may lead to incomplete cement hydration. In traditional curing processes, cement hydration is a gradual process that takes a long time to complete. Microwave heating accelerates the hydration process by rapidly raising the temperature, thereby quickly increasing early strength [35]. However, it may also cause some cement particles to lose activity due to excessive temperature, affecting subsequent hydration. Therefore, as curing time extends, the strength growth of the CTB slows down or even stops. Secondly, during microwave heating, the temperature distribution inside the CTB may be uneven, leading to incomplete hydration in some areas. Although microwave heating is characterized by uniform heating, in practice, due to the material properties and the multifield coupling stress distribution of the microwave field, temperature gradients may exist inside the fill material, resulting in weaker hydration in certain regions [9,36]. These regions may show higher strength in the early stages but experience slow or even declining strength growth during later curing due to incomplete hydration. Thirdly, microwave heating may increase microcracks and pores inside the CTB, affecting its long-term strength. During microwave heating, rapid temperature rise and moisture evaporation may cause uneven stress distribution within the CTB, leading to the formation of microcracks. These microcracks may have little impact on early strength, but, over time, their expansion and connection can lead to strength reduction [16]. Microwave heating can affect the microstructure of CTB, primarily manifested as changes in its pore structure. Under microwave heating, the moisture within the CTB can evaporate more uniformly, reducing the uneven distribution of moisture and improving the material's density. Microwave heating can also induce grain growth and phase transitions of minerals within CTB to some extent, further enhancing its strength. Additionally, rapid moisture evaporation during microwave heating may create pores within the CTB, weakening the overall strength of the material, especially evident during long-term curing.

As shown in Figure 5, the CTB without microwave heating treatment had an average strength growth rate of 188.08% after 3 and 7 days of curing. In contrast, Figure 3 shows that CTB subjected to microwave heating treatment had an average strength growth rate of 234.15% after 3 days of curing, significantly enhancing the early strength of CTB, further confirming the hypotheses and conclusions of this study.

3.4. *Effect of Univariate on UCS*

To visually illustrate the impact of factors other than the cement-tailings ratio and slurry concentration on the UCS, an analysis was conducted on the effects of MHT, MDT, and MP on the UCS.

3.4.1. Microwave Heating Time

Figure 6 illustrates the effect of MHT (4, 6, 8, and 10 min) on UCS of CTB samples cured for different durations (3 and 7 days). Figure 6 also shows that during the early stage (0–3 days), UCS of CTB increases with MHT from 0 to 8 min; however, from 8 to 10 min, UCS of CTB decreases with further heating time. CTB heated for 8 min shows the highest UCS value. These results indicate that MHT longer than 8 min can reduce the early strength enhancement effect on CTB samples. These findings suggest that microwave treatment can significantly enhance the mechanical properties of CTB within an appropriate MHT range.

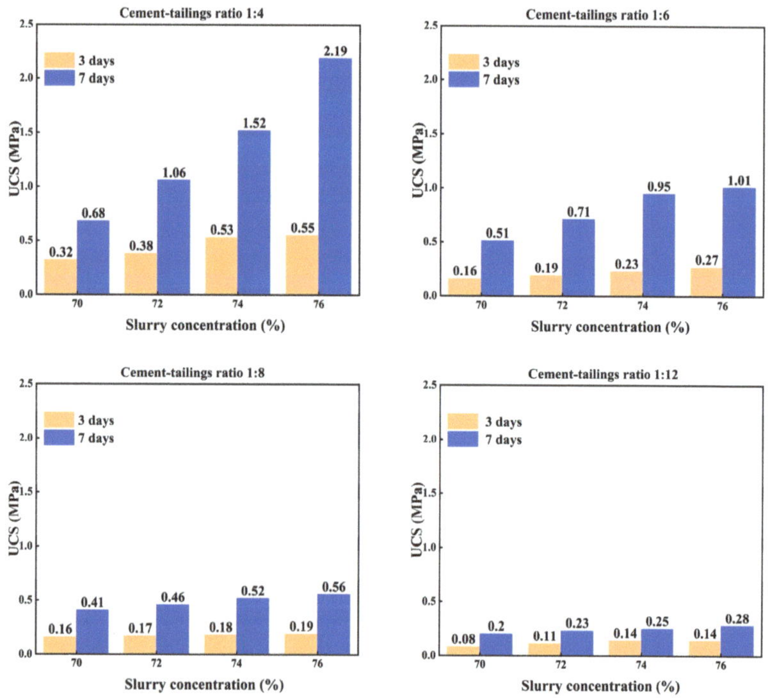

Figure 5. UCS comparison of CTB with a normal curing.

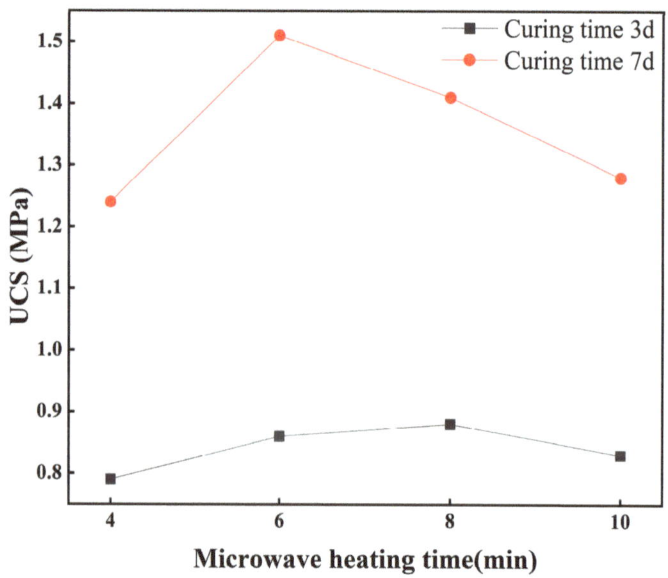

Figure 6. Effect of microwave heating time on UCS of CTB.

Figure 6 also shows that the effect of MHT on CTB cured for 7 days is different from that cured for 3 days. From 0 to 6 min, the UCS of CTB increases with longer heating time, but from 6 to 10 min, UCS decreases as MHT increases. In CTB cured for 7 days, the hydration of the binder continues, but at a slower rate compared to early-stage CTB (0–3 days). MHT for

6 min can appropriately increase the temperature for the formation of hydration products, thereby enhancing the strength of CTB. However, when the MHT exceeds 6 min, excessive heat is generated, causing thermal damage that disrupts the internal microstructure, leading to adverse effects. Based on the above results, it is concluded that microwave heating can significantly enhance UCS of early-stage CTB (0–7 days), but the MHT should not be too long (6–8 min in this study). As previously mentioned, the high early strength growth of CTB is crucial for mining operations. Therefore, it is expected that microwave technology will be practically incorporated into the mining filling schedule to enhance the early strength of CTB.

3.4.2. Microwave Delay Time

The effect of MDT on the timing of microwave heating should be another important consideration in the application of microwave technology in mining fill operations. However, previous studies [9] did not take this factor into account. Therefore, MDTs of 0, 2, 4, and 6 h were used to reveal the impact of MDT on the development of UCS in CTB. Figure 7 shows the effect of MDT on UCS of CTB samples at different curing ages (3 days and 7 days). From Figure 7, it can be noted that regardless of MDT, the higher UCS gains in CTB are associated with longer curing ages. The study also found that if the MDT is 0 h, microwave heating (for 6 min) has a significant promoting effect on the development of UCS in CTB. From Figure 7 it can be observed that the optimal MDT differs between the 3-day and 7-day curing periods. For the 3-day curing period, the UCS gradually increases as the MDT increases from 0 to 4 h; beyond 4 h, UCS begins to decrease. For the 7-day curing period, the UCS gradually increases as the MDT increases from 0 to 2 h; beyond 2 h, the UCS begins to decrease.

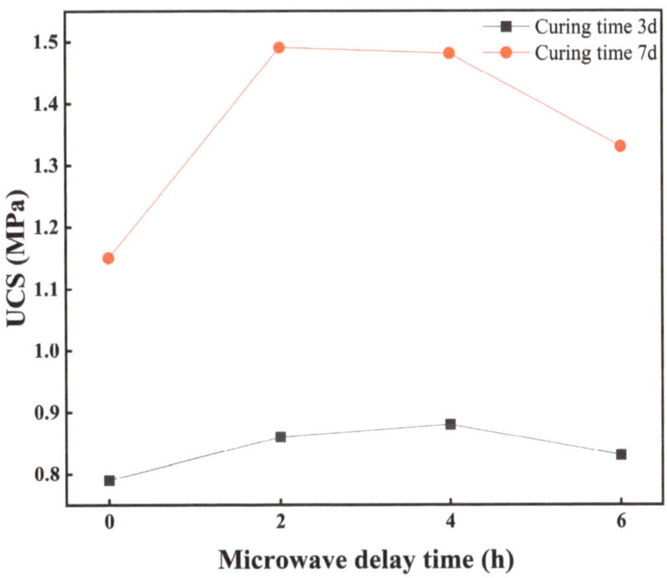

Figure 7. Effect of microwave delay time on UCS of CTB.

The positive impact of shorter MDT for microwave heating can be explained as follows: microwave radiation provides thermal energy that accelerates the early hydration reactions of the binder, producing more hydration products and thereby improving the UCS. In contrast, the negative impact of longer MDT for microwave heating on UCS of 7-day cured CTB may be due to the removal of water, a reactant in binder hydration, by microwave radiation. Consequently, fewer hydration products are formed, which reduces the later strength of CTB. Additionally, the CTB structure has already been set and hardened before microwave heating. The application of microwaves generates significant heat, which inevitably causes thermal expansion, potentially leading to cracks and weakening the structure of CTB. Therefore, the shorter the MDT, the better the mechanical properties of CTB.

3.4.3. Microwave Power

MP, which affects the temperature during microwave heating, should be the most important consideration in the application of microwave technology in mining fill operations. However, previous studies failed to consider this important factor, and the findings of this study differ from those of previous research. Therefore, MPs of 60, 180, 300, and 420 W were used to reveal the impact of MP on the development of UCS in CTB. Figure 8 shows the effect of MP on UCS of CTB samples at different curing ages (3 days and 7 days). From Figure 8, it can be noted that regardless of MP, UCS of CTB increases in both curing periods. The study also found that when the MP is 420 W, microwave heating significantly promotes the development of the UCS in the CTB. From Figure 8, it can be observed that the optimal MP is the same for both 3-day and 7-day curing periods. For a 3-day curing period, UCS of CTB gradually increases as the MP increases from 0 to 420 W. For the 7-day curing period, UCS of CTB gradually increases as the MP increases from 0 to 420 W. Since the maximum MP in this experiment is set at 420 W, and UCS of CTB samples is still increasing, optimal MP parameters above 420 W remain to be explored.

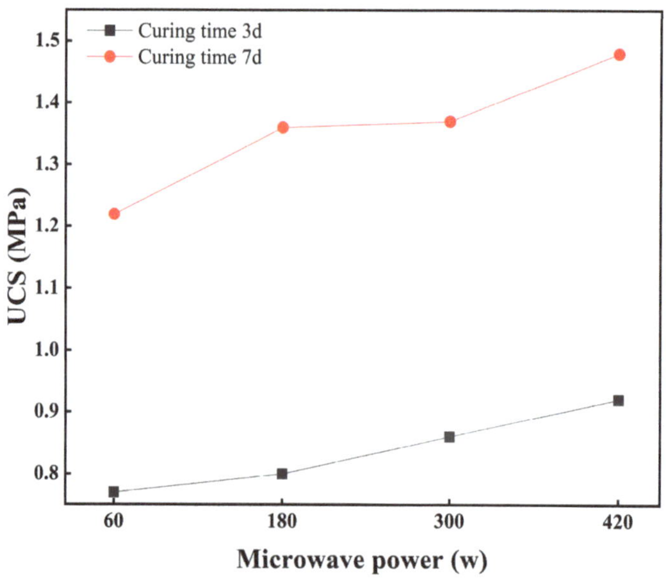

Figure 8. Effect of microwave power on UCS of CTB.

The beneficial effects of MP on UCS can be attributed to the following: microwave radiation delivers thermal energy, and the level of MP determines the amount of thermal energy provided. Increasing the MP appropriately enhances the thermal energy, producing more hydration products and thereby improving the UCS. Conversely, excessively high MP can have a negative impact, possibly due to the excessive thermal energy provided by microwave radiation, which can damage the internal structure of CTB and thereby reduce its mechanical properties. Additionally, the CTB structure has already been set and hardened before microwave heating. The application of microwaves generates significant heat, which inevitably causes thermal expansion, potentially leading to cracks and weakening the structure of CTB. Therefore, selecting the optimal MP parameters for microwave heating is crucial for enhancing the early strength of CTB.

3.5. Effect of Univariate on the Modulus of Elasticity

3.5.1. Microwave Heating Time

The elastic modulus reflects the CTB's resistance to deformation and is a key indicator of its stiffness and stress-strain relationship [12,37,38]. The influence of MHT on the modulus of elasticity of the CTB cured with different ages is illustrated in Figure 9. The figure shows that, regardless of the curing time, the elastic modulus of the CTB is significantly affected by the MHT. It is also noteworthy that, as the curing time increases, the variation in the elastic modulus differs from that observed at the 3-day curing period. In the early 3 days, the addition of microwave heating during conventional curing significantly enhances the elastic modulus. This finding is consistent with the effect of heating time on the UCS at the same age. However, for CTB cured for 7 days, excessive heating time reduces the elastic modulus, especially when the heating time exceeds 8 min.

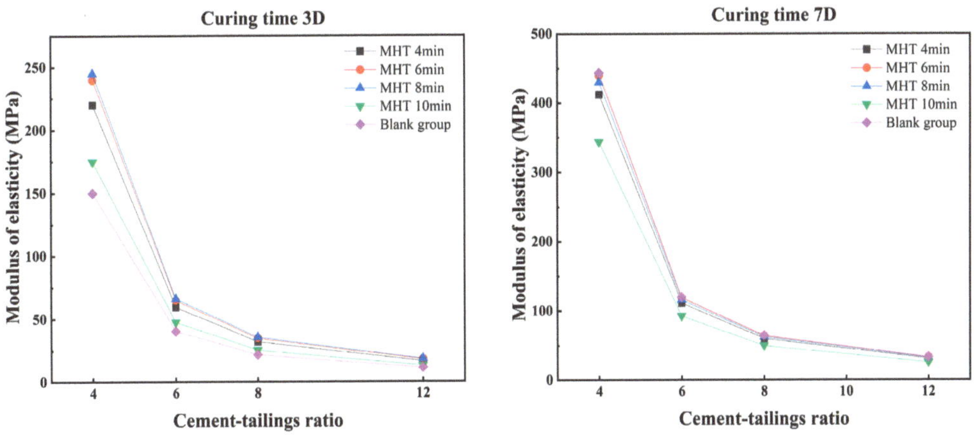

Figure 9. Effect of microwave heating time on the modulus of elasticity of CTB.

During the early 3-day curing stage, the cement hydration reaction in CTB is not yet complete, and the early strength and elastic modulus mainly rely on the formation of early hydration products and the increase in filling density. Microwave heating rapidly and uniformly increases the temperature around the cement particles, enhancing the reaction rate between cement and water, thereby accelerating the formation of early hydration products. The generated products fill the pores in the CTB, improving the structure's density and early rigidity, thus significantly enhancing the elastic modulus. On the other hand, microwave heating can reduce porosity through thermal effects, causing closer contact between particles and improving the early UCS and elastic modulus of CTB. For CTB cured for 7 days, microwave heating, especially when the heating time is too long (e.g., exceeding 8 min), can cause a decrease in elastic modulus. This may be due to excessive internal temperature from prolonged heating, resulting in thermal stress and thermal expansion effects. This uneven thermal expansion may lead to the formation and propagation of microcracks within the material, weakening the overall elastic modulus. The nonuniformity of microwave heating during the process may lead to the formation of cracks or voids in certain regions of the material, which directly affects its elastic modulus.

An increase in voids or the expansion of cracks can reduce the overall stiffness and strength of the material, thereby affecting its elastic modulus.

3.5.2. Microwave Delay Time

Figure 10 shows the influence of delay time on the modulus of elasticity of the CTB at the curing ages of 3 and 7 d. When the curing time reaches 3 days, microwave heating can only increase the elastic modulus with appropriate delay times (2 and 4 h), which differs from the previous studies [12], where the delay times were 0 and 3 h. The elastic modulus value with a delay time of 0 h is unexpectedly lower than that at 2 h. It should be noted that experimental errors may lead to unexpected results. When the curing time is extended to 7 days, adding microwave heating during conventional curing reduces the elastic modulus.

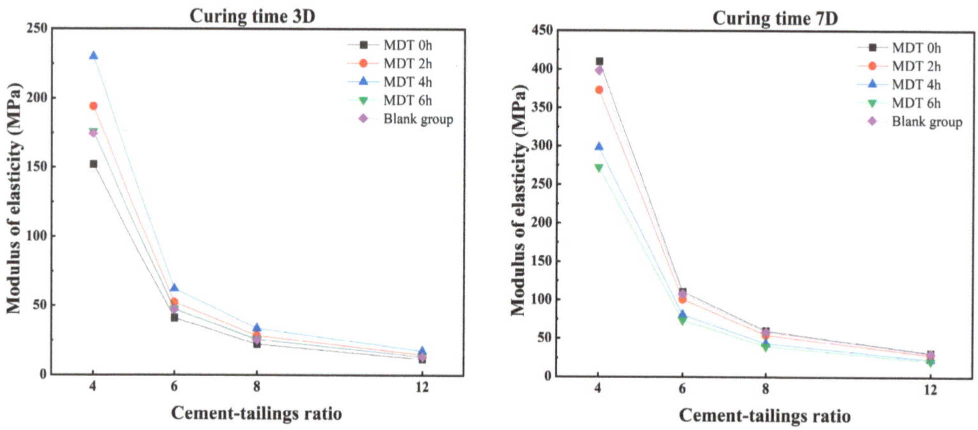

Figure 10. Effect of microwave delay time on the modulus of elasticity of CTB.

The reason for this result has been discussed in Section 3.4.1. MDT of 2 and 4 h may coincide with the efficient stage of cement hydration, promoting the densification of the cement paste and thereby increasing the elastic modulus. If the MDT is too early (e.g., 0 h), before the cement paste has fully begun hydration, microwave heating may cause rapid evaporation of moisture, inhibit the hydration reaction, and increase voids, leading to a reduction in the elastic modulus. On the other hand, during the 7-day curing process, cement hydration has reached a more mature stage, and any rapid thermal expansion or uneven heating may induce cracks or voids within the material, leading to a reduction in the overall elastic modulus. Microwave heating may also affect the interfaces within the material, such as the bonding or interaction forces between particles. When the material is heated, the interactions between particles or molecules may change, which in turn affects the elastic modulus.

3.5.3. Microwave Power

The effect of MP on the elastic modulus at different curing ages is shown in Figure 11. Similarly, a previous study [12] did not consider the impact of MP, the most important microwave parameter, on the elastic modulus of CTB. It can be observed from Figure 11 that regardless of curing time, an increase in MP positively influences the elastic modulus. It is also noteworthy that as the curing time extends to seven days, the variation in elastic modulus is reduced compared to that observed during the 3-day curing period. During the early 3-day curing period, incorporating microwave heating into conventional curing significantly enhances the elastic modulus, with the highest value observed at a power of

420 W. For CTB cured for 7 days, microwave heating also improves the elastic modulus, particularly when the microwave power exceeds 180 W.

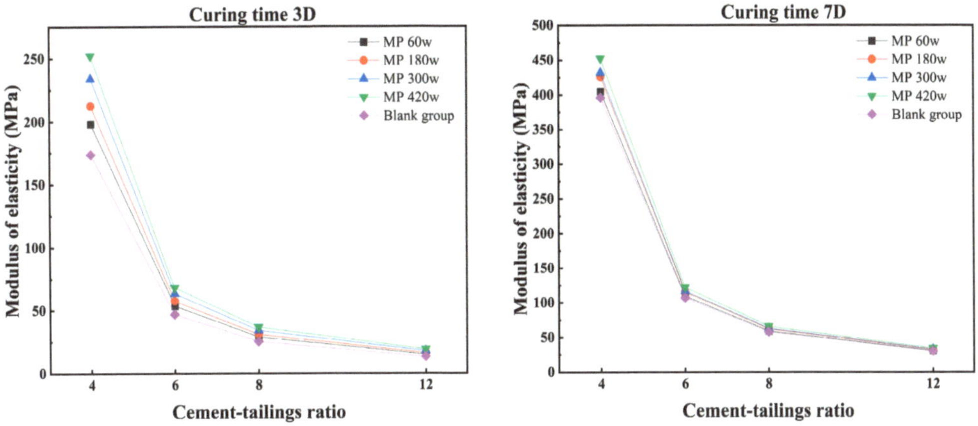

Figure 11. Effect of microwave power on the modulus of elasticity of CTB.

When the MP is moderate, the hydration reaction of the cement particles becomes more uniform and rapid, thereby improving the mechanical properties of the material. Microwave heating can transfer heat more uniformly, avoiding issues such as local overheating or uneven heating that may occur with traditional methods, thereby improving the material's uniformity and further enhancing its elastic modulus. By the seventh day, the cement hydration is nearly complete, but appropriate MP can still accelerate the reaction of partially unhydrated cement particles, continuing the formation of hydration products and further strengthening the structure of CTB. Higher MP (e.g., above 300 W) can promote the later-stage hydration reaction in a shorter period, allowing for a more complete formation of hydration products, which enhances the strength and elastic modulus.

4. Conclusions

This experimental study explored the use of microwave radiation to boost the early strength of CTB. The study specifically examined how microwave heating influences the mechanical properties of the CTB. The findings lead to the following conclusions.

1. Internal factors of CTB have a greater impact on its strength than microwave heating does. However, selecting appropriate MHT and MP, followed by conventional curing, helps improve the early mechanical properties of the CTB. Within an appropriate time frame, extending the MHT can accelerate the hydration rate of the cementitious materials, producing more hydration products, which thus contribute to higher UCS and elasticity modulus in the CTB. However, if the MHT is too long, thermal damage can negatively affect the mechanical properties of the CTB.
2. The MDT is also crucial, hence the consideration of MDT between placing the CTB and microwave heating. Shorter microwave delay times are associated with better mechanical properties in the early-stage CTB. Microwave radiation significantly affects the strength and elasticity modulus of the CTB.
3. An appropriate microwave radiational setting can be established: 420 W of MP, 6 min of MHT, and an MDT of 4 h. The optimal MP parameters above 420 W remain to be explored. The mechanism by which microwave radiation enhances the CTB can be explained through studies of its microstructure.

4. The enhancement of early strength in CTB through microwave treatment has significant practical implications in mining applications. In terms of economic feasibility, although the initial investment in microwave treatment equipment is high, its advantages—such as improved tailings recovery, reduced operational costs, and energy savings—can lead to significant economic benefits in the long run.

Author Contributions: Conceptualization, X.C.; methodology, X.C.; software, X.C.; validation, X.C.; K.Z. and Z.P.; formal analysis, X.C.; investigation, X.C.; resources, X.C.; data curation, X.C.; writing—original draft preparation, X.C.; writing—review and editing, Z.P.; visualization, X.C.; supervision, Z.P.; project administration, K.Z.; funding acquisition, X.C. All authors have read and agreed to the published version of the manuscript.

Funding: This study was funded by the National Natural Science Foundation of China (Project No. 52204167), the Natural Science Foundation of Hunan Province (Project No. 2024JJ6504), and the Fundamental Research Funds for the Central Universities of Central South University (Project No. 2023ZZTS0715).

Data Availability Statement: Data are contained within the article.

Conflicts of Interest: The authors declare that they have no known competing financial interests or personal relationships that could have appeared to influence the work reported in this paper.

References

1. Jin, A.-b.; Wang, J.; Chen, S.-j.; Li, H. Strength and damage characteristics of tailings filling body with different particle size distributions. *Rock Soil Mech.* **2022**, *43*, 3083–3093.
2. Zhang, Q.; Yin, G.; Fan, X.; Li, X.; Wang, W.; Geng, W.; Liu, H. A damage constitutive model of tailings based on the analysis of elastic-plastic and sliding of skeleton grains. *Disaster Adv.* **2012**, *5*, 730–735.
3. Hou, Y.; Kong, P.; Yang, K.; Yin, S.; Yu, X.; Wang, Y.; Kou, P. Enhancing performance and structural integrity of cemented paste backfill: Rheological behavior, strength characteristics, and microstructural dynamics. *Case Stud. Constr. Mat.* **2024**, *20*, e03367. [CrossRef]
4. He, Y.; Chen, Q.; Qi, C.; Zhang, Q.; Xiao, C. Lithium slag and fly ash-based binder for cemented fine tailings backfill. *J. Environ. Manag.* **2019**, *248*, 109282. [CrossRef]
5. Chen, Q.; Chen, H.; Wang, P.; Chen, X.; Chen, J. Effect of Sulphur-Containing Tailings Content and Curing Temperature on the Properties of M32.5 Cement Mortar. *Materials* **2021**, *14*, 5751. [CrossRef]
6. Chindaprasirt, P.; Rattanasak, U.; Taebuanhuad, S. Role of microwave radiation in curing the fly ash geopolymer. *Adv. Powder Technol.* **2013**, *24*, 703–707. [CrossRef]
7. Fall, M.; Benzaazoua, M. Modeling the effect of sulphate on strength development of paste backfill and binder mixture optimization. *Cem. Concr. Res.* **2005**, *35*, 301–314. [CrossRef]
8. Fall, M.; Samb, S.S. Effect of high temperature on strength and microstructural properties of cemented paste backfill. *Fire Saf. J.* **2009**, *44*, 642–651. [CrossRef]
9. Ding, P.; Yan, S.; Guo, Q.; Chang, L.; Li, Z.; Zhou, C.; Han, D.; Yang, J. Study on the Effects of Microwave Heating Time and Power on the Mechanical Properties of Cemented Tailings Backfill. *Minerals* **2024**, *14*, 944. [CrossRef]
10. Kong, Y.; Wang, P.; Liu, S. Microwave pre-curing of Portland cement-steel slag powder composite for its hydration properties. *Constr. Build. Mater.* **2018**, *189*, 1093–1104. [CrossRef]
11. Tang, T.; Cai, L.; You, K.; Liu, M.; Han, W. Effect of microwave pre-curing technology on carbide slag-fly ash autoclaved aerated concrete (CS-FA AAC): Porosity rough body formation, pore characteristics and hydration products. *Constr. Build. Mater.* **2020**, *263*, 120112. [CrossRef]
12. Wu, D.; Sun, W.; Liu, S.; Qu, C. Effect of microwave heating on thermo-mechanical behavior of cemented tailings backfill. *Constr. Build. Mater.* **2021**, *266*, 121180. [CrossRef]
13. Wei, W.; Shao, Z.; Zhang, P.; Chen, W.; Qiao, R.; Yuan, Y. Experimental assessment of microwave heating assisted aggregate recycling from dried and saturated concrete. *Mater. Struct.* **2021**, *54*, 142. [CrossRef]
14. Wang, Y.; Luo, S.; Yang, L.; Ding, Y. Microwave curing cement-fly ash blended paste. *Constr. Build. Mater.* **2021**, *282*, 122685. [CrossRef]
15. Zhou, F.; Pan, G.; Meng, H.; Mi, R. Effect of secondary curing on the performance of microwave cured concrete. *Constr. Build. Mater.* **2022**, *330*, 127256. [CrossRef]

16. Wu, D.; Zhao, P.; Cheng, W.; Hao, Z.; Zhang, Y. Effect of microwave on coupled rheological and mechanical properties of cemented tailings backfill. *Constr. Build. Mater.* **2024**, *441*, 137558. [CrossRef]
17. Haddad, R. Characterization and Deterioration Detection of Portland Cement Concrete Using Electromagnetic Waves over a Wideband of Frequency. 1996. Available online: https://vtechworks.lib.vt.edu/items/984c4c4b-1763-4193-83b5-2ca5ccab357c (accessed on 1 January 2025).
18. Somaratna, J.; Ravikumar, D.; Neithalath, N. Response of alkali activated fly ash mortars to microwave curing. *Cem. Concr. Res.* **2010**, *40*, 1688–1696. [CrossRef]
19. Xu, H.; Jaynes, J.; Ding, X. Combining Two-Level and Three-Level Orthogonal Arrays for Factor Screening and Response Surface Exploration. *Stat. Sin.* **2014**, *24*, 269–289. [CrossRef]
20. Voelkel, J.G.; Gallagher, K.P. The design and analysis of order-of-addition experiments: An introduction and case study. *Qual. Eng.* **2019**, *31*, 627–638. [CrossRef]
21. Georgiou, S.D. Orthogonal designs for computer experiments. *J. Stat. Plann. Inference* **2011**, *141*, 1519–1525. [CrossRef]
22. Georgiou, S.D.; Stylianou, S.; Drosou, K.; Koukouvinos, C. Construction of orthogonal and nearly orthogonal designs for computer experiments. *Biometrika* **2014**, *101*, 741–747. [CrossRef]
23. Huynh, T.V. Orthogonal Array Experiment in Systems Engineering and Architecting. *Syst. Eng.* **2011**, *14*, 208–222. [CrossRef]
24. Simonsen, A.M.T.; Solismaa, S.; Hansen, H.K.; Jensen, P.E. Evaluation of mine tailings' potential as supplementary cementitious materials based on chemical, mineralogical and physical characteristics. *Waste Manag.* **2020**, *102*, 710–721. [CrossRef]
25. Zhu, P.; Wang, L.; Hong, D.; Zhou, M. A Study of Cordierite Ceramics Synthesis From Serpentine Tailing and Kaolin Tailing. *Sci. Sinter.* **2012**, *44*, 129–134. [CrossRef]
26. Lei, G.; Guo, Y.G.; Zhu, J.G.; Chen, X.M.; Xu, W.; Shao, K.R. Sequential Subspace Optimization Method for Electromagnetic Devices Design with Orthogonal Design Technique. *IEEE Trans. Magn.* **2012**, *48*, 479–482. [CrossRef]
27. Cao, S.; Yilmaz, E.; Song, W. Fiber type effect on strength, toughness and microstructure of early age cemented tailings backfill. *Constr. Build. Mater.* **2019**, *223*, 44–54. [CrossRef]
28. Jiang, H.; Qi, Z.; Yilmaz, E.; Han, J.; Qiu, J.; Dong, C. Effectiveness of alkali-activated slag as alternative binder on workability and early age compressive strength of cemented paste backfills. *Constr. Build. Mater.* **2019**, *218*, 689–700. [CrossRef]
29. Kou, Y.; Deng, Y.; Tan, Y.; Han, C.; Song, W. Hydration Characteristics and Early Strength Evolution of Classified Fine Tailings Cemented Backfill. *Materials* **2023**, *16*, 963. [CrossRef]
30. Wang, Y.; Wu, J.; Ma, D.; Pu, H.; Yin, Q.; Chen, W. Study on macro-meso mechanical properties of cemented tailings backfill with high fly ash content. *Environ. Sci. Pollut. Res.* **2023**, *30*, 2904–2917. [CrossRef]
31. Qiu, H.; Zhang, F.; Sun, W.; Liu, L.; Zhao, Y.; Huan, C. Experimental Study on Strength and Permeability Characteristics of Cemented Rock-Tailings Backfill. *Front. Earth Sci.* **2022**, *10*, 802818. [CrossRef]
32. Yin, Z.; Li, R.; Lin, H.; Chen, Y.; Wang, Y.; Zhao, Y. Analysis of Influencing Factors of Cementitious Material Properties of Lead–Zinc Tailings Based on Orthogonal Tests. *Materials* **2023**, *16*, 361. [CrossRef]
33. Cong, M.; Zhang, S.; Sun, D.; Zhou, K. Optimization of preparation of foamed concrete based on orthogonal experiment and range analysis. *Front. Mater.* **2021**, *8*, 778173. [CrossRef]
34. Yin, S.; Hou, Y.; Yang, S.; Chen, X. Study on static and dynamic flocculation settlement characteristics of fine tailings slurry and influence of flocculant on strength of fine tailings backfill. *Case Stud. Constr. Mater.* **2022**, *17*, e01525. [CrossRef]
35. Makul, N.; Rattanadecho, P.; Agrawal, D.K. Applications of microwave energy in cement and concrete—A review. *Renew. Sustain. Energy Rev.* **2014**, *37*, 715–733. [CrossRef]
36. Sun, J.; Wang, W.; Yue, Q. Review on Microwave-Matter Interaction Fundamentals and Efficient Microwave-Associated Heating Strategies. *Materials* **2016**, *9*, 231. [CrossRef] [PubMed]
37. Pan, Z.; Zhou, K.; Wang, Y.; Lin, Y.; Saleem, F. Comparative Analysis of Strength and Deformation Behavior of Cemented Tailings Backfill under Curing Temperature Effect. *Materials* **2022**, *15*, 3491. [CrossRef]
38. Chen, S.; Yilmaz, E.; Xiang, Z.; Wang, Y. Curing conditions effect on pore structure, compressive strength and elastic modulus of cementitious tailings backfills. *Powder Technol.* **2023**, *422*, 118458. [CrossRef]

minerals

Article

Optimization Model for Mine Backfill Scheduling Under Multi-Resource Constraints

Yuhang Liu [1,2], Guoqing Li [1,*], Jie Hou [1], Chunchao Fan [1,3], Chuan Tong [1,3] and Panzhi Wang [3]

1 School of Civil and Resource Engineering, University of Science and Technology Beijing, Beijing 100083, China
2 Department of Civil and Environmental Engineering, National University of Singapore, Singapore 119077, Singapore
3 Jiaojia Gold Mine, Shandong Gold Mining (Laizhou) Company Limited, Yantai 261441, China
* Correspondence: qqlee@ustb.edu.cn

Abstract: Addressing the resource constraints, such as manpower and equipment, faced by mine backfilling operations, this study proposed an optimization model for backfill scheduling based on the Resource-Constrained Project Scheduling Problem (RCPSP). The model considered backfilling's multi-process, multi-task, and multi-resource characteristics, aiming to minimize total delay time. Constraints included operational limits, resource requirements, and availability. The goal was to determine optimal resource configurations for each stope's backfilling steps. A heuristic genetic algorithm (GA) was employed for solution. To handle equipment unavailability, a new encoding/decoding algorithm ensured resource availability and continuous operations. Case verification using real mine data highlights the advantages of the model, showing a 20.6% decrease in completion time, an 8 percentage point improvement in resource utilization, and a 47.4% reduction in overall backfilling delay time compared to traditional methods. This work provides a reference for backfilling scheduling in similar mines and promotes intelligent mining practices.

Keywords: RCPSP; mine backfilling; GA; scheduling optimization; intelligent mine

check for updates

Citation: Liu, Y.; Li, G.; Hou, J.; Fan, C.; Tong, C.; Wang, P. Optimization Model for Mine Backfill Scheduling Under Multi-Resource Constraints. *Minerals* **2024**, *14*, 1183. https://doi.org/10.3390/min14121183

Academic Editor: Bekir Genc

Received: 31 October 2024
Revised: 18 November 2024
Accepted: 19 November 2024
Published: 21 November 2024

1. Introduction

Backfilling allows for substantial disposal of mine waste into stopes, contributing significantly to safety, environmental protection, cost efficiency, and operational effectiveness [1–4]. As a crucial method for achieving green mining, zero-tailing operations, and sustainable resource extraction, backfill mining is progressively becoming the preferred practice in underground mining [5–8].

The operational efficiency of backfill mining largely depends on the coordinated management of multiple stopes, production stages, and resource allocations. Current scheduling and management techniques are typically inflexible, leading to lower resource utilization and production efficiency. Mines employing the backfill mining method handle several backfill tasks in numerous stopes on monthly and weekly schedules [9–11], with each stope requiring several sequential backfill steps involving multiple resources [12]. The backfill scheduling problem is characterized by high flexibility, allowing each stage to select among various parallel resources [13]. Enhancing backfill efficiency requires optimized resource coordination across projects, stages, and constraints. However, the core resource—the sand silo—is commonly managed in an "one operational, one standby" configuration, which imposes a rigid and inefficient framework.

In the era of Mining 4.0 and the development of intelligent mines, significant advances in backfill technologies have been achieved worldwide [14]. Increased investment has led to improved automation, enabling real-time adjustments of slurry concentration and composition as well as centralized control of material feeding, slurry mixing, and transportation processes [15–17]. Nonetheless, substantial opportunities remain to enhance the allocation of key resources, such as manpower, equipment, and materials, to further optimize backfill operations.

These challenges are fundamentally optimization problems. Applying advanced scheduling optimization tools to backfill operations presents a promising research direction. Optimized workflows enhance control over the entire production process [18,19], aligning production with higher-level schedules and facilitating efficient resource allocation and task management.

A substantial body of research in production scheduling optimization, particularly in the manufacturing sector, centers on the Job Shop Scheduling Problem (JSP) [20]. The JSP, a foundational challenge in production scheduling, was introduced by Johnson in 1954 [21]. Later, in 1990, Bruker and Schlie [22] expanded the JSP to the Flexible Job Shop Scheduling Problem (FJSP), broadening equipment options from single to multiple machines, which enhanced flexibility and complexity, thus extending its applicability. Gong [23] further incorporated workforce flexibility into the FJSP framework, enabling improved human resource allocation. Key optimization metrics in FJSP include makespan, delivery delays, machine load, and energy consumption [24–26]. Additionally, preliminary studies have explored the application of these scheduling principles to mine backfill operations [27,28].

In model construction, certain assumptions are typically applied, with most models presuming continuous availability of manpower, materials, and equipment. However, real-world production environments are dynamic, complex, and subject to uncertainty, with frequent instances of equipment unavailability [29–32]. The Resource-Constrained Project Scheduling Problem (RCPSP), first introduced by Johnson and Russell in 1967 [33], addresses resource availability limitations by optimizing project planning and scheduling under resource constraints. RCPSP is widely used to optimize scheduling in vehicle fleets and production lines. In backfill operations, limitations stem from factors such as manpower constraints, equipment reliability issues, and unavailability due to sand silo loading and slurry preparation requirements, as well as equipment occupancy. Adapting RCPSP to the backfill context enables rational scheduling of backfill tasks according to equipment status, facilitating planned maintenance, optimizing timing for sand loading and unloading, and reducing extended delays.

In summary, while the adoption of backfill mining continues to expand, challenges related to rigid resource management and suboptimal utilization persist. Although automation and centralized control have advanced as part of intelligent mining development, these improvements have yet to fully inform scheduling decision-making. To address these issues, this study integrates resource constraints into a backfill scheduling optimization model, accounting for task parallelism and complex workflows. By systematically scheduling tasks with respect to both time and resource allocation, the model aims to fulfill target objectives while maximizing resource utilization.

The methodology of this study includes: first, identifying resource requirements for each backfill step and determining possible instances of resource unavailability; next, establishing model assumptions, defining input and output data, and setting model parameters, objectives, and constraints; subsequently, developing an algorithmic solution; and finally, validating the model and algorithm using actual mine data and analyzing scheduling results to evaluate the model's effectiveness.

2. Constraint Analysis

Backfill resources are categorized into three types: manpower, equipment, and materials. Among these, materials include tailings and cementitious components. Metal mines generate substantial amounts of tailings daily from mineral processing, with less than half typically allocated for backfilling under normal conditions [34]. The supply of cementitious materials is highly accessible, ensuring an adequate supply at any given time. Therefore, this study does not consider constraints on material resources.

Both manpower and equipment resources face limitations in production. The number of stopes that a backfill team can service simultaneously is capped, requiring the model to account for constraints on manpower occupancy.

Equipment encompasses various machinery used in backfilling operations. Equipment reliability is a critical factor for resource availability. Maintenance needs or equipment malfunctions can lead to unavailability and delays in tasks. Reliability degradation of key equipment components can impact the quality, speed, or energy consumption of backfilling. Therefore, the primary resource constraints in backfill operations include occupancy constraints, sandloading constraints, and reliability constraints.

2.1. Equipment Reliability Constraints

Analyzing equipment reliability allows for the quantification of its capability to perform designated functions, enabling the development of preventive maintenance plans. This approach also optimizes the allocation of available resources, ensuring the smooth progression of backfill operations.

If the failure behavior of backfill equipment follows a Weibull distribution, its probability density function is given by Equation (1), where t denotes the equipment's service age, and β, η and r are the shape, scale, and location parameters. These parameters influence the curve's shape, horizontal displacement, and scale. The cumulative distribution function (unreliability function) and reliability function are provided in Equations (2) and (3), respectively.

$$f(t) = \frac{\beta}{\eta}\left(\frac{t-r}{\eta}\right)^{\beta-1} e^{-(\frac{t-r}{\eta})^{\beta}} \tag{1}$$

$$F(t) = \int_0^t f(t)dt = 1 - e^{-(\frac{t-r}{\eta})^{\beta}} \tag{2}$$

$$R(t) = 1 - F(t) = e^{-(\frac{t-r}{\eta})^{\beta}} \tag{3}$$

When $\beta < 1$, $\beta = 1$ and $\beta > 1$, these values respectively describe the early failure phase, random failure phase, and wear-out phase of the equipment, corresponding to the left, bottom, and right segments of the failure probability density curve, also known as the bathtub curve. The parameter η represents the characteristic life of the equipment, while $r > 0$ indicates the time at which failures begin to occur, representing the minimum service age.

Figure 1 illustrates the reliability variation of backfill equipment, where R represents equipment reliability. Equipment reliability decreases with increasing service age. Prior to each backfill task, the system evaluates whether the equipment's reliability at the task's completion would fall below the critical threshold RT. If the predicted reliability is below RT, maintenance is initiated. Complete maintenance restores the equipment to its original condition ("as good as new"), while partial maintenance improves reliability to an initial level, R^{ini}.

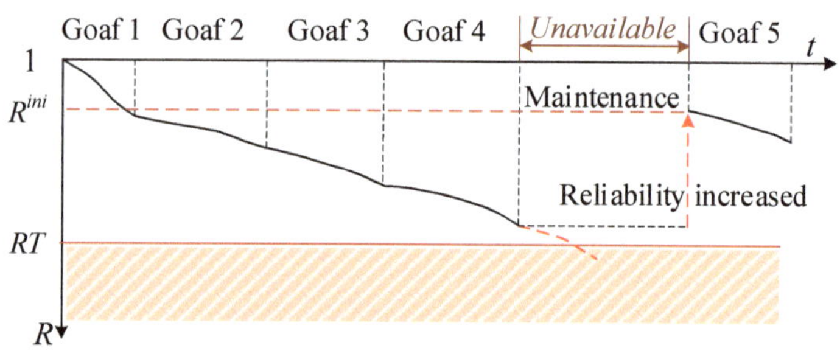

Figure 1. Schematic Diagram of Equipment Reliability and Availability.

2.2. Characteristics of Sand Silo Feeding and Slurry Preparation

The sand silo, a core component of the backfill system, thickens low-concentration tailings (about 20%, mass concentration) from the processing plant to a high concentration (around 80%, mass concentration) for use in mine backfilling after activation and slurry preparation [35].

Apart from reliability-related failures, the sand silo is also unavailable for slurry preparation at certain times due to its operational characteristics. Before discharging slurry, the silo must first undergo feeding and settling processes, during which it cannot support any backfilling operations. This is the primary reason mines adopt a "one operational, one standby" mode.

The sand silo feeding and settling process is illustrated in Figure 2. The red lettering represent the process of feeding, sedimentation and filling the sand. Initially, low-concentration tailings slurry is pumped into the empty silo. After a period of settling, the slurry separates into two layers: a supernatant on top and high -concentration tailings at the bottom. Subsequently, more low-concentration slurry is pumped into the silo from the bottom, pushing the supernatant out of the silo. Overflow of the supernatant usually carries a small amount of tailings. This feeding and settling cycle is repeated as needed until the silo is nearly filled with high-concentration tailings, depending on operational requirements.

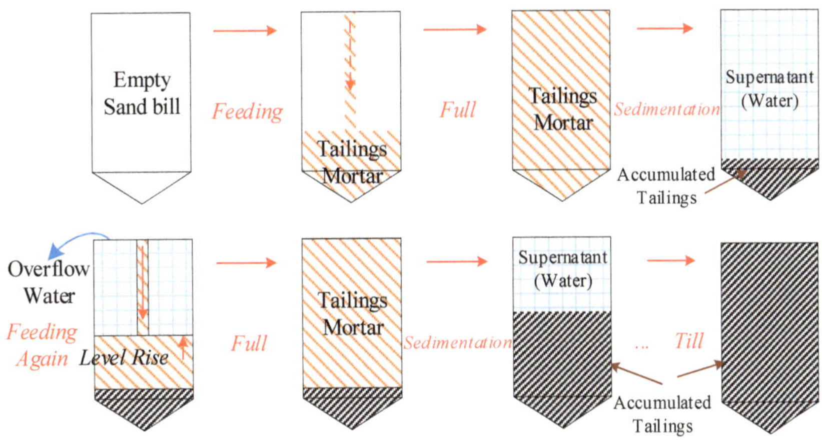

Figure 2. Schematic Diagram of Vertical Sand Silo Feeding and Settling.

The changes in tailings mass within the sand silo during normal operation are illustrated in Figure 3. Initially, the silo is filled with high-concentration tailings, with a mass of *MA*. As it supports backfilling tasks, the tailings mass inside gradually decreases. When the remaining tailings are deemed insufficient to meet backfilling needs, as stated in the red section, the feeding process is initiated. This process consists of multiple cycles of tailings feeding and settling, leading to intermittent increases in mass within the silo. During these cycles, the silo is unavailable for slurry preparation or backfilling, as stated in the brown section. Once feeding is completed, the silo can again serve backfilling tasks.

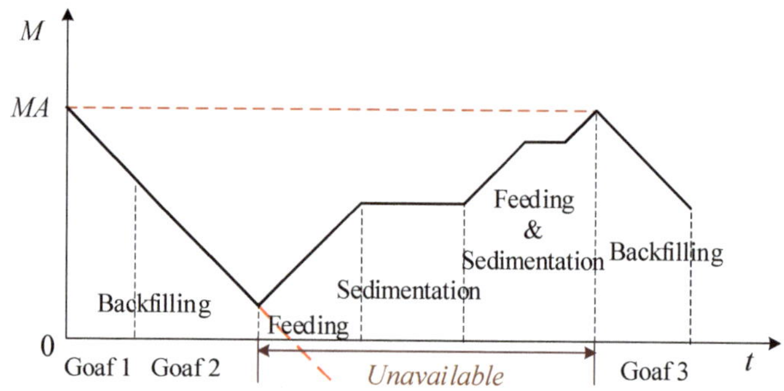

Figure 3. Variation of Tailings Mass and Availability in the Sand Silo.

2.3. Resource Occupancy and Delay

An essential factor in optimizing scheduling is managing resource occupancy. When conflicts arise in equipment allocation, this study prioritizes assigning equipment to the stope ready for filling first. Stopes are sorted by allowed start time, with resources allocated to earlier tasks, while later tasks are scheduled at the next available time slot.

Delays in stope backfilling may result from factors such as equipment reliability issues, sand silo refilling, or conflicting equipment usage schedules. Once backfilling starts, each step must proceed without interruption to prevent incidents such as pipe blockages or bursts caused by slurry solidification. If a delay is anticipated in any step, the start time for backfilling that stope is adjusted, and resources are rescheduled to ensure all required resources are available for each step throughout the process.

2.4. Backfilling Process and Resource Constraints

This study focuses on equipment reliability and preventive maintenance, as well as unavailability due to sand silo feeding and resource occupancy. The resources required for each step in the backfilling process, along with their respective constraints, are shown in Table 1. The backfilling operation consists of four steps: preparation, water flow induction, layered backfilling, and pipeline cleaning. Resources include manpower, water storage, slurry mixing equipment, cementitious material management equipment, and conveying equipment. The abbreviations in the table are explained below. A slash indicates that the process does not require the corresponding resource, so resource constraints are not considered.

Table 1. Resource Constraints for Each Step in the Backfilling Process.

Operation Step	Manpower	Water Storage	Slurry Mixing Equipment	Cementitious Material Management Equipment	Conveying Equipment
Preparation	O	\	\	\	\
Water Flow Induction	O	O, R	\	\	O, R
Layered Backfilling	O	O, R	O, R, S	O, R	O, R
Pipeline Cleaning	O	O, R	\	\	O, R

Abbreviations: O: Occupancy Constraint, R: Reliability Constraint, S: Sand Silo Feeding Constraint, \: Not applicable.

3. Optimization Model Construction

3.1. Model Assumptions, Inputs, and Outputs

The mine backfill scheduling optimization problem under resource constraints can be described as follows: suppose there are A levels in the mining area where mining and backfilling operations are being conducted, with each level containing B_a stopes. Each stope requires C steps. The mining area has D types of resources, with each step requiring at least one type, and the total available resources amount to E_d. Each task has different resource options, leading to variations in resource occupancy and maintenance, which in turn result in differing delay times.

The model assumptions are as follows:

1. The preceding process for backfilling is mining, and backfilling can only begin once a stope has been mined out. Consequently, each stope has an earliest possible start time for backfilling.
2. At time 0, all slurry mixing equipment is fully stocked with tailings.
3. At any given time, each resource can only serve one backfilling task.
4. Each step may require multiple types of resources, but each step is executed only once and utilizes only one resource from each required type.
5. Backfilling within a stope must proceed continuously; if any resource becomes unavailable, all subsequent steps are delayed.
6. Preparation times between different tasks for the same resource are not considered.
7. The duration of a given backfill task is consistent across all resources.

3.2. Symbols and Definitions

Based on the above assumptions, the symbols used in this model are defined as follows:

Subscripts

- t, Scheduling time index, $t \in (1,2,\ldots,T)$, where T is the total scheduling period?
- a, Level index, $a \in (1,2,\ldots,A)$, where A is the total number of levels in the mining area.
- b, Stope index, $b \in (1,2,\ldots,B_a)$, where B_a is the total number of stopes to be backfilled.
- c, Step index, $c \in (1,2,\ldots,C)$, where C is the maximum number of steps.
- d, Resource type index, $d \in (1,2,\ldots,D)$, where D is the total number of resource types.
- e, Resource unit index, $e \in (1,2,\ldots,E_d)$, where E_d is the total number of units for a specific type of resource.

Parameters

- $BF_{a,b}$, Earliest feasible backfilling time, in hours (h)
- $DT_{a,b,c,d,e}$, Duration of the backfilling task, in hours (h)
- $TB_{a,b,c}$, Tailings consumption for backfilling, in tonnes (t)
- MA_e, Total capacity of tailings held by the slurry mixing equipment, in tonnes (t)
- M_e^t, Mass of tailings in the slurry mixing equipment at time t, in tonnes (t)
- CM_e, Critical mass of tailings in the slurry mixing equipment, in tonnes (t); below this value, additional feeding is required.
- FT_e, Time required for tailings feeding, in hours (h)
- $RT_{d,e}$, Reliability threshold for equipment; maintenance is required if reliability falls below this level
- $MDT_{d,e}$, Duration of equipment maintenance, in hours (h)
- $R_{d,e}^t$, Equipment reliability at time t
- SeT, Equipment service age post-maintenance, in hours (h)
- PN_a, Number of pipelines leading to level a
- FS_0, Number of primary units for backfilling operations
- N, A large positive integer
- $DeT_{a,b}$, Backfilling delay time, in hours (h)

Decision Variables

- $X_{a,b,c,d,e}^t$, An integer variable, equal to 1 if a resource of type d and unit e is used for the c-th step in stope b at level a at time t; otherwise, it is 0.
- $Y_{d,e}^t$, An integer variable, equal to 1 if the d-th resource of type e is available at time t; otherwise, it is 0.
- $Z_{a,b,c}^t$, An integer variable, equal to 1 if the backfilling task begins at time t; otherwise, it is 0.

Auxiliary Decision Variables

- $RA_{a,b,c,d}$, An integer variable, equal to 1 if a resource of type d can serve the c-th step in stope b at level a; otherwise, it is 0.
- $ST^0_{a,b,c}$, Expected start time for the backfilling in stope b at level a in hours (h).
- $ST_{a,b,c}$, Actual start time for the c-th step in stope b at level a in hours (h).
- $ET_{a,b,c}$, Expected end time for the c-th step in stope b at level a in hours (h).
- $ET_{a,b,c}$, Actual end time for the c-th step in stope b at level a in hours (h).
- $MS^t_{d,e}$, An integer variable, equal to 1 if the backfilling equipment is under maintenance at time t; otherwise, it is 0.
- S^t_e, An integer variable, equal to 1 if the slurry mixing equipment is undergoing feeding at time t; otherwise, it is 0.
- *delay, delay_m, delay_t, delay_on*, represent total delay time, delay time caused by resource maintenance, sand silo feeding, and equipment occupancy, respectively.

3.3. Optimization Objective

The goal of backfill scheduling is to organize manpower, equipment, and other resources to carry out operations as soon as the backfilling conditions are met in the stope [36]. The benefits are:

1. Minimizing backfill delay time, as the backfilling time for the goaf remains unchanged, the total completion time will be minimized, leading to higher equipment utilization.
2. In the coordinated mining and backfilling mode, timely backfilling enables earlier commencement of the next mining stage, improving overall production efficiency.
3. Prompt backfilling can prevent long-term exposure of the goaf, reducing the risk of roof collapse and side wall failure incidents.

Thus, the optimization goal of the model (Equation (4)) is to minimize the total delay time in backfilling, thereby maximizing production continuity, reducing stope exposure time, and enhancing safety in underground mining.

$$F(x) = min\sum_{a=1}^{A}\sum_{b=1}^{B}(ET_{a,b,C} - BF_{a,b}) = min\sum_{a=1}^{A}\sum_{b=1}^{B}(DeT_{a,b}) \quad (4)$$

3.4. Constraints

The model's constraints are divided into two categories: backfilling operation constraints and resource availability constraints.

3.4.1. Backfilling Operation Constraints

The following business constraints are implemented to ensure the effective and efficient scheduling of the backfilling process. These constraints govern the logical flow, resource allocation, and timing of various tasks involved in backfilling operations.

1. Each backfilling step c in stope b at level a is performed only once:

$$\sum_{t=1}^{T} Z^t_{a,b,c} = 1, \forall a, b, c \quad (5)$$

2. Only one resource of each type is selected to serve a specific backfilling task:

$$\sum_{t=1}^{T}\sum_{e=1}^{E} X^t_{a,b,c,d,e} = RA_{a,b,c,d}, \forall a, b, c, d \quad (6)$$

3. Earliest feasible backfill time constraint:

$$ST_{a,b,1} \geq BF_{a,b}, \forall a, b \quad (7)$$

4. Continuity constraint for backfilling steps, where the start time of the next step is equal to the end time of the current step. When applying this constraint, it is necessary to account for steps that require multiple resource types:

$$ET_{a,b,c} = ST_{a,b,c} + \frac{\sum_{d=1}^{D}\sum_{e=1}^{E}\left(DT_{a,b,c,d,e} \cdot \sum_{t=1}^{T} X^{t}_{a,b,c,d,e}\right)}{\sum_{d=1}^{D}\sum_{e=1}^{E}\sum_{t=1}^{T} X^{t}_{a,b,c,d,e}} \tag{8}$$

where:

$$\sum_{d=1}^{D}\sum_{e=1}^{E}\sum_{t=1}^{T} X^{t}_{a,b,c,d,e} \geq 1ST_{a,b,c+1} = ET_{a,b,c}, \forall a,b,c \in [1, C-1] \tag{9}$$

3.4.2. Resource Availability Constraints

Equations (11) and (12) correspond to the equipment reliability constraints and maintenance strategies described in Section 2.1. Equations (13) and (14) pertain to the sand silo tailings level assessment and feeding/discharging operational mode discussed in Section 2.2. Equations (10) and (15) correspond to the equipment occupancy constraints described in Section 2.3.

5. Pipeline Occupancy Constraint: The number of stopes being backfilled at level a and below must not exceed the number of pipelines leading to level a.

$$\sum_{i=a}^{A}\sum_{b=1}^{B}\sum_{c=1}^{C} Z^{t}_{a,b,c} \leq PN_{a}, \forall t \tag{10}$$

6. Equipment Reliability Constraint: Before a task begins, the equipment's reliability at the task's end must meet the minimum requirement, as indicated by Equation (11).

$$R_{d,e}^{ET_{a,b,c}} \geq RT_{d,e}, \ \forall e, d \tag{11}$$

If this requirement is not met, as specified in Equation (12), maintenance is required. The resource is marked as unavailable, the task start time is adjusted, and the equipment reliability is updated post-maintenance.

$$if\ R_{d,e}^{ET^{0}_{a,b,c}} < RT_{d,e},\ then \begin{cases} MS_{d,e}^{ST^{0}_{a,b,c}}, \ldots, MS_{d,e}^{ST^{0}_{a,b,c}+MDT_{d,e}} = 1 \\ ST_{a,b,1} = ST^{0}_{a,b,1} + MDT_{d,e} & \forall e, d \\ R_{d,e}^{ST_{a,b,1}} = R_{d,e}^{ini} \end{cases} \tag{12}$$

7. Slurry Mixing Equipment Feeding Constraint: Before a task begins, the tailings level in the slurry mixing equipment must meet the required amount, as specified by Equation (13).

$$M_{e}^{ET_{a,b,c}} \geq CM_{e}, \forall e \tag{13}$$

If this requirement is not satisfied, as indicated in Equation (14), or if the tailings level in the slurry mixing equipment has fallen below the critical value before the task starts, then feeding is initiated. The equipment is marked as unavailable, the backfilling task start time is adjusted, and the tailings level in the slurry mixing equipment is updated.

$$\left.\begin{matrix} M_{e}^{ST^{0}_{a,b,c}} \leq CM_{e} \\ or\ M_{e}^{ET^{0}_{a,b,c}} \leq 0 \end{matrix}\right\} then, \begin{cases} S_{e}^{ST^{0}_{a,b,c}}, \ldots, S_{e}^{ST^{0}_{a,b,c}+FT_{e}} = 1 \\ ST_{a,b,1} = ST^{0}_{a,b,1} + FT_{e} & \forall e \\ M_{e}^{ST_{a,b,1}} = MA_{e} \end{cases} \tag{14}$$

8. Manpower Resource Constraint: The number of stopes being backfilled at any given time must not exceed the available number of backfilling units.

$$\sum_{d=1}^{D}\sum_{e=1}^{E} MS_{d,e}^{t} \leq MS_0, \forall t \tag{15}$$

3.4.3. Parameter Constraints

9. Equations (16) and (17) establish the assignment relationships between $X_{a,b,c,d,e}^{t}$ and $Z_{a,b,c}^{t}$. Both variables are binary, and when $X_{a,b,c,d,e}^{t}$ = 1, the corresponding $Z_{a,b,c}^{t}$ is also ensured to be 1. Equation (18) defines the assignment relationship between $ST_{a,b,c}$ and $Z_{a,b,c}^{t}$, where $ST_{a,b,c}$ is equal to the t value for which $Z_{a,b,c}^{t}$ = 1.

$$\sum_{d=1}^{D}\sum_{e=1}^{E} X_{a,b,c,d,e}^{t} \leq N \cdot Z_{a,b,c}^{t}, \forall a,b,c,t \tag{16}$$

$$\sum_{d=1}^{D}\sum_{e=1}^{E} X_{a,b,c,d,e}^{t} \geq Z_{a,b,c}^{t}, \forall a,b,c,t \tag{17}$$

$$ST_{a,b,c} = \sum_{t=1}^{T} t \cdot Z_{a,b,c}^{t}, \forall a,b,c \tag{18}$$

10. When backfilling equipment is under maintenance or the slurry mixing equipment is in a feeding state, the corresponding resource is unavailable:

$$\text{if } d=3, S_e^t = 1, \forall e \text{ or } MS_{d,e}^{t} = 1, \forall d, e Y_{d,e}^{t} = 0 \tag{19}$$

4. Algorithm Design

The RCPSP has been proven to be an NP-hard (Non-deterministic Polynomial-time hard) problem [37,38]. The discrete nature of equipment reliability constraints and tailings constraints for slurry mixing equipment further increases the complexity of the solution. Intelligent optimization methods, such as GA, Simulated Annealing (SA), Ant Colony Optimization (ACO), and Particle Swarm Optimization (PSO), are effective for quickly obtaining approximate solutions to large-scale NP-hard problems, offering an advantage in solution efficiency [39]. In this study, the model is implemented using the Python programming language, with GA applied for solving.

4.1. Encoding and Decoding

4.1.1. Encoding Rules

The backfill scheduling problem with multi-resource availability involves assigning the e-th resource of type d to the c backfilling step in stope b at level a, requiring consideration of five dimensions of data. As shown in Equation (20), a chromosome is composed of five dimensions: A, B, C, D, and E, which correspond to level index, stope index, backfilling step index, resource type, and resource unit, respectively. Resource types D (1, 2, 3, 4, 5) represent manpower, water storage, slurry mixing equipment, cementitious material management equipment, and conveying equipment.

$$I = \begin{bmatrix} A \\ B \\ C \\ D \\ E \end{bmatrix} = \begin{bmatrix} 1,1,1,\ldots,2,\ldots,A \\ 1,1,1,\ldots,1,\ldots,B_A \\ 1,2,2,\ldots,1,\ldots,C_{BA} \\ 1,1,2,\ldots,2,\ldots,D_{CBA} \\ 1,1,2,\ldots,1,\ldots,E_{DCBA} \end{bmatrix} \tag{20}$$

The logic of the encoding algorithm is illustrated in Figure 4. Based on the above rules, an initial population of individuals is generated with specified population size. The encoded data includes level information, stope details, backfilling steps, and backfilling resources, forming the initial chromosomes for the algorithm.

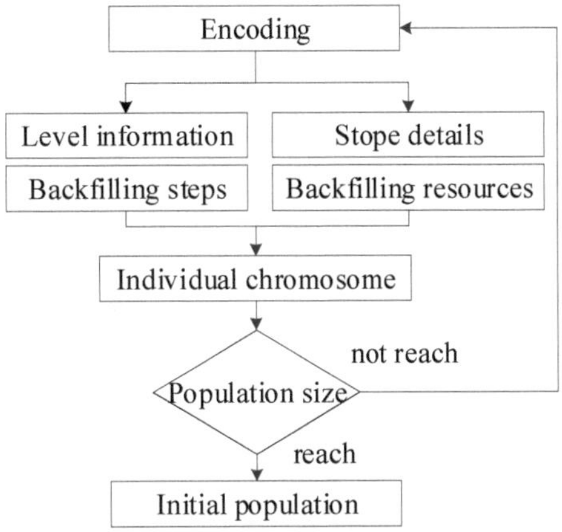

Figure 4. Encoding Algorithm Logic.

The algorithm's solution process must address two key issues: first, that each backfilling task requires multiple resources to operate simultaneously; and second, the timing of when each resource begins servicing the task.

To resolve the first issue, a data dictionary was created to map each backfill task to its required resource types, providing the basis for encoding. Unique identifiers are assigned to differentiate resource types during this process.

For the second issue, precise scheduling times cannot be determined during encoding due to potential delays in the optimization process. To manage this, delay times are calculated and recorded during optimization. In the decoding phase, a detailed schedule—incorporating timing—is generated using these delay times and the earliest possible start time.

4.1.2. Decoding Algorithm

The logic of the decoding algorithm is illustrated in Figure 5.

The detailed decoding steps are as follows:

1. With known level indices, stopes, backfilling step durations, allowed start times, and tailings consumption, stopes are sorted based on the allowed start time.
2. According to Equation (3), calculate the resource reliability R after completing the c-th step in stope b at level a. If $R_{d,e}^{ET_{a,b,c}} \geq RT_{d,e}$, the resource does not require maintenance. Otherwise, maintenance is performed, rendering the resource unavailable. A maintenance record is created, and the R is reset. The delay caused by maintenance is equal to the maintenance duration, $delay\_m = MDT_{d,e}$, and the resource's service age is updated accordingly.
3. Check whether the tailings weight can meet the requirements for backfilling stope b at level a. If $M_e^{ET_{a,b,c}} \geq CM$, the equipment has sufficient tailings and no additional feeding is needed. Otherwise, feeding is performed, create a feeding record, and update the new tailings level. The delay caused by insufficient tailings is equal to the feeding time, $delay\_t = FT_e$.
4. Calculate the occupancy times for all resources required to determine whether each resource is within its available time. If yes, no further action is needed; otherwise, calculate the delay caused by resource occupancy, $delay\_on$.
5. Calculate the overall delay for backfilling using Equation (21). The actual start time is obtained by adding this delay to the allowed start time of the stope. Determine the

timing for each task, create the backfilling schedule for the stope, and update resource occupancy accordingly.

$$DeT_{ab} = delay\_on + delay\_m + delay\_t \tag{21}$$

6. Return to Step 2 and proceed with calculations for the next stope until scheduling plans are completed for all stopes.
7. Calculate the total delay across all stopes, using it as the fitness value for the GA. Return the detailed schedule, fitness value, maintenance records, and feeding records.

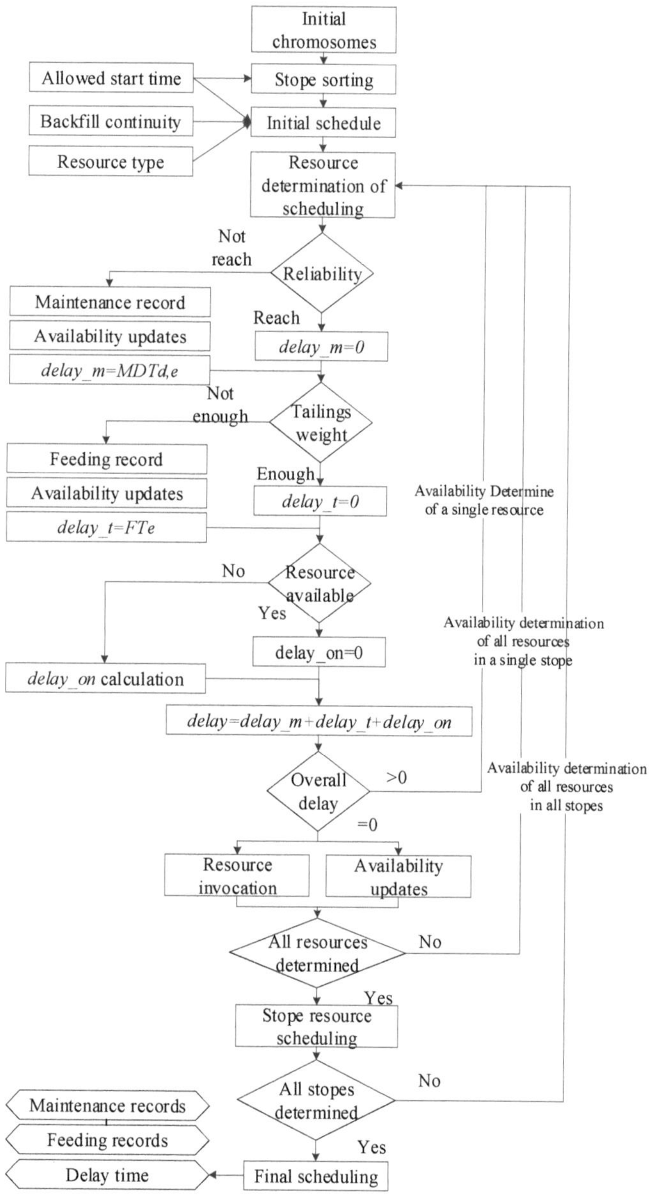

Figure 5. Decoding Algorithm Logic.

4.2. Execution Process of the GA

The execution process of the GA includes population initialization, evaluation and selection, crossover and mutation, and fitness calculation. The steps for implementing the GA in this model are as follows:

1. Define parameters such as population size, number of iterations, and mutation probability.
2. Initialize the first generation of the population according to the encoding rules. Calculate fitness values through decoding, copy the best individual to replace the worst individual to ensure elitism, and update the global best individual. Elitism ensures that the best individual in each generation is passed to the next generation.
3. Select two parent individuals using a roulette-wheel selection method and apply two-point crossover to generate offspring. Roulette wheel selection determines the probability of an individual being chosen as a parent based on its fitness. Two-point crossover selects two crossover points in two parent genes and exchanges sequences to produce new individuals.
4. Define mutation rules, allowing mutation only within the same resource type. Mutate resource numbers according to the mutation probability to create new offspring. Mutation randomly alters an individual's genes to increase diversity and prevent premature convergence.
5. Decode the new offspring and calculate their fitness values, retaining the two best individuals and updating the global best.
6. Repeat steps (3), (4), and (5) until the specified number of iterations is reached.
7. Return the global best individual.

In a simple GA, suppose we have individuals and fitness values as shown in Table 2:

Table 2. Example gene of GA.

Individual	Gene	Fitness
1	101010	20
2	110011	30
3	111000	50

Execute the following steps:

- Selection: Use the roulette wheel selection method to select two individuals, such as individual 2 and individual 3.
- Crossover: After performing the two-point crossover, two new offspring individuals are generated, such as the new individuals "110000" and "111011" produced by the crossover between individual 2 and individual 3.
- Mutation: Apply mutation to the offspring "110000", which could change to "110001".
- Elitism: Retain the current best individual (for example, the individual with a fitness of 50) and directly pass it to the next generation.

The individuals after executing the GA are: "110001", "111011", and "111000".

The search logic of the algorithm involves assigning resources of the same type but with different identifiers to tasks. To ensure that all generated solutions are valid, crossover and mutation operations are applied exclusively to the E dimension of the chromosome. Crossover is permitted only when the D values of the two chromosomes are identical, indicating the use of the same type of equipment. Similarly, the mutation range is restricted to the same category of resources. For instance, if the E value is 5, representing human resources with an identifier range of (1, 6) the mutation is confined to this range.

5. Case Study, Results Analysis, and Discussion

To verify the effectiveness of the model and algorithm, an empirical analysis is conducted using production data from a large underground metal mine in Shandong, China. The mine operates at a production scale of 8000 t/d and employs an upward layered backfilling method. The backfill structure is divided into two layers vertically, with five sequential backfilling steps: preliminary preparation, water flow induction, bottom filling, top filling, and pipeline cleaning, where C is set to 5.

5.1. Case Data and Inputs

The mine establishes a monthly backfilling plan, with adjustments and optimizations approximately every ten days. During a specific optimization instance, 27 stopes in 11 levels were scheduled for backfilling; detailed stope information is provided in Appendix A. The backfilling station includes six tailings silos, each equipped with corresponding mixing facilities, collectively forming a slurry mixing system. Three backfilling teams are available, each capable of handling two tasks simultaneously. Additionally, there are two water storage units and two cementitious material silos, each independently able to meet the water and material demands of all the slurry mixing systems. A total of six pipelines connect to the underground backfill main network.

The equipment reliability data is sourced from the ledger and management system of the backfilling center at the mine, covering equipment operation and failure data since 2020. The dataset includes various metrics such as equipment usage time, failure records, maintenance history, and more, encompassing different types of equipment and operating conditions. The Pareto principle was applied to analyze equipment failure modes. The backfilling station at the case has been in service for several years, with equipment now in the wear-out phase, corresponding to the third stage of the bathtub curve, where $\beta > 1$. This process follows a Non-homogeneous Poisson Process (NHPP). The equipment failure rate function adheres to a Weibull distribution, with fitted parameters for the Weibull function shown in Table 3. The primary backfilling units remain continuously available, with a reliability of 1. When the model is applied to other mines, it will need to be adapted according to actual data and select different failure functions based on the specific failure modes observed.

Table 3. Reliability Parameters for Backfilling Equipment.

Type	β	η	r	Service Age (h)	D	E
Manpower	\	\	\	\	1	1, 2, 3, 4, 5, 6
Water tank	2	1920	0	550, 550, 550, 420, 420, 420	2	6, 7, 8, 9, 10, 11
Pulping & mixing	2	969	0	490, 210, 430, 150, 480, 270	3	12, 13, 14, 15, 16, 17
Cementitious material	2	1983	0	460, 460, 460, 390, 390, 390	4	18, 19, 20, 21, 22, 23
Slurry conveying	2	2064	0	480, 510, 530, 550, 570, 590	5	24, 25, 26, 27, 28, 29

As shown in Table 3, when the value range of E is (1, 6), it represents human resources, and its reliability is set to 1 by default, so no other parameters need to be assigned.

Calculating the reliability of the water tank and cementitious material management equipment is relatively complex, as these two systems operate in parallel. The combined reliability of the parallel configuration is determined using Equation (22). This parallel setup is then represented as six virtual units with uniform reliability, with their service age cumulatively tracked across tasks.

$$R_p(t) = 1 - [1 - R_1(t)] \cdot [1 - R_2(t)] \tag{22}$$

In summary, additional data required for the model are listed in Table 4, including the values or ranges for A, B, C, D, and E; the tailings capacity and feeding time for the slurry mixing equipment; and the equipment reliability threshold.

Table 4. Additional Input Data for the Model.

Parameter	Value	Parameter	Value
MA (t)	1424	*A*	11
FT (h)	10	*B*	1–3
RT	0.8	*C*	5
MDT (h)	12	*D*	5
SeT (h)	80	*E*	1–30
FS_0	6	*N*	10^5

5.2. Optimization Results

The model is constructed using Python on the Anaconda platform, and the computations are performed on a Lenovo Thinkpad notebook equipped with an Intel Core i7-8550 processor, that was purchased in Shanghai, China. The parameters of the GA, such as crossover rate, mutation rate, and population size, significantly affect the solving speed and efficiency.

A higher crossover rate and mutation rate can accelerate convergence but may reduce the diversity of solutions, potentially leading to the loss of the optimal solution. Conversely, when the crossover rate is too low, the convergence speed decreases, and the probability of obtaining suboptimal solutions increases. However, under the elitism strategy, the loss of the optimal solution does not occur. To reduce computational intensity and the number of iterations, this case study adopts relatively high crossover and mutation rates, both set at 0.2.

A larger population size requires more computational resources and time but provides a broader range of solutions. In this case study, multiple simulations were conducted with the same number of iterations (50). When the population size was 20, only 2 out of 10 simulations achieved an approximate minimum value (825 h). As the population size increased, the number of times the optimal fitness value was achieved also increased. When the population size reached 50, all 10 simulations achieved the optimal fitness value.

Regarding computation time and efficiency, the time per iteration varied between 6 and 9 min. However, the number of iterations required to reach the optimal fitness value varied. Subsequently, a set of data was randomly selected to represent the iterative process. Figure 6 shows the trend of fitness value changes during the iterations, displaying a gradient descent and reaching the minimum value at the 27th generation.

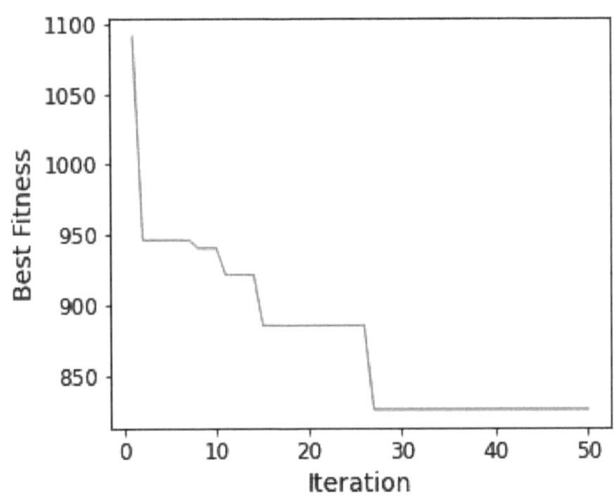

Figure 6. GA Iteration Chart.

The model solution results are shown in Table 5, presenting the start, end, and delay times for stope backfilling. Due to space constraints, only data for 2 out of the 5 dimensions is displayed in the table. Using the first stope in the first level as an example, the resource allocation for each backfilling step is detailed in Table 6.

Table 5. Model Solution Results.

Level/*A*	Stope/*B*	Start (h)	End (h)	Delay (h)
1	1	0	13	0
1	2	48	59	0
2	1	34	48	10
2	2	34	49	10
2	3	14	33	14
3	1	61	79	13
3	2	10	32	10
3	3	51	71	51
4	1	80	102	32
4	2	43	59	19
4	3	94	112	70
5	1	112	135	88
5	2	96	111	0
6	1	130	150	10
7	1	41	63	41
7	2	73	96	25
7	3	152	173	56
7	4	113	137	65
8	1	61	77	13
8	2	175	191	31
8	3	103	125	55
9	1	147	166	75
9	2	158	182	14
10	1	193	212	73
10	2	88	101	16
11	1	121	138	25
11	2	129	146	9

Table 6. Backfilling Resource Allocation (Example: First Stope (*B*) in the First Level).

Level/*A*	Stope/*B*	Step/*C*	Type/*D*	No./*E*	Start (h)	End (h)
1	1	1	1	5	0	8
1	1	2	1	5	8	8.5
1	1	2	2	10	8	8.5
1	1	2	5	28	8	8.5
1	1	3	1	5	8.5	11.5
1	1	3	2	10	8.5	11.5
1	1	3	3	16	8.5	11.5
1	1	3	4	19	8.5	11.5
1	1	3	5	28	8.5	11.5
1	1	4	1	5	11.5	12.5
1	1	4	2	10	11.5	12.5
1	1	4	3	16	11.5	12.5
1	1	4	4	19	11.5	12.5
1	1	4	5	28	11.5	12.5
1	1	5	1	5	12.5	13
1	1	5	2	10	12.5	13
1	1	5	5	28	12.5	13

Table 6 provides a clear view of the resource IDs (*E*) and types (*D*) used in each backfilling step (*C*), along with the start and end times for resource allocation.

Figure 7 illustrates the optimal Gantt chart for the stope backfilling schedule. The horizontal axis denotes the scheduling time in hours, while the vertical axis displays the level numbers and stope identification codes. Five different colors are used in the chart to distinguish the backfilling steps, with each color representing a specific operation phase.

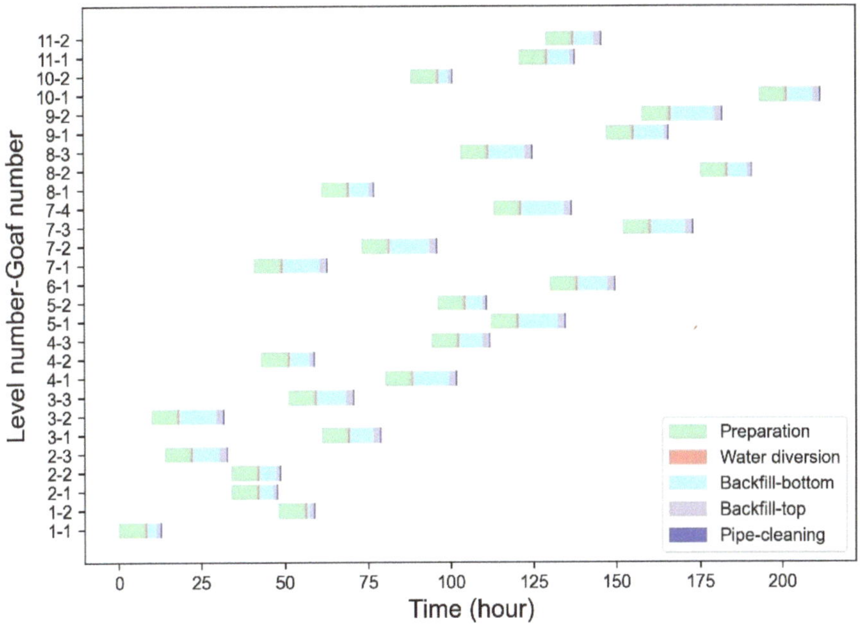

Figure 7. Stope Backfilling Schedule.

Figure 8 demonstrates the utilization status of five backfilling resources, where the horizontal axis represents time and the vertical axis indicates resource types (resource codes are detailed in Table 2). As in the case description, each number in PM and SC represents a resource. the data in the box represents the level No.- stope No. Seven colors are employed in the chart: the first five colors correspond to the basic backfilling resources, while the sixth color (yellow) indicates the sand feeding status of the slurry mixing equipment, and the seventh color (purple) represents equipment maintenance status. During this scheduling period, 13 sand feeding operations and 2 equipment maintenance sessions were recorded. Notably, human resources, water tanks, and cementitious material management equipment can simultaneously serve multiple backfilling processes; thus, their specific service allocation to individual stops is not marked in the figure.

As shown in Figure 8, pumping equipment PM-5 exhibits the lowest utilization rate, primarily due to its extended service age. The frequent maintenance requirements of this equipment would increase scheduling delays and consequently raise the fitness value. Regarding slurry transportation equipment, pipelines SC-3 and SC-4 show significantly higher utilization rates compared to other facilities. This can be attributed to their dedicated function in serving deeper levels, as the backfilling slurry for deep stopes can only be transported through these two specific pipeline routes.

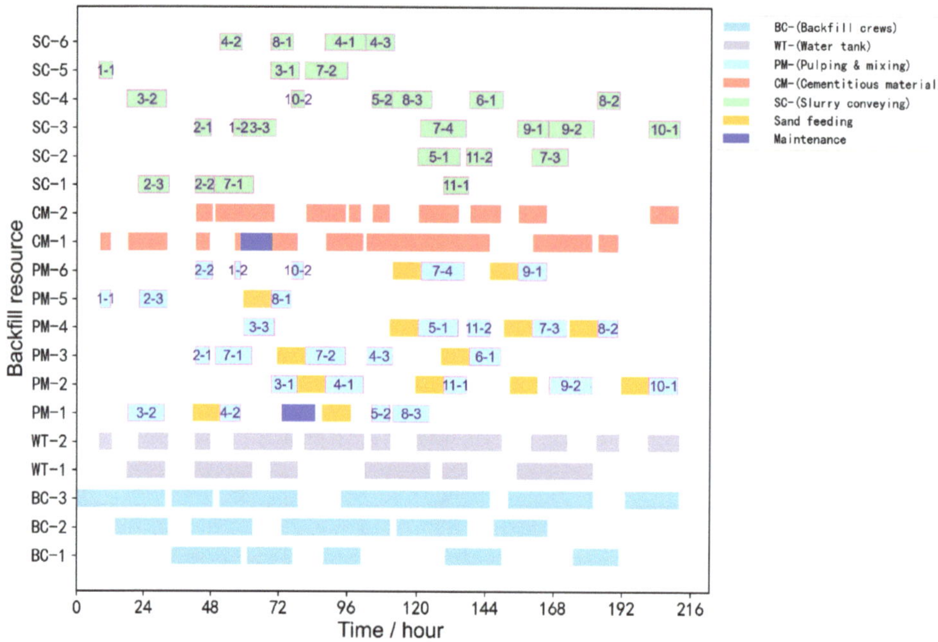

Figure 8. Resource Allocation Distribution.

5.3. Advantages over Traditional Mode

This model implements precise scheduling by comprehensively considering the operational characteristics and reliability of slurry mixing equipment. Compared to the current 'one-operating-one-standby' mode used in mines, where only 50% of total resources are available at any given time, this model optimizes the synergistic utilization of all system resources, thereby significantly improving operational efficiency. Table 7 presents a comparative summary of scheduling results between the traditional mode and the optimized solution.

Table 7. Backfilling Resource Allocation (Example: First Stope in the First Level).

	Traditional	Optimization
Makespan (h)	267	212
Utilization	31.07%	39.13%
Total delay (h)	1568	825

Compared to the traditional model, the optimized solution offers significant advantages, primarily in the following areas:

1. Completion time was reduced by 20.6%, from 267 h to 212 h.
2. Resource utilization increased by 8 percentage points, from 31.07% to 39.13%.
3. Total delay time decreased by 47.4%, from 1568 h to 825 h. This reduction significantly minimizes goaf exposure time, allowing for more timely ground pressure management and reducing the risk of ground pressure incidents.

5.4. Limitation and Future Research

Despite the model's effectiveness in the current mining context, its generalizability to diverse equipment types and complex operational scenarios requires further investigation.

The model validation was primarily conducted under specific operational conditions, which may limit its broader applicability.

Future research directions include:

1. Extending the model to accommodate diverse equipment types and environmental factors, particularly non-linear failure modes.
2. Investigating the model's real-time implementation in large-scale mining operations, with emphasis on computational efficiency.
3. Integrating advanced technologies such as artificial intelligence and big data analytics to enhance the model's adaptability and robustness in dynamic mining environments.

The proposed extensions would significantly advance the practical application of backfill scheduling optimization in modern mining operations.

6. Conclusions

This study develops a scheduling optimization model for backfilling operations under resource constraints, addressing the complex requirements of coordinating multiple resources, sequential steps, stopes, and specific process-related needs. To achieve effective scheduling, a tailored algorithm has been designed to solve the model, and its effectiveness has been validated through application to a case study mine, optimizing the assignment of backfilling personnel, equipment, and tasks. Comparative analysis with conventional scheduling approaches confirms the model's effectiveness. The main findings are as follows:

1. Resource-Constrained Scheduling Optimization Model: This study systematically identifies the various resource requirements in mine backfilling processes and examines operational procedures when resources are unavailable. By incorporating constraints on resource availability and setting total backfilling delay time as the optimization objective, the model provides a comprehensive approach to scheduling that accounts for multiple resource limitations.
2. Heuristic Solution Algorithm for Complex Processes: Given the NP-hard nature of the problem, a heuristic algorithm has been applied, featuring encoding methods adapted to process constraints and decoding methods aligned with resource limitations. The algorithm calculates delay times as fitness values and integrates specific procedures for managing resource unavailability, ensuring a structured and efficient solution process.
3. Model Validation and Performance Analysis: Detailed data processing has been conducted for the case study mine, producing a precise scheduling solution that identifies start and end times for each task across various resources. Compared to the traditional model, the optimized scheduling solution shows clear advantages, with a 20.6% reduction in completion time, an 8-percentage point increase in resource utilization, and a 47.4% reduction in total backfilling delay time.

In conclusion, this study presents a novel optimization model for mine backfill scheduling, along with its solution methodology and validation through a real-world case study. The model demonstrates both theoretical significance and practical value in mining operations. By implementing intelligent scheduling optimization in mine backfilling, the model not only enhances production efficiency and underground mining safety but also advances mining intelligentization. The model's capability for dynamic decision-making and optimal resource allocation provides valuable support for intelligent mining operations, particularly in addressing complex scheduling challenges in modern mining environments.

Author Contributions: Methodology, G.L.; Model, J.H.; Validation, Y.L.; Resources, C.F., C.T. and P.W.; Result analysis, J.H. and Y.L.; Writing—original draft, Y.L.; Writing—review and editing, G.L. and J.H.; Visualization, Y.L.; Supervision, G.L. All authors have read and agreed to the published version of the manuscript.

Funding: This research was funded by the National Key Research and Development Program of China–2023 Key Special Project (No. 2023YFC2907403), the National Natural Science Foundation of China (No. 52404161), and the China Scholarship Council (202206460036).

Data Availability Statement: Data are contained within the article.

Acknowledgments: The authors would like to express their sincere gratitude to Justin, Qixuan Song, and Chu Chu for their valuable insights and constructive suggestions on the solution methodology. The authors also appreciate the anonymous reviewers for their helpful comments that improved the quality of this paper.

Conflicts of Interest: Panzhi Wang is employee of Jiaojia Gold Mine, Shandong Gold Group Mining (Laizhou) Co., Ltd., Yantai, China. The paper reflects the views of the scientists and not the company.

Appendix A. Input Data, Stopes Information

No.	Level/*A*	Goaf/*B*	Earliest Time/*BF* (h)	Tailings/*TB* (t)		Backfill Duration/*DT* (h)				
				Bottom	Top	1	2	3	4	5
1	1	1	0	289.5	47.8	8	0.5	3	1	0.5
2	1	2	48	90.1	11	8	0.5	1	1	0.5
3	2	1	24	349.8	33	8	0.5	4	1	0.5
4	2	2	24	440.4	75.7	8	0.5	5	1	0.5
5	2	3	0	778.1	96.8	8	0.5	8	2	0.5
6	3	1	48	778.1	117	8	0.5	7	2	0.5
7	3	2	0	981.8	108.4	8	0.5	11	2	0.5
8	3	3	0	852.4	115.3	8	0.5	9	2	0.5
9	4	1	48	1018.3	121	8	0.5	11	2	0.5
10	4	2	24	550.3	65.1	8	0.5	6	1	0.5
11	4	3	24	646.6	96.8	8	0.5	7	2	0.5
12	5	1	24	1106.3	134.7	8	0.5	12	2	0.5
13	5	2	96	476.3	57.3	8	0.5	5	1	0.5
14	6	1	120	879.2	113.3	8	0.5	9	2	0.5
15	7	1	0	1093	116.6	8	0.5	11	2	0.5
16	7	2	48	1129.4	120.2	8	0.5	12	2	0.5
17	7	3	96	1087.3	112	8	0.5	10	2	0.5
18	7	4	48	1044.7	126.8	8	0.5	13	2	0.5
19	8	1	48	536.5	73.3	8	0.5	6	1	0.5
20	8	2	144	559.8	71.6	8	0.5	6	1	0.5
21	8	3	48	1051.2	129.1	8	0.5	11	2	0.5
22	9	1	72	833.5	79.5	8	0.5	9	1	0.5
23	9	2	144	1278.6	143.8	8	0.5	13	2	0.5
24	10	1	120	742.4	119.6	8	0.5	8	2	0.5
25	10	2	72	251.8	30.7	8	0.5	3	1	0.5
26	11	1	96	651.8	61.4	8	0.5	7	1	0.5
27	11	2	120	536.4	92.1	8	0.5	6	2	0.5

References

1. Behera, S.K.; Mishra, D.P.; Singh, P.; Mishra, K.; Mandal, S.K.; Ghosh, C.N.; Kumar, R.; Mandal, P.K. Utilization of mill tailings, fly ash and slag as mine paste backfill material: Review and future perspective. *Constr. Build. Mater.* **2021**, *309*, 125120. [CrossRef]
2. Kasap, T.; Yilmaz, E.; Sari, M. Physico-chemical and micro-structural behavior of cemented mine backfill: Effect of pH in dam tailings. *J. Environ. Manag.* **2022**, *314*, 115034. [CrossRef] [PubMed]
3. Guo, M.; Tan, Y.; Chen, D.; Song, W.; Cao, S. Optimization and Stability of the Bottom Structure Parameters of the Deep Sublevel Stope with Delayed Backfilling. *Minerals* **2022**, *12*, 709. [CrossRef]
4. Zhang, C.; Taheri, A.; Du, C.; Xia, W.; Tan, Y. Mechanical Characteristics and Macro–Microscopic Response Mechanisms of Cemented Paste Backfill under Different Curing Temperatures. *Minerals* **2024**, *14*, 433. [CrossRef]
5. Ji, X.; Gu, X.; Wang, Z.; Xu, S.; Jiang, H.; Yilmaz, E. Admixture Effects on the Rheological/Mechanical Behavior and Micro-Structure Evolution of Alkali-Activated Slag Backfills. *Minerals* **2022**, *13*, 30. [CrossRef]
6. Wang, B.; Gan, S.; Yang, L.; Zhao, Z.; Wei, Z.; Wang, J. Additivity Effect on Properties of Cemented Ultra-Fine Tailings Backfill Containing Sodium Silicate and Calcium Chloride. *Minerals* **2024**, *14*, 154. [CrossRef]
7. Wang, Z.; Jiang, H.; Fu, Y.; Ma, Z.; Wang, X. Calcined alunite-modified alkali-sulphate-activated slag as a novel binder for high-performance cemented paste backfill. *J. Build. Eng.* **2024**, *91*, 109687. [CrossRef]
8. Chen, X.; Guo, L.; Zhou, Y.; Xu, W.; Zhao, Y. Remediation of grassland subsidence and reduction of land occupation with tailings backfill technology: A case study of lead-zinc mine in Inner Mongolia, China. *Front. Environ. Sci.* **2023**, *11*, 1183945. [CrossRef]

9. Lu, H.; Qi, C.; Chen, Q.; Gan, D.; Xue, Z.; Hu, Y. A new procedure for recycling waste tailings as cemented paste backfill to underground stopes and open pits. *J. Clean. Prod.* **2018**, *188*, 601–612. [CrossRef]
10. Mashifana, T.; Sithole, T. Clean production of sustainable backfill material from waste gold tailings and slag. *J. Clean. Prod.* **2021**, *308*, 127357. [CrossRef]
11. Qi, C.; Fourie, A. Cemented paste backfill for mineral tailings management: Review and future perspectives. *Miner. Eng.* **2019**, *144*, 106025. [CrossRef]
12. Liu, Y.; Li, G.; Hou, J.; Guo, G.; Pan, D.; Yu, Q. An Underground Mine Safety-Oriented Optimization Model for Mine Tailings Backfill Scheduling Considering Multi-Process and Multi-Cycle Issues. *Minerals* **2023**, *13*, 1409. [CrossRef]
13. Liu, K.; Mei, B.; Li, Q.; Sun, S.; Zhang, Q. Collaborative Production Planning Based on an Intelligent Unmanned Mining System for Open-Pit Mines in the Industry 4.0 Era. *Machines* **2024**, *12*, 419. [CrossRef]
14. Wu, A.; Wang, Y.; Ruan, Z.E.; Xiao, B.; Wang, J.; Wang, L. Key theory and technology of cemented paste backfill for green mining of metal mines. *Green Smart Min. Eng.* **2024**, *1*, 27–39. [CrossRef]
15. Wang, Z.; Wu, A.; Ruan, Z.; Bürger, R.; Wang, S.; Mo, Y. Flocculation behavior, mechanics, and optimization of tailings based on multi-objective: Insight into the concentration and time-dependent floc size. *Powder Technol.* **2024**, *439*, 119718. [CrossRef]
16. Lööw, J.; Abrahamsson, L.; Johansson, J. Mining 4.0—The Impact of New Technology from a Work Place Perspective. *Min. Metall. Explor.* **2019**, *36*, 701–707. [CrossRef]
17. Zhironkina, O.; Zhironkin, S. Technological and Intellectual Transition to Mining 4.0: A Review. *Energies* **2023**, *16*, 1427. [CrossRef]
18. Hou, J.; Wang, H.; Li, G.; Sheng, B.; Wang, Q. Multistage dynamic optimisation of ore flow for underground metal mines. *Int. J. Min. Reclam. Environ.* **2024**, *38*, 407–423. [CrossRef]
19. Jiang, S.; Lian, M.; Lu, C.; Gu, Q.; Ruan, S.; Xie, X. Ensemble Prediction Algorithm of Anomaly Monitoring Based on Big Data Analysis Platform of Open-Pit Mine Slope. *Complexity* **2018**, *2018*, 1–13. [CrossRef]
20. Xu, X.-c.; Gu, X.-w.; Wang, Q.; Gao, X.-w.; Liu, J.-p.; Wang, Z.-k.; Wang, X.-h. Production scheduling optimization considering ecological costs for open pit metal mines. *J. Clean. Prod.* **2018**, *180*, 210–221. [CrossRef]
21. Johnson, S.M. Optimal two- and three-stage production schedules with setup times included. *Nav. Res. Logist. Q.* **1954**, *1*, 61–68. [CrossRef]
22. Brucker, P.; Schlie, R. Job-shop scheduling with multi-purpose machines. *Computing* **1990**, *45*, 369–375. [CrossRef]
23. Gong, G.; Chiong, R.; Deng, Q.; Gong, X. A hybrid artificial bee colony algorithm for flexible job shop scheduling with worker flexibility. *Int. J. Prod. Res.* **2020**, *58*, 4406–4420. [CrossRef]
24. Soltani Khaboushan, A.; Osanloo, M.; Esfahanipour, A. Optimization of open pit to underground transition depth: An idea for reducing waste rock contamination while maximizing economic benefits. *J. Clean. Prod.* **2020**, *277*, 123530. [CrossRef]
25. Upadhyay, S.P.; Askari-Nasab, H. Simulation and optimization approach for uncertainty-based short-term planning in open pit mines. *Int. J. Min. Sci. Technol.* **2018**, *28*, 153–166. [CrossRef]
26. Vaziri, V.; Sayadi, A.R.; Mousavi, A.; Parbhakar-Fox, A.; Monjezi, M. Mathematical modeling for optimized mine waste rock disposal: Establishing more effective acid rock drainage management. *J. Clean. Prod.* **2021**, *288*, 125124. [CrossRef]
27. Guo, L.; Funari, V.; Li, M. Editorial: Advances in sustainable mine tailings management. *Front. Earth Sci.* **2023**, *11*, 1269955. [CrossRef]
28. Kou, Y.; Liu, Y.; Li, G.; Hou, J.; Luan, L.; Wang, H.; Guo, L. Design and Implementation of an Integrated Management System for Backfill Experimental Data. *Adv. Civ. Eng.* **2022**, *2022*, 1–9. [CrossRef]
29. Bahroun, Z.; Tanash, M.; As'ad, R.; Alnajar, M. Artificial Intelligence Applications in Project Scheduling: A Systematic Review, Bibliometric Analysis, and Prospects for Future Research. *Manag. Syst. Prod. Eng.* **2023**, *31*, 144–161. [CrossRef]
30. Chimunhu, P.; Topal, E.; Ajak, A.D.; Asad, W. A review of machine learning applications for underground mine planning and scheduling. *Resour. Policy* **2022**, *77*, 102693. [CrossRef]
31. Ding, H.; Zhuang, C.; Liu, J. Extensions of the resource-constrained project scheduling problem. *Autom. Constr.* **2023**, *153*, 104958. [CrossRef]
32. Nesbitt, P.; Blake, L.R.; Lamas, P.; Goycoolea, M.; Pagnoncelli, B.K.; Newman, A.; Brickey, A. Underground mine scheduling under uncertainty. *Eur. J. Oper. Res.* **2021**, *294*, 340–352. [CrossRef]
33. Johnson, R.T.J. An Algorithm for the Resource Constrained Project Scheduling Problem. Ph.D. Thesis, Massachusetts Institute of Technology, Cambridge, MA, USA, 1967.
34. Yin, S.; Shao, Y.; Wu, A.; Wang, H.; Liu, X.; Wang, Y. A systematic review of paste technology in metal mines for cleaner production in China. *J. Clean. Prod.* **2020**, *247*, 119590. [CrossRef]
35. Qi, C.; Fourie, A.; Chen, Q.; Zhang, Q. A strength prediction model using artificial intelligence for recycling waste tailings as cemented paste backfill. *J. Clean. Prod.* **2018**, *183*, 566–578. [CrossRef]
36. Huang, S.; Li, G.; Ben-Awuah, E.; Afum, B.O.; Hu, N. A robust mixed integer linear programming framework for underground cut-and-fill mining production scheduling. *Int. J. Min. Reclam. Environ.* **2019**, *34*, 397–414. [CrossRef]
37. Golab, A.; Gooya, E.S.; Falou, A.A.; Cabon, M. Review of conventional metaheuristic techniques for resource-constrained project scheduling problem. *J. Proj. Manag.* **2022**, *7*, 95–110. [CrossRef]

38. Hartmann, S.; Briskorn, D. A survey of variants and extensions of the resource-constrained project scheduling problem. *Eur. J. Oper. Res.* **2010**, *207*, 1–14. [CrossRef]
39. Pellerin, R.; Perrier, N.; Berthaut, F. A survey of hybrid metaheuristics for the resource-constrained project scheduling problem. *Eur. J. Oper. Res.* **2020**, *280*, 395–416. [CrossRef]

minerals

MDPI

Article

Evolution of Pore Structure and Mechanical Characteristics of Red Sandstone Under Drying–Wetting Cycles

Hongwei Deng [1], Shiyu Zhou [1], Songtao Yu [2,*], Yao Liu [1] and Jingbo Xu [1]

[1] School of Resources and Safety Engineering, Central South University, Changsha 410083, China; denghw208@126.com (H.D.); 225512098@csu.edu.cn (S.Z.); 225501021@csu.edu.cn (Y.L.)

[2] School of Emergency Management and Safety Engineering, Jiangxi University of Science and Technology, Ganzhou 341000, China

* Correspondence: 9120210003@jxust.edu.cn

Abstract: Red sandstone is widely distributed in southern China. Due to the significant difference in mechanical properties before and after hydration and its poor water stability, red sandstone often triggers landslide accidents. In this paper, red sandstone from an open pit slope in Jiangxi Province was taken as the research object. Two variables, namely the initial saturation degree (25%, 50%, 75%, and 100%) and the number of wetting–drying cycles (0, 10, 20, 30, and 40), were set. With the help of nuclear magnetic resonance, the Brazilian disc test, and fractal theory, the relationships among its meso-structure, macroscopic fracture mechanics characteristics, and deterioration mechanism were analyzed. The research results are as follows: (1) Wetting–drying cycles have a significant impact on the pore structure and fracture mechanics characteristics of red sandstone. Moreover, the higher the initial saturation degree, the more obvious the deterioration effect of the wetting–drying cycles on the rock mass. (2) After further subdividing the pores according to their size for research, it was found that sandstone is mainly composed of mesopores, and the deterioration laws of different types of pores after the wetting–drying cycles are different. The porosities of total pores and macropores increase, while the proportions of mesopores and micropores decrease. The fractal dimensions of macropores and total pores of each group of rock samples are all within the range of 2–3, and the fractal dimension value increases with the increase in the number of wetting–drying cycles, showing significant and regular fractal characteristics. Micropores and some mesopores do not possess fractal characteristics. The fractal dimension of rock samples basically satisfies the rule that the larger the pore diameter, the larger the fractal dimension and the more complex the pore structure. (3) Both the type I and type II fracture toughness of rock samples decrease with the increase in the number of cycles, and the decrease is the most significant when the initial saturation degree is 100%. After 40 cycles, the decreases in type I and type II fracture toughness reach 23.578% and 30.642%, respectively. The fracture toughness is closely related to the pore structure. The porosity and fractal dimension of rock samples and their internal macropores are linearly negatively correlated with the type II fracture toughness. The development of the macropore structure is the key factor affecting its fracture mechanics performance. (4) After the wetting–drying cycles, the internal pores of red sandstone continue to develop. The number of pores increases, the pore diameter enlarges, and the proportion of macropores rises, resulting in internal damage to the rock mass. When bearing loads, the expansion and connection of internal cracks intensify, ultimately leading to the failure of the rock mass. The research results can provide important reference for the stability analysis of sandstone slope engineering.

Academic Editors: Yuye Tan, Xun Chen, Yuan Li and Elsabe Kearsley

Received: 21 November 2024
Revised: 24 January 2025
Accepted: 5 February 2025
Published: 7 February 2025

Citation: Deng, H.; Zhou, S.; Yu, S.; Liu, Y.; Xu, J. Evolution of Pore Structure and Mechanical Characteristics of Red Sandstone Under Drying–Wetting Cycles. *Minerals* **2025**, *15*, 158. https://doi.org/10.3390/min15020158

Keywords: red sandstone; wetting–drying cycle; initial saturation; pore structure characteristics; fracture mechanics; fractal dimension

1. Introduction

Red sandstone is widely distributed in the southern regions of China. Owing to its favourable mechanical properties, durability, heat resistance, and its ease of cutting and processing in the natural state, it has wide applications [1]. However, when it is exposed to water, it expands, disintegrates, and softens quickly. This leads to a significant reduction in the mechanical strength. The increase in the water content in the rock mass not only weakens the rock strength but also changes its failure mode. The cumulative damage caused by wetting–drying cycles on the sandstone accelerates the weathering rate of the rock mass, promotes the development of the pore structure, and reduces the mechanical properties [2]. Red sandstone is often exposed to complex natural environments. It is affected by various climatic and hydrogeological conditions, such as precipitation and water level fluctuations. These conditions have a serious impact on the safety and service level of its engineering applications [3].

Therefore, studying the influence of different initial saturations and the number of wetting–drying cycles on the pore structure and mechanical characteristics of red sandstone is extremely important when conducting a stability analysis of sandstone slope engineering.

Studying the variation in the microscopic pore structure is a crucial approach to revealing the failure mechanism of red sandstone under the effect of wet–dry cycling [4]. Nuclear magnetic resonance spectroscopy can be utilized to rapidly and non-destructively detect the pore structure of rock masses. Ye Wanjun et al. [5,6] employed NMR testing technology to explore the evolution law of the pore structure of soil after different wetting–drying cycles. Due to the characteristics of water absorption and expansion as well as the water loss and contraction of cementing minerals, the mesoscopic structure of the rock mass deteriorates remarkably under the wet–dry cycling, which is generally manifested as an increase in porosity, a gradual coarsening and connection of tiny pores, and an evolution towards macropores [7–9]. When the rock mass has different saturation degrees, significant differences also exist in its internal micro-mesoscopic structure and evolution law. Yang et al. [10,11] observed through electron microscopy (SEM) that after the wet–dry cycling, the particle characteristics and pore distribution on the rock surface changed. Water molecules infiltrated into the interior of the rock, water-soluble particles were dissolved by water, and fine rock fragments were detached, leaving behind pore cracks. Li et al. [12] conducted an exploration on medium-sized sandstone after wet–dry cycling tests using NMR technology, and the results indicate that the medium-grained sandstone of the rock is directly related to its mechanical properties. Most of the existing research on sandstones mainly aims to use nuclear magnetic resonance (NMR) techniques to detect mesoscopic structures. However, there are relatively few studies that introduce geometric theories to conduct qualitative and quantitative analyses on the pore structures of sandstones by using the mathematical features contained in the T_2 spectrum.

Fractal theory can be applied to characterize the complexity and homogeneity of the pore structure of porous materials, providing more detailed and comprehensive information about the pore structure [13–15]. Gao Lirong et al. [16] integrated fractal theory with the cluster analysis approach to carry out in-depth research on the characteristics and classification methods of reservoir pore structures. Zhang et al. [17] quantified the degree of damage to the microstructure of rocks under wet–dry cycling conditions through fractal dimensional fitting and established the relationship between the microstructure and mechanical parameters.

Wet–dry cycling represents a significant factor that affects the mechanical strength of rock masses. A multitude of scholars have conducted conventional and dynamic mechanical property tests in this regard. Yu Yue et al. [18,19] utilized uniaxial compression experiments to study the impact of wet–dry cycling on compressive strength. Their findings reveal that as the number of wetting–drying cycles escalates, the compressive strength of the rock progressively declines. Liu Mu [20] investigated the influence of different rock inclinations on the uniaxial compressive strength under wet–dry cycling conditions. It was demonstrated that the larger the rock inclination, the smaller the compressive strength of the rock. Fu Yan et al. [21,22] indirectly measured the tensile strength of rocks using the Brazilian disc splitting test. Zhao et al. [23] conducted uniaxial compression tests and triaxial compression tests on red sandstone under wet–dry cycling conditions to reveal the relationship between wetting–drying cycles and mechanical properties. Wang et al. [24] examined the influence of wetting–drying cycles on the creep characteristics of sandstone, analyzing the macroscopic and microscopic morphologies of the rock after destruction to disclose the deterioration mechanism. In general, wet–dry cycling exhibits a negative correlation with the strength of soft rock, that is, as the number of wetting–drying cycles increases, the strength of soft rock diminishes. Moreover, a higher water content corresponds to a lower rock strength. Jiang Jingdong et al. [25] conducted triaxial compression tests on mudstones with different water contents to study their mechanical properties and energy mechanisms. Cherblanc et al. [26] discovered that the loss of mechanical strength in rocks due to the water content is linearly related to the amount of water adsorbed by clay minerals. Nevertheless, the mechanism underlying the effect of wetting–drying cycles on the mechanical strength of rock masses remains elusive.

To elucidate the mechanism of wet–dry cycling, this study takes the red sandstone of an open slope in Jiangxi as the research object, with different initial saturation degrees and different numbers of wetting–drying cycles as variable factors. Through NMR and Brazilian splitting experiments and the incorporation of fractal theory, the pore structure characteristics and the variation rules of type I and type II fracture toughness characteristics of red sandstone in different water-saturated environments after wet–dry cycling are analyzed. The results of this analysis are expected to provide a valuable reference for slope stabilization and maintenance.

2. Test Programme

2.1. Specimen Preparation

Red sandstone from the slope of an open pit mine in Jiangxi Province was taken as the research object, and its mineral composition mainly consists of detrital minerals and clay minerals. The detrital minerals mainly include quartz, feldspar, calcite, and a small amount of hematite; the clay minerals are montmorillonite, illite, kaolinite, and chlorite. Under natural conditions, the material is brownish red, with uniform particles and no obvious texture on the surface.

The rock was processed into standard cylinders with diameter × height dimensions of 50 mm × 50 mm. Specimens of the Brazilian disc with a central crack (CSTBD) [27,28] were fabricated by mechanical cutting. The length of the central through crack was 15 mm, and the thickness was 1 mm.

To determine the time required for the specimen to reach a predetermined saturation level and the corresponding mass, the sandstone was saturated using a vacuum saturation device, with the saturation pressure set at 0.1 MPa for 48 h. The specimens were taken out and weighed at 10 min intervals for the first 2 h and then at 30 min intervals after that. Curve fitting was performed on the obtained data to obtain the time required for the red sandstone to reach 25%, 50%, 75%, and 100% saturation, which were 4 min and 16 s, 14 min and 47 s, 1 h and 8 min, and 2 days, respectively. To ensure uniform water distribution in the specimen and sufficient water–rock reaction in the specimens, the specimens were wrapped in cling film for 24 h after reaching the predetermined saturation time.

2.2. Tests

(1) Dry and wet cycle test

Given the alternating dry and wet environment of the test rock samples and the operability of the test, the "wetting–drying" cycle of the sandstone samples was realized by means of natural water absorption at room temperature and oven drying. In accordance with the standard specification of GB/T 50266—2013 [29], the red sandstone samples were placed in a water container to absorb water freely for 24 h and then put into an oven to be dried for 24 h (in order to minimize the impact of temperature on the mechanical properties of the rocks as much as possible, the oven temperature was set at 60 °C). This constituted one wetting–drying cycle. In this experiment, four initial saturation levels of 25%, 50%, 75%, and 100% were designed, together with five wetting–drying cycles performed for 0, 10, 20, 30, and 40 times, and each treatment was carried out in triplicate to ensure the reliability of the results.

(2) Nuclear magnetic resonance (NMR) test

The NMR system was used to detect the internal structure of the rock samples and obtain data on the total porosity, the T_2 spectral curve, pore size distribution, etc. The specific parameters were set as follows: the sampling frequency of the Carr–Purcell–Meiboom–Gill (CPMG) sequence was 200 kHz, the echo time (TE) was 0.15 ms, and the number of echo counts (NECH) was 8000. The waiting time (TW) was 2000 ms, and 16 scans were carried out. T_2 spectra were analyzed for each group of specimens, and the pore size as well as the pore distribution were calculated by relaxation time inversion.

(3) Fracture mechanical strength test

The Brazilian disc test and fracture toughness test were conducted in accordance with the recommendations of the International Society of Rock Mechanics standards (ISRM) [30]. The tests were performed by displacement-controlled pressurization: the displacement loading rate was 0.1 mm/min, and the angles between the fissure inclination and the direction of cleavage loading were set to be 0° and 27.2°, respectively, in order to obtain the Type I and Type II fracture toughness of the specimens [31]. The flow chart of the test is shown in Figure 1 below.

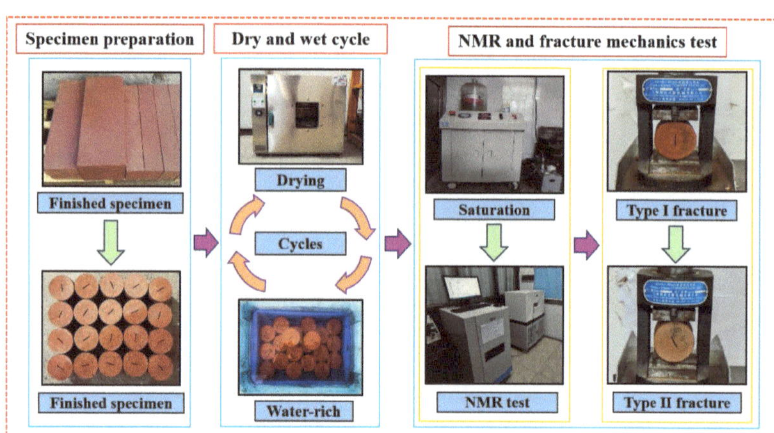

Figure 1. The flowchart of the test.

2.3. NMR Aperture Conversion and Fractal Calculation

2.3.1. NMR Aperture Conversion

The relaxation time (T_2) distribution curve of rock samples can be obtained using the NMR technique, and it has been shown that the NMR transverse relaxation time T_2 has a good linear correspondence with sandstone pore properties. Therefore, the size of the distributed pores of the rock can be approximated by the transverse relaxation time T_2, and the relationship between T_2 and pore size, r, can be expressed as follows [32,33]:

$$r = \rho_2 F_s T_2 \tag{1}$$

where r is the diameter of the pore structure, μm; ρ_2 is the surface relaxation strength of the material, μm/ms; F_s is the geometrical shape factor of the pore structure, which takes the value of 3 for spherical pores and 2 for columnar pores; and T_2 is the transverse relaxation time in ms.

In this paper, by referring to similar studies [34,35], the following values of relaxation strength are chosen: ρ_2 = 0.01 μm/ms; F_s = 3. Therefore, Equation (1) can be expressed as follows:

$$r = 0.03T_2 \tag{2}$$

2.3.2. NMR Fractal Dimension Calculation and Pore Structure Delineation

Red sandstone is a complex porous medium with complex and unevenly distributed pore structures. It is difficult to comprehensively characterize its pore structure and its evolution using traditional descriptive methods. Fractal theory, as a geometric method for studying the structural characteristics of irregular objects with self-similarity properties, can be used to quantitatively analyze the complexity, similarity, or heterogeneity of the pore structure of sandstone [36,37].

The parameter used to quantitatively describe this self-similarity is called the fractal dimension (D), which is related to the complexity of pores and the roughness of the surface. The effective range of the fractal dimension of rock pores is from 2 to 3 [38–40]. When the fractal dimension of a certain type of pore is less than 2, it indicates that the pore has poor fractal properties or lacks fractal characteristics. The closer the value is to 3.0, the more complex the pore structure is and the stronger the heterogeneity is; conversely, the simpler the pore structure is, the weaker the heterogeneity is.

When pores have fractal characteristics, the relationship between the number of pores and the pore diameter can be expressed by a power law function [41,42], and its mathematical expression is shown in Equation (3):

$$N(r) = \propto r^{-D} \tag{3}$$

In the above formula, $N(r)$ represents the number of pores with a pore diameter greater than r in the rock mass, the symbol r refers to the pore diameter of the internal pore structure of the rock mass, and D is the fractal dimension of the rock mass.

According to the relevant methods of calculating the dimension, the number of pores with a pore diameter greater than r in the rock mass with fractal characteristics can be further expressed by Equation (4) [43,44].

$$N(r) = \int_r^{r_{\max}} g(r)dr = Ar^{-D} \tag{4}$$

where $N(r)$ is the number of pores with a pore diameter greater than r; r is the pore radius; $r_{\max}$ the maximum pore radius in the rock; A is the fractal factor; $g(r)$ is the density function of pore size distribution in the rock; and D is the fractal dimension of the rock. Assuming that the pore space of the rock is composed of a series of capillary bundles, according to the Brooks–Corey [45] model, it is known that the cumulative volume fraction, S_v, with a pore radius less than r is expressed by Equation (5):

$$S_v = \left(\frac{r_{max}}{r}\right)^{D-3} = \left(\frac{T_{2max}}{T_2}\right)^{D-3} \quad (5)$$

The derivation of both sides of the above equation leads to Equation (6) for the NMR fractal dimension:

$$\lg(S_v) = (3-D)\lg(T_2) - (3-D)\lg(T_{2max}) \quad (6)$$

By plotting the curve using the above equation, with lg (S_v) representing the vertical axis, lg (T_2) representing the horizontal axis, and (3-D) lg (T_{2max}) representing the constant, the slope (3-D) can be determined. Subsequently, the fractal dimension of sandstone pores is calculated from this slope.

Considering the complexity of the internal pore structure of red sandstone, it is beneficial to further classify the pores to analyze the pore structure in more detail. In this paper, the classification of the pore size is mainly based on the study of Yan et al. [35,36]. The pore sizes of sandstones are classified into three categories: micropores (r < 0.1 μm), mesopores (0.1–1 μm), and macropores (r ≥ 1 μm). The porosity sizes of micropores, mesopores, and macropores in the specimens were calculated and noted as Φ_{mi}, Φ_{me}, and Φ_{ma}, respectively.

The classification basis shown in Figure 2a yielded the results of pore structure division, and the fractal dimension of each pore was calculated as shown in Figure 2b; the fractal dimensions of the total pore, micropore, mesopore, and macropore are characterized by D_{total}, D_{mi}, D_{me}, and D_{ma}.

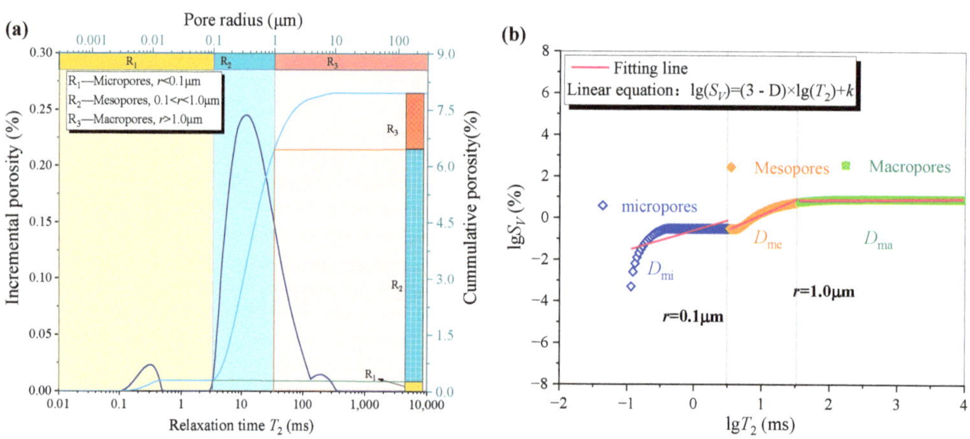

Figure 2. (**a**) Pore division basis. (**b**) Example diagram of fractal dimension calculation.

3. Analysis of Experimental Results

3.1. Characteristics of T_2 Spectrum Distribution and Evolution Law

The NMR T_2 spectrum curves adeptly reveal the information of pores of diverse sizes within the tested rock specimens. Notably, the T_2 relaxation time positively correlates with the pore aperture size, r; specifically, a larger T_2 value indicates a larger pore aperture. Moreover, the amplitude of the T_2 curve serves as a proxy for the quantity of pores within a specific aperture range such that a greater amplitude corresponds to a larger number of such pores. Figure 3 presents the T_2 spectra of each sandstone specimen group subjected to varied treatments.

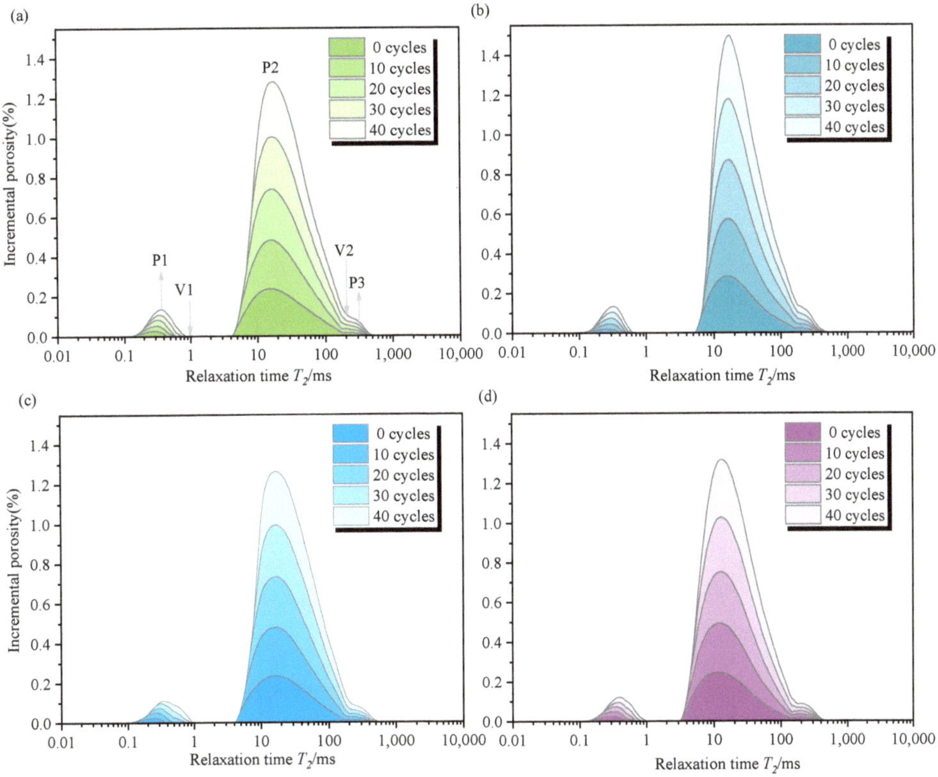

Figure 3. T_2 spectra of pore characteristics of sandstones with different initial saturations undergoing different numbers of wet and dry cycles. (**a**) Initial saturation of 25%; (**b**) Initial saturation of 50%; (**c**) Initial saturation of 75%; (**d**) Initial saturation of 100%.

From Figure 3, it is observed that the pore structure development characteristics of each group of red sandstone under dry and wet cycling exhibit relative consistency. The T_2 pore increment curves primarily exhibit three spectral peaks, namely P1, P2, and P3, along with two spectral valleys, V1 and V2. The relaxation time of P1 spans from 0.061 to 0.977 ms. P2 is concentrated within the range of 3.181–580.523 ms, and P3 within the range of 580.523–4055.461 ms. These three peaks correspond to three distinct pore clusters. Notably, the P2 peak is the tallest, and its relaxation time range is the broadest, indicating that the specimens witness the most favourable mesopore development, followed by micropore growth. The boundary separating the macropore and mesopore spectral clusters is somewhat indistinct. In contrast, a more conspicuous trough lies between the small and medium pore spectral peaks, signifying that the pore throats among connected pores are relatively narrow, and the pore connectivity is deficient [46].

Under the influence of dry and wet cycling, all three peaks experience concurrent growth. This implies that as the cycle number increases, the porosity increment surges. Concurrently, the relaxation time of each pore shifts rightward, denoting an augmentation in pore size and, consequently, a rise in cumulative porosity. Moreover, the relaxation time of the two troughs undergoes an overall reduction, demonstrating that dry and wet cycling not only expedites pore size expansion but also bolsters inter-pore connectivity.

3.2. *Characteristics of Sandstone Pore Structure*

3.2.1. Change Rule of Total Porosity of Mortar

To quantitatively analyze the pore structure characteristics of sandstone, the results of the NMR test on sandstone porosity are presented in Table 1.

Table 1. Sandstone porosity.

Number of Drying–Wetting Cycles (n)	25% Saturation	50% Saturation	75% Saturation	100% Saturation
0	8.032	8.103	8.001	7.956
10	8.198	8.192	8.359	8.280
20	8.365	8.370	8.445	8.442
30	8.416	8.487	8.560	8.726
40	8.436	8.674	8.739	8.875

The porosity of the rock mass with the same initial saturation increases with the increase in the number of wet and dry cycles. For example, for the rock sample with an initial saturation of 50%, the average porosity after 0 wet and dry cycles was 8.103%. After 10 cycles, the porosity was 8.192%, and the growth rate was 1.108%. After 20 cycles, the porosity was 8.370% with a growth rate of 3.305%. After 30 cycles, the porosity reached 8.487%, and the growth rate was 4.739%. After 40 cycles, the porosity was 8.674%, and the growth rate was 7.049%.

It is observed that in the first 10 cycles, the porosity growth rate of 50% saturation rock samples was slow. After 10 cycles, the growth rate accelerated. There are two main reasons for this. First, in the 50% initial saturation rock samples, fewer pores formed at the start of the cycle. Second, the micropore size led to poor porosity connectivity and permeability. The damage and pore structure change during the wet and dry cycle is a progressive and cumulative process. More porosity provides better conditions for later wet and dry cycles. So, the porosity grows faster in the later stage than in the earlier stage.

Among the four groups of initial saturation samples, the rock sample with 100% saturation had the most significant porosity growth. From 0 to 40 wet and dry cycles, the growth rates were 4.077%, 6.111%, 9.686%, and 11.554%. There are two main reasons for this. First, in the NMR test with 0 cycles, the 100% saturated rock samples were wet. More water filled the pores, making the porosity seem smaller before cycling. Second, these samples were saturated the longest. They fully reacted with water. Soluble salts in the rock reacted with water [47]. Some ions dissolved. Clay minerals expanded and generated force. The macropores in red sandstone released this force. The reaction of salts, clay, and water enlarged the pore space. Also, carbonate hydrolysis and clay mineral changes made quartz and feldspar lose adhesion. So, the rock lost cementation and had many pores.

In the later wet and dry cycle tests, water molecules enter into the sandstone via the structural surfaces. When drying, the water in the rock evaporates, and soluble substances move to the rock surface. When these substances dissolve, the sandstone forms new pores. So, the pore growth rate is significant.

3.2.2. Mortar Division of Pore Structure Porosity

The results of pore structure division are presented in Figure 4, where (a)–(d) represent the porosity distribution of micropores, mesopores, and macropores of sandstone with initial saturation ranging from 25% to 100% after wet and dry cycles, respectively.

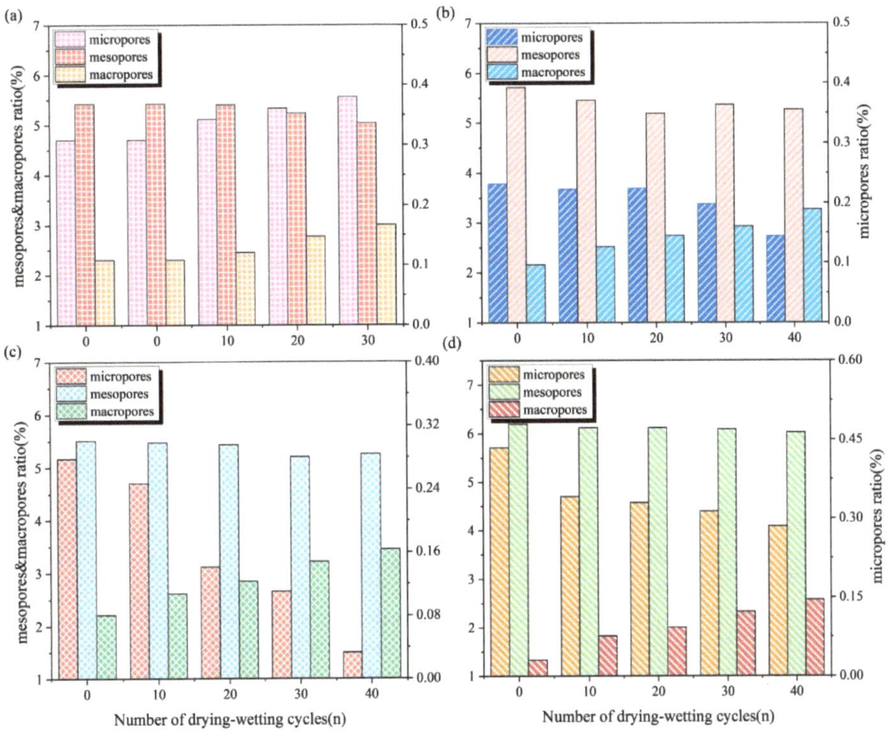

Figure 4. Pore distribution of sandstones with different initial saturation after wet and dry cycles. (**a**) Initial saturation of 25%; (**b**) Initial saturation of 50%; (**c**) Initial saturation of 75%; (**d**) Initial saturation of 100%.

As shown in Figure 4, each group of sandstone samples is mainly mesoporous prior to wet and dry cycles. The development of different pore types changes with the increase in cycle numbers.

The Φ_{mi} distribution of each specimen ranges from 0.033 to 0.443. The Φ_{mi} of 25% saturated sandstone shows an overall increasing trend with more cycles, while that of the 50%, 75%, and 100% saturated rock samples decreases. It can be found that when the initial water saturation time is shorter, the damage to the rock samples during the dry and wet cycling process is minor. Although the micropore size shows growth and development, it does not transform a large number of micropores into mesopores and macropores. With the increase in the initial water saturation time, the water–rock action is intensified, which promotes the growth of the size of the micropores and their conversion into medium pores during dry and wet cycling.

From the perspective of mesopores, the distribution of Φ_{me} within each saturated sandstone does not display obvious growth or reduction trends under the influence of wet and dry cycling but rather exhibits a fluctuating state. This is attributed to two aspects. First, the original mesopore size increases under the effect of wet and dry cycling, and some micropores further develop into mesopores during this process. Second, a portion of mesopores further evolve under the action of wet and dry cycling and enlarge to become macropores. The expansion of this part of the mesopore gaps leads to the reduction in the mesopore spectral area. Under the combined effect of these two factors, the spectral area of the mesopores in the rock samples shows a fluctuating pattern of change.

The Φ_{ma} value of rock samples increases remarkably with wet and dry cycling according to each initial saturation degree. Among them, the Φ_{ma} value of the rock samples with an initial saturation degree of 100% exhibits the most significant increase during wet and dry cycling. This is because macroporosity represents the final stage of pore development. The transformation of existing small- and medium-sized pores to macroporosity and the development of macroporosity itself result in an increase in the spectral area of macroporosity. Moreover, the damage caused to the rock body is a cumulative process. With the deepening of wet and dry cycling, the internal structure of the sample is more susceptible to damage due to the accumulation of damage, and thus, the growth rate of Φ_{ma} shows a trend of rapid increase.

3.3. The Characteristics of the Correlation Between the Pore Structure and Fractal Dimension

3.3.1. Fractal Characterization of Pore Structure

The fractal dimensions of the samples and each pore are presented in Table 2. The overall fractal dimension values of the rock samples range from 2.454 to 2.521, demonstrating a satisfactory level of self-similarity. Strictly speaking, this self-similarity is of a statistical nature. Due to the complex composition and disordered structure of actual rocks, the pore networks therein do not exhibit the strict self-similarity as that of the theoretical fractals [48–50]. After 0 wet and dry cycles, the total fractal dimension (DTotal) of the rock samples is less than 2.5. This implies that the pore geometry of the original rock is relatively simple and homogeneous. As the number of wet and dry cycles increases, the internal pores changes due to dissolution and expansion, which makes the rock structure more complex and less homogeneous. Consequently, the fractal dimension of the rock samples increases.

Table 2. Fractal dimensions of sandstone pores.

Cycles	Initial Saturation	Fractal Dimension			
		D_{total}	D_{mi}	D_{me}	D_{ma}
0	25% Saturation	2.494	−0.930	2.531	2.897
10		2.501	−0.480	2.562	2.923
20		2.499	−1.071	2.554	2.931
30		2.506	−0.800	2.593	2.938
40		2.514	−0.547	2.672	2.947
0	50% Saturation	2.454	0.082	2.263	2.865
10		2.486	0.011	2.359	2.871
20		2.499	0.202	2.396	2.915
30		2.506	0.409	2.424	2.931
40		2.508	0.488	2.434	2.942
0	75% Saturation	2.454	−0.786	2.238	2.891
10		2.500	−0.749	2.305	2.913
20		2.499	0.051	2.292	2.927
30		2.509	−0.101	2.348	2.924
40		2.507	−0.606	2.402	2.935
0	100% Saturation	2.483	−0.644	1.856	2.927
10		2.521	0.113	1.898	2.913
20		2.502	0.387	1.836	2.937
30		2.513	0.376	1.807	2.944
40		2.520	0.074	1.879	2.948

The relationship among different pore sizes, namely $D_{mi} < D_{me} < D_{ma}$, indicates that macropores are more complex and possess stronger fractal characteristics. Non-homogeneity increases as the number of cycles and the initial saturation rise, which is most likely due to the formation of macropores. As the saturation and the number of cycles increase, the walls of small and medium pores start to dissolve from their weak points. Subsequently, some of these pores connect with each other to form macropores. These macropores retain some characteristics of small and medium pores, so they possess specific fractal features and have high spatial complexity and strong self-similarity (Figure 5). D_{mi} does not exceed 0.488 and is outside of the normal range (2–3) [38,51]. Moreover, its values change irregularly. Therefore, it can be considered that the micropores within the rock samples do not possess fractal characteristics, as for the fractal dimension of micropores, this phenomenon is quite common [52,53]. This is because when the pore diameter is smaller than a certain range, the spatial structure of the pores will change, manifested as smaller and less connected pores [54], and they are no longer similar to the spatial structure of the rock mass [55]. This also suggests that micropores may be nearly linear [56], mainly originating from microfractures caused by the local expansion of clay cement. The values of D_{me} range from 1.807 to 2.672, showing obvious fractal features compared to micropores. When the initial saturation is 100%, D_{me} decreases significantly, and non-homogeneity weakens. This may be because mesopores grow due to erosion when fully saturated, and many of them turn into macropores.

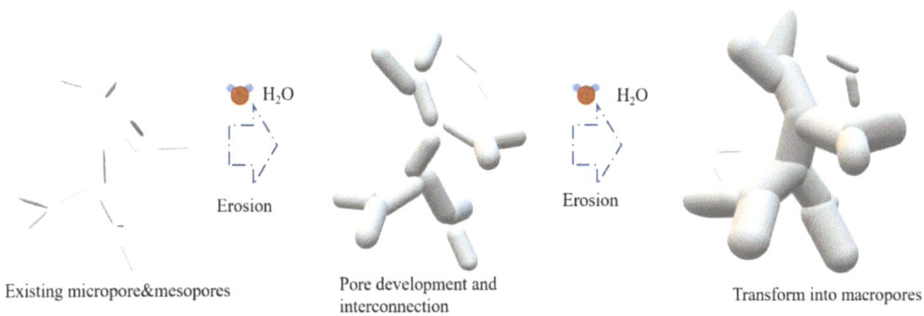

Figure 5. The process of macroporosity formation.

3.3.2. Correlation Between Fractal Dimension and Porosity

As the pore fractal dimension indicates the pore structure quality, it is necessary to further explore its relationship with porosity variation. Hence, it is necessary to further determine the relationship between the variation in the fractal dimension and porosity. Based on the porosity and fractal dimension features of different pore structures in sandstone samples with various saturation and cycle numbers, a linear relationship between the fractal dimension and porosity of the sandstone was set up and then graphically fitted. Since the fractal dimensions of micropores are all less than 2, only the fractal dimensions of mesopores, macropores and total pores are fitted together with porosity.The results are shown in Figure 6.

According to the fitting results, the D_{Total} and Φ_{Total} values do not show an obvious correlation. Only the porosity of the sample with 75% saturation has a better linear fitting relationship with the fractal dimension. This might be due to the following reasons: Firstly, the fractal dimension of the total porosity [47] mainly characterizes the complexity of the overall pore structure of the rock samples. Secondly, according to the pore division, it is known that the complexity of the total pore structure is influenced not only by the total porosity of the sample but also by the composition and proportion of various types of pores in the sample. These factors together lead to the fluctuation in the fractal dimension of total porosity. Consistent with the research findings of Shi et al. [49,57], an increase in porosity does not necessarily cause an increase in the complexity of the pore structure.

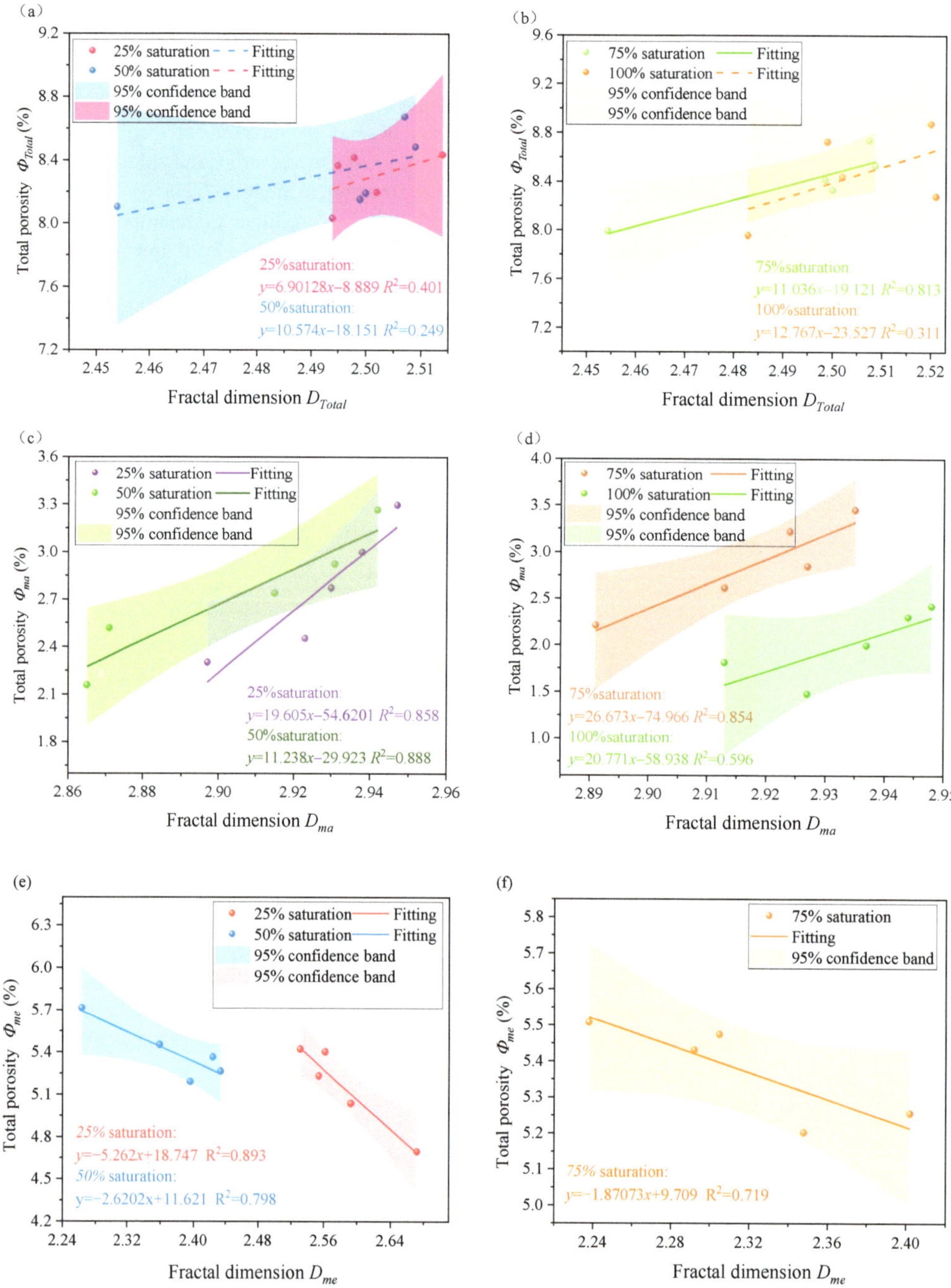

Figure 6. Plot of fractal dimension versus porosity fit. (**a**) Fitting conditions of D_{Total} and Φ_{Total} for rock samples with initial saturations of 25% and 50%; (**b**) Fitting conditions of D_{Total} and Φ_{Total} for rock samples with initial saturations of 75% and 100%; (**c**) Fitting conditions of D_{ma} and Φ_{ma} for rock samples with initial saturations of 25% and 50%; (**d**) Fitting conditions of D_{ma} and Φ_{ma} for rock samples with initial saturations of 75% and 100%; (**e**) Fitting conditions of D_{me} and Φ_{me} for rock samples with initial saturations of 25% and 50%; (**f**) Fitting conditions of D_{me} and Φ_{me} for rock samples with initial saturations of 75% and 100%.

The correlation between D_{ma} and Φ_{ma} and between D_{me} and Φ_{me} is good (R^2: 0.719–0.893). This means the porosity of red sandstone's macropores and mesopores is closely tied to their pore structure complexity. Specifically, D_{me} and Φ_{me} are positively related. As the original macropores grow, some meso- and meso-small pores connect and turn into macropores, making the macropore structure more complex. But D_{me} and Φ_{me} also have a negative linear correlation. In dry and wet cycles, although the Φ_{me} value decreases overall, the mesopore's internal structure becomes more complex. This may be because cycling makes small- and medium-sized pores develop and connect more. So, mesopores keep turning into macropores, and porosity drops. Thus, the negative correlation between porosity and structure complexity appears.

3.4. Characteristics of Fracture Mechanics Changes in Red Sandstone

3.4.1. Changing Law of Fracture Toughness

Figure 7 shows the changes in the type I and II fracture toughness after different initial saturation tests and dry and wet cycling tests. As the number of cycles rises, the fracture toughness of both type I and type II red sandstone drops notably. This is because the growth in the pore space damages the clay cement structure between sand grains, and this affects the shear strength. It is known that the volume, size, and shape of the pore space are the main reasons for the fall in the rock body's fracture performance.

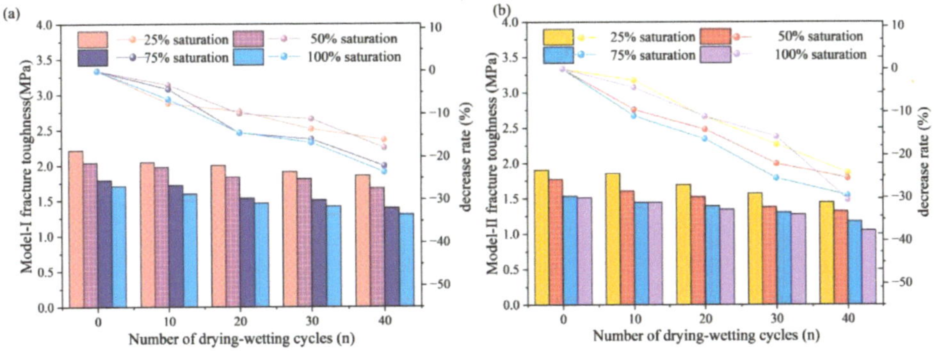

Figure 7. Type I (**a**) and II (**b**) fracture toughness of sandstones with different initial saturations.

Dry and wet cycling impacts the type II fracture toughness of rock samples more than type I. After 40 cycles, the type I fracture toughness decreases by 24.13% from its initial value at 0 cycles, while the type II fracture toughness drops by 31.35%. This is due to different fracture modes in the stress concentration area. Type I fracture is perpendicular to the crack surface and mainly affected by tensile force. Type II is parallel to it and mainly affected by shear force [58,59].

The higher the initial saturation, the larger the decrease in the rock samples' fracture toughness. This means the higher the water content, the more obvious the effect of cycling on fracture toughness. From Section 3.2's analysis (showing that more water in the rock body leads to more water dissolution and erosion and more pores), we can guess that during cycling, the initial saturation mainly affects the pore space and indirectly worsens the rock body's anti-shear ability.

The fracture toughness of rock samples without cycling also decreases as the initial saturation rises. This shows that other factors work when water fills pores. Some researchers think substances like calcite, dolomite, and gypsum in the rock mass dissolve in water or expand after absorbing water, thus breaking the rock mass structure [60]. Another idea is that water molecules with high polarity can form weak hydrogen bonds with oxides like

SiO_2. This makes the close bond between sand and clay cement worse, causing micro-cracks and local loosening. This process builds up at the macroscopic level and weakens the rock's resistance to slippage and fracture [61]. When loaded, rock samples are more likely to have internal slippage and fractures. The cyclic suction–drying effect makes the damage from wet and dry cycling worse and leads to more serious internal damage. So, the fracture strength of rock samples with high initial saturation changes more with the number of dry and wet cycles.

3.4.2. Characteristics of Correlation Between Pore Structure and Fracture Mechanics

In order to further investigate the correlation between porosity, pore structure evolution, and fracture mechanical strength of red sandstone under dry and wet cycling, the porosity, fractal dimension, and fracture toughness were fitted, respectively. According to the literature, there may exist linear [39,62], exponential [47,63], and power function [64] relationships between the pore indexes of the rock body and its mechanical properties. These three functions were fitted separately, and the correlation coefficients (R^2) are presented in Table 3. The linear correlation degree between the porosity, fractal dimension, and fracture toughness (with an R^2 mean value of 0.799) is the highest. This might be attributed to the fact that the maximum porosity of the rock mass after undergoing 40 cycles of wetting and drying is only 8.9%, indicating that it is still in the primary stage of pore development. Moreover, the results of the study by Amiri M [65] also demonstrate that for porosity below 10%, there is a good linear relationship with fracture mechanical strength.

Table 3. Correlation coefficients for different fitted lines.

Fracture Toughness	Saturation	Related Coefficient					
		Porosity			Fractal Dimension		
		Linear Function	Exponential Function	Power Function	Linear Function	Exponential Function	Power Function
Type I	25%	0.914	0.495	0.876	0.812	0.442	0.836
	50%	0.927	0.736	0.931	0.783	0.638	0.689
	75%	0.894	0.628	0.833	0.610	0.702	0.601
	100%	0.923	0.764	0.918	0.331	0.340	0.382
Type II	25%	0.937	0.759	0.895	0.789	0.441	0.702
	50%	0.932	0.642	0.931	0.881	0.738	0.858
	75%	0.957	0.661	0.911	0.688	0.502	0.435
	100%	0.942	0.834	0.839	0.456	0.490	0.268

The mean value of the coefficient of correlation R^2 for the linear fit of porosity and fractal dimension to type II fracture toughness is 0.823, which is higher than that of R^2 with type I fracture toughness (0.774). This corresponds to the conclusion that the shear resistance of the rock mass is more sensitive to porosity.

Different sizes of pores exert different degrees of influence on the fracture toughness of the rock body. In combination with the analyses in Sections 3.2 and 3.3, micropores account for a relatively small proportion, and under the effect of dry and wet cycling, the porosity of micropores generally decreases, and structural changes are not obvious. The porosity of mesopores fluctuates in its changing trend, while its structural complexity increases. The porosity and geometric complexity of macropores increase regularly. For this reason, it is hypothesized that macropores have the maximum effect on fracture toughness. Taking type II fracture toughness as an example, the correlation between the porosity and fractal dimension of macropores and type II fracture toughness was analyzed by linear fitting, with the total pores used for comparison. The results are shown in Figures 8 and 9. The fitting results reveal that the porosity and fractal dimension of macropores and total pores were negatively correlated with type II fracture toughness. Additionally, there was a general tendency for the R^2 value to decrease in the equation of type II fracture toughness versus Φ_{ma} and D_{ma} as the saturation degree increased. This is because different initial saturations lead to different outcomes even after undergoing the same dry and wet cycling process, which also affects the complexity of the internal pore structure of rock samples. From a macroscopic perspective, during the migration of water inside the rock, a large number of soluble minerals and other components are migrated and dissolved along with the water, resulting in the formation of more microstructures within the rock body. These micropores continue to develop, weakening the strength of the rock and causing the regularity between the fractal dimension and mechanical strength to no longer be strongly correlated.

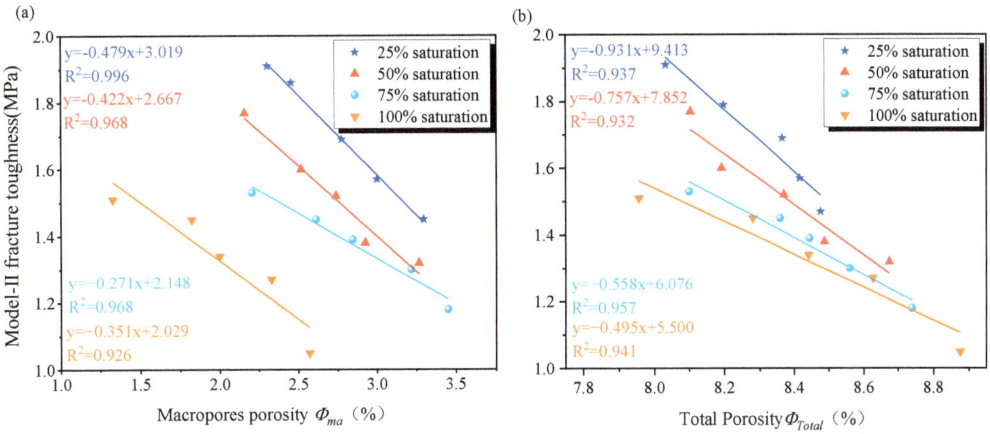

Figure 8. Comparison of Fitting Relationships between Porosity and Type II Fracture Toughness. (**a**) Fitting results between Φ_{ma} and Type II fracture toughness; (**b**) Fitting results between Φ_{Total} and Type II fracture toughness.

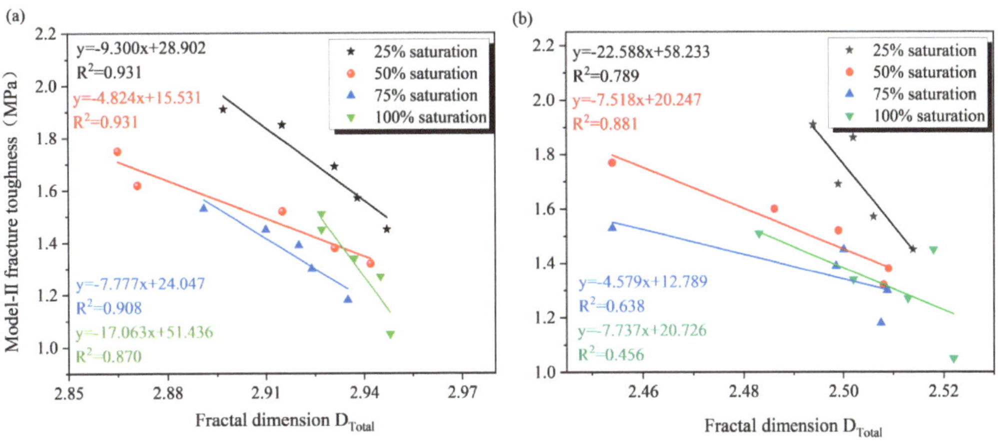

Figure 9. Comparison of the Fitting Relationships between the Fractal Dimension of Sandstone and Type II Fracture Toughness. (**a**) Fitting results between D_{ma} and Type II fracture toughness; (**b**) Fitting results between D_{Tota} and Type II fracture toughness.

Figure 8 shows that the linear fit of type II fracture toughness to the Φ_{ma} and Φ_{Total} values of rock samples with each amount of saturation is better, with the fitting coefficient R^2

ranging from 0.876 to 0.966, which basically shows a more direct negative linear correlation. That is, with the growth of Φ_{ma} and Φ_{Total} in the pore structure, the macroscopic fracture toughness of the rock mass decreases. This also indicates that the porosity of macropores and total pores has a significant effect on the mechanical strength of rocks.

On the contrary, in Figure 9, the R^2 variation interval of the fractal dimension of macropores and type II fracture toughness is located in the range of 0.927–0.944, which is overall much higher than that of the total porosity (R^2 at 0.456–0.881). This is attributed to the fact that the total porosity in the pore structure also contains micropores and mesopores, and the trend of their fractal dimension is not in line with the fracture toughness, which suggests that macropores are the main cause of the effect on the rock's shear-resistant performance. The development of mesopores into macropores is the main mechanism for the deterioration of macroscopic fracture mechanical properties of the rock mass by dry and wet cycling.

4. Discussion

In order to explore the evolution laws of the pore structure and mechanical properties of red sandstone under the action of drying–wetting cycles, the red sandstone from an open pit slope in Jiangxi was taken as the research object. Two variables, the initial saturation and the number of drying–wetting cycles, were set. With the help of NMR, CSTBD tests, and fractal theory, the changes in the microscopic pore structure and macroscopic fracture mechanical properties of red sandstone were explored based on aspects such as its porosity, pore proportion, the geometric complexity of different internal pores, and the type I and type II fracture toughness. Furthermore, a fitting analysis of the correlation between the two was carried out, which can, to a certain extent, reveal the damage mechanism of red sandstone under the action of drying–wetting cycles. However, the degradation of the microstructure and macro-mechanical properties of red sandstone is usually a multi-level phenomenon resulting from the combined action of physical and chemical factors, which is worthy of an in-depth exploration. Therefore, in this section, combined with the existing experimental results and the relevant research theories on the water–rock interaction of red sandstone, the degradation process of red sandstone under the action of drying–wetting cycles is analyzed and discussed. The damage mechanism is shown in Figure 10.

(1) Red sandstone is primarily composed of detrital minerals and clay minerals, with the clay minerals serving as the cementing agents. The structural arrangement of mineral particles is relatively loose, and there are numerous minute pores in the original red sandstone, providing convenient channels for water to penetrate the interior of the rock. Among the cementing substances, montmorillonite and calcite are readily soluble in water [66]. The dissolution of clay minerals leads to weakening of the cementation strength. Illite exhibits remarkable hydrophilicity and water absorption capacity and is prone to undergoing hydration and expansion due to chemical reactions upon contact with water [67]. When kaolinite clay particles encounter water, the bound water film on the particle surface thickens, resulting in intergranular expansion [68]. The expansion of clay particles contributes to increasing the distance between clay particles and between particles and pores. After the volume expansion of mineral particles, the structure of the sample becomes increasingly loose. Therefore, red sandstone has a relatively high degree of water sensitivity [69].

(2) At the initial stage of water saturation, water molecules gradually penetrate into the interior of the rock from the surface. Under the combined effects of physical softening and chemical erosion, such as the action of pore water pressure, the expansion of mineral particles, the hydration reactions of mineral components, and dissolution [70], the structure tends to become loose, further expanding the water flow channels. During the wetting–

drying cycle, in the drying stage, the water molecules in the pores evaporate, which is similar to the "reverse reaction" during water saturation: the pore water pressure decreases, the expanded particles contract, and some dissolved substances migrate with the water. The water saturation test shows that the water molecule content and the degree of penetration inside the red sandstone increase with the increase in the water saturation time. That is, an increase in saturation implies that the pores are filled with a large amount of water, and both the physical and chemical reactions of the internal components are more thorough. For the sample with an initial saturation of 25%, due to the relatively short water saturation time, water molecules have difficulty penetrating deep into the structure, and the mineral components do not react fully. After 40 cycles, the total porosity only increases by 5.030%. For red sandstone with a saturation in the range of 50%–100%, with the increase in soaking time, the increase in porosity under the wetting–drying cycle continues to rise. For the fully saturated rock sample, the increase in porosity after 40 cycles reaches 11.551%. The increase in the fractal dimension value, which characterizes the geometric complexity of the pores, is twice that of the rock sample with a 25% saturation after 40 cycles. The changes in the mesoscopic structure of the sandstone increase the internal defects, leading to damage and failure. The damage spreads to the weak structural planes, and the increase in the complexity of the internal structure and the tendency towards heterogeneity, to a certain extent, result in the deterioration of macroscopic mechanical properties. After 40 wetting–drying cycles, the type I fracture toughness of the 100% saturation rock sample decreases by 23.578%, and the type II fracture toughness decreases by as much as 30.642%.

(3) Red sandstone with different initial saturations continuously undergoes water-absorption and drying during the wetting–drying cycle. During drying, the water inside the rock evaporates, and the hydrated reaction products dissolved during water saturation are carried out. The dissolution of soluble substances generates new micropores and microcracks. The development and connection of the original pore sizes accelerate, the porosity increases, and medium-sized and small pores transform into large pores. The porosity of large pores and the fractal dimension value continue to increase, the rock structure becomes loose, and the cementation strength and cohesion weaken. Eventually, the water–rock interaction causes the internal cracks in the red sandstone sample to develop and expand fully, resulting in microstructural damage. The framework minerals are more likely to peel off and slide under load [71]. During the Brazilian disc test, when loaded, the internal cracks expand and penetrate, leading to the destruction of macroscopic mechanical properties.

It is worth noting that this study has certain limitations. On the one hand, the research was not carried out directly from the perspective of various components of red sandstone. The exploration of the complexity of mineral compositions, such as clay minerals and detrital minerals, as well as their specific evolution mechanisms and reaction sensitivities under the water–rock interaction, is not deep enough. Therefore, the specific degree of their influence on the long-term stability of red sandstone under drying–wetting cycle conditions remains unclear. On the other hand, in the research on the correlation between the pore structure and fracture mechanics, although a linear relationship among porosity, fractal dimension, and fracture toughness was found, the heterogeneity of the rock was not analyzed in depth in combination with the characteristics of different clay minerals. This may affect the accuracy of the established models and relationships, as well as the understanding of the sensitivity of each component to the cracking mechanism and the reaction mechanisms. Based on the above-mentioned deficiencies, future research should include the following aspects:

(1) Research on the migration of clay minerals at the micro level: More experimental techniques should be introduced to directly observe the migration of clay minerals at the

micro level. For example, in situ monitoring equipment can be used to track the movement of clay particles during the drying–wetting cycle process. Additionally, micro-imaging techniques can be used to analyze the morphological changes in clay minerals before and after phenomena such as water-induced swelling, hydration, and dissolution so as to clarify the specific role of clay migration in the degradation mechanism of red sandstone.

(2) An expanded research scope for clay minerals: The research scope should be expanded to the fields of interlayer clay and clay roughness. With the help of material characterization methods such as X-ray diffraction and scanning electron microscopy, the structural characteristics of different types of clay minerals in red sandstone and their responses to water–rock interaction should be analyzed. These factors should be incorporated into a theoretical model to improve accuracy when predicting the degradation process of red sandstone.

(3) An in-depth exploration of the correlation between the pore structure and fracture mechanics: The influence of factors such as the pore shape and connectivity on the correlation between the pore structure and fracture mechanics should be further explored. Numerical simulation methods such as a finite element analysis should be used to establish a more comprehensive mechanical model while considering the heterogeneity of the rock. This model should be verified by a large amount of experimental data to provide a more reliable theoretical basis for the engineering stability analysis of red sandstone.

Overall, this study lays a certain foundation for understanding the evolution of the pore structure and mechanical properties of red sandstone under drying–wetting cycles, but there are still many aspects that need to be improved upon and explored in depth. Future research should follow the above-mentioned directions to continuously deepen the understanding of the behaviour of red sandstone in complex environments.

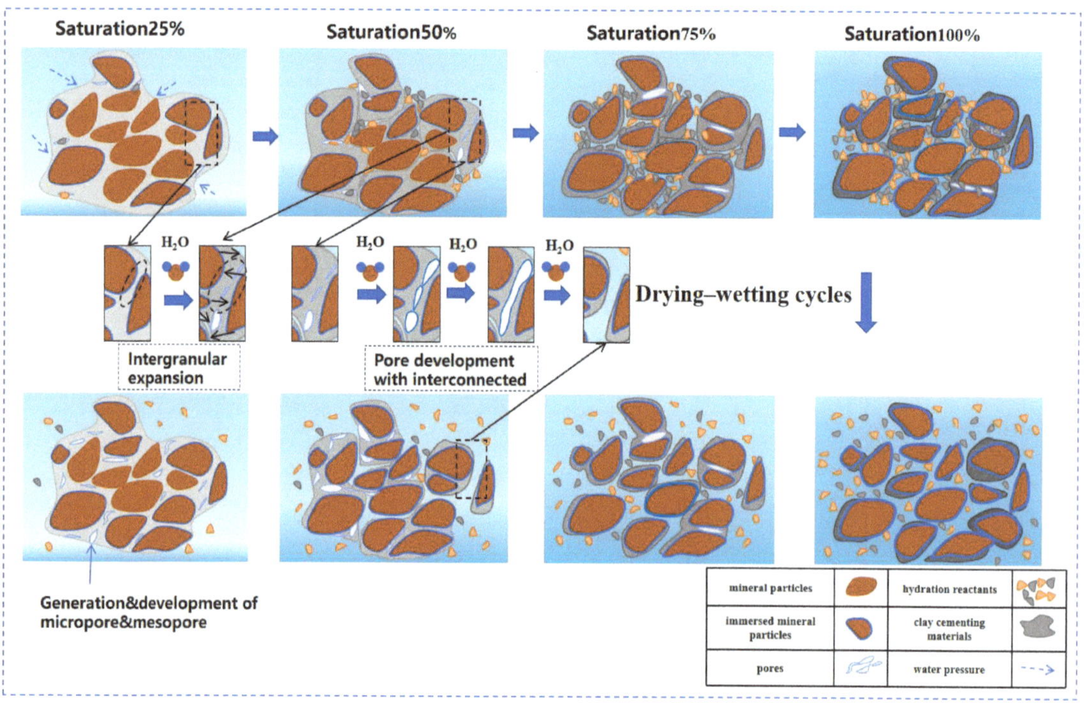

Figure 10. Pore structure damage mechanism of red sandstone.

5. Conclusions

(1) Red sandstone's porosity and fractal dimension increase continuously under wet–dry cycling, especially when fully saturated. After 40 cycles, the porosity increases by 11.554% compared to 0 cycles with an accelerated rate. Concurrently, the fractal dimension value rises from 2.483 to 2.520. These changes can be ascribed to the dissolution and expansion of clay minerals within the rock body.

(2) The NMR T_2 spectrum of red sandstone specimens shows three spectral peaks. Based on the pore size, the pores are further categorized into Φ_{ma}, Φ_{me}, and Φ_{mi}, and the fractal dimension is computed to comprehend the distribution traits of each pore type. During wetting and drying cycles, the porosity of micropores generally decreases. Owing to their small size and simple shape, they lack fractal characteristics. The structural complexity of mesopores experiences a slight increase. Although the porosity proportion of mesopores remains the highest throughout the experiment, it does not change remarkably. This can be attributed to the progressive development of micropores and mesopores into macropores. The porosity and geometric complexity of macropores increase conspicuously and display a strong correlation. Since macropores are mainly formed by the interconnection of small and medium pores, they possess certain fractal features. As the porosity of macropores changes, the fractal dimension also increases.

(3) The type I and type II fracture toughness of the rock samples both decrease with the increase in the number of wetting and drying cycles. Notably, the fracture toughness of the rock samples in a fully saturated water state exhibits the most significant reduction under the influence of wetting and drying cycles. This indicates that wetting and drying cycles cause tensile and compressive fracture damage to the rocks to varying extents, and such damage is aggravated by the increase in the initial water saturation degree of the rock body. The Φ_{total} and Φ_{ma} are both negatively correlated with the two types of fracture mechanical toughness of the sandstone, and the linear fitting effect is satisfactory. The growth in the total porosity and macroporosity has a relatively direct relationship with the attenuation of fracture toughness. Under various variable conditions, the fractal dimension and porosity of macropores are linearly related to the two mechanical strengths (R^2: 0.876–0.996), suggesting that the porosity and structural complexity of macropores are keys in the shift from microscopic pore damage to the macroscopic level in red sandstone.

(4) Red sandstone mainly exists in the form of cementation between clay minerals and detrital minerals. As the initial water saturation time increases, its internal components react fully with water. While the degree of cementation weakens, the pore space further expands. With the increase in the number of cycles, the porosity continues to grow. Pores keep growing, developing, and connecting with each other, eventually evolving into the form of large pores, and the macroscopic fracture toughness also keeps decreasing. With the increase in large pores and the pore structure becoming more complex, the rock structure becomes increasingly loose, causing microstructural damage to the rock mass. Eventually, after the water–rock interaction, the internal pores of the red sandstone sample are fully developed and expanded, and the deterioration of the microstructure is aggravated, resulting in a decline in macroscopic mechanical properties.

Author Contributions: Conceptualization, H.D. and S.Z.; methodology, S.Z.; validation, S.Z., Y.L. and S.Y.; formal analysis, Y.L.; investigation, S.Z.; resources, S.Y. and H.D; writing—original draft preparation, S.Z.; writing—review and editing, Y.L.; visualization, J.X.; supervision, Y.L.; project administration, H.D.; funding acquisition, S.Z. All authors have read and agreed to the published version of the manuscript.

Funding: This work was financially supported by the National Natural Science Foundation of China under grant number 51874352.

Data Availability Statement: Data are contained within the article.

Conflicts of Interest: The authors declare no conflict of interest.

References

1. Du, K.; Liu, M.; Li, X.; Wang, L.; Zhou, J.; Yang, T.; Li, P. Mechanical properties of hard red sandstone in conventional triaxial compression tests incorporating acoustic emission. *Trans. Nonferrous Met. Soc. China* **2013**, 1–23.
2. Shen, S.W.; Wu, F.; Gan, L.; Jiang, M. Physical and mechanical properties of red sandstone with different initial saturation after freezing and thawing. *Drill. Proj.* **2021**, *48*, 120–128.
3. Chen, J.; Tong, J.; Rui, Y.C.; Cui, Y.; Pu, Y.; Du, J.; Apel, D. Step-path failure mechanism and stability analysis of water-bearing rock slopes based on particle flow simulation. *Theor. Appl. Fract. Mech.* **2024**, *131*, 104370. [CrossRef]
4. Miao, F.; Wu, Y.; Li, L.; Tang, H.; Xiong, F. Weakening Laws of Slip Zone Soils during Wetting–drying Cycles based on Fractal Theory. A Case Study in the Three Gorges Reservoir (China). *Acta Geotech.* **2020**, *15*, 1909–1923. [CrossRef]
5. Ye, W.J.; Wu, Y.; Yang, G.; Jing, H.; Chang, S.; Chen, M. Study on the change rule of fine microstructure and macro-mechanical properties of ancient soil under dry and wet cycling. *J. Rock Mech. Eng.* **2019**, *38*, 2126–2137.
6. Cai, X.; Zhou, Z.L.; Tan, L.H.; Zang, H.Z.; Song, Z.Y. Fracture behavior and damage mechanisms of sandstone subjected to wetting-drying cycles. *Eng. Fract. Mech.* **2020**, *234*, 107109. [CrossRef]
7. Li, X.; Deng, L.; Zhang, G.; Yi, Q.; Deng, M.; Shang, M.; Li, J.; Zhang, Y.; Yang, J. Mechanism of fine structure evolution and macro-mechanical multi-scale study of soil under loaded wet and dry cycles in the downward sliding zone. *Water Conserv. Hydropower Technol.* **2024**, *55*, 180–192.
8. Jiang, Q.; Xu, Y.; Wang, H.; Yang, L. Effects of dry and wet cycles on suction and shear strength of slip belt soil matrix. *J. Earth Sci. Environ.* **2022**, *44*, 959–970.
9. Liu, K.; Ye, W.J.; Gao, H.J.; Dong, Q. Multi-scale effects of mechanical property degradation of expansive soils in dry and wet environments. *J. Rock Mech. Eng.* **2020**, *39*, 2148–2159.
10. Yang, X.; Wang, J.; Hou, D.; Zhu, C.; He, M. Effect of dry-wet cycling on the mechanical properties of rocks, a laboratory-scale experimental study. *Processes* **2018**, *6*, 199. [CrossRef]
11. Zhang, F.; Jiang, A.; Yang, X. Shear creep experiments and modeling of granite under dry-wet cycling. *Eng. Geol. Environ.* **2021**, *80*, 5897–5908. [CrossRef]
12. Li, X.; Peng, K.; Peng, J.; Hou, D. Experimental investigation of cyclic wetting-drying effect on mechanical behavior of a medium-grained sandstone. *Eng. Geol.* **2021**, *293*, 106335. [CrossRef]
13. Zang, Q.; Liu, C.; Awan, R.S. Occurrence characteristics of the movable fluid in heterogeneous sandstone reservoir based on fractal analysis of NMR data: A case study of the Chang 7 Member of Ansai Block, Ordos Basin, China. *J. Pet. Sci. Eng.* **2022**, *214*, 110499. [CrossRef]
14. Liu, M.; Bi, R.; Luo, X.; Du, K. Natural joint effect on mechanical characteristics and fracture evolution of In-Site rocks under uniaxial compression. *Eng. Fail. Anal.* **2024**, *157*, 107880. [CrossRef]
15. Liu, Y.; Deng, H. Study on permeability performance of cemented tailings backfill based on fractal characteristics of pore structure. *Constr. Build Mater.* **2023**, *365*, 130035. [CrossRef]
16. Gao, L.; Zhang, J. Evaluation of reservoir pore classification based on fractal features and cluster analysis. *Adv. Fine Pertrochem.* **2024**, *25*, 27–34.
17. Zhang, H.; Lu, K.; Zhang, W.; Li, D.; Yang, G. Quantification and acoustic emission characteristics of sandstone damage evolution under dry–wet cycles. *J. Build. Eng.* **2022**, *48*, 103996. [CrossRef]
18. Yu, Y.; Li, C.; Hong, W.; Meng, J.; Fu, G. Strength characteristics and structural damage of red bedded sandstone of Xiaoba Formation in Baihetan under dry and wet cycling. *Saf. Environ. Eng.* **2022**, *29*, 24–32.
19. Yuan, P.; Qi, S.; Bit, N. Strength deterioration and energy evolution law of deep sandstone under dry and wet cycling. *Sci. Technol. Eng.* **2022**, *22*, 337–343.
20. Liu, M. Experimental study on strength damage of dolomite under dry and wet cycling. *Roadbed Eng.* **2022**, *5*, 117–121. [CrossRef]
21. Fu, Y.; Wang, Z.; Liu, X.; Yuan, W.; Miao, L.; Liu, J.; Deng, Z. Study on fine-scale damage evolution and macroscopic deterioration of sandstone under dry and wet cycling. *J. Geotech. Eng.* **2017**, *39*, 1653–1661.
22. Chen, Z.; Lan, H.; Liu, S.; Du, K. Tensile strength deterioration mechanism and damage mode of grotto temple sandstone under dry and wet cycling. *Earth Sci.* **2024**, *49*, 612–624.
23. Zhao, B.; Li, Y.; Huang, W.; Yang, J.; Sun, J.; Li, W.; Zhang, L.; Zhang, L. Mechanical characteristics of red sandstone under cyclic wetting and drying. *Environ. Earth Sci.* **2021**, *80*, 738. [CrossRef]
24. Wang, X.; Lian, B.; Wang, J.; Feng, W.; Gu, T. Creep damage properties of sandstone under dry-wet cycles. *J. Mt. Sci.* **2020**, *17*, 3112–3122. [CrossRef]

25. Jiang, J.; Chen, S.; Xu, J.; Liu, Q. Mechanical properties and energy characteristics of mudstone in different water-bearing states. *J. Coal* **2018**, *43*, 2217–2224.
26. Cherblanc, F.; Berthonneau, J.; Bromblet, P.; Huon, V. Influence of water content on the mechanical behaviour of limestone, role of the clay minerals content. *Rock Mech. Rock Eng.* **2016**, *49*, 2033–2042. [CrossRef]
27. Jiang, Y.; Zhou, L.; Zhu, Z.; Ma, L.; Chen, J.; Li, Y. Research on dynamic cracking properties of cracked rock mass under the effect of thermal treatment. *Theor. Appl. Fract. Mech.* **2022**, *122*, 103580. [CrossRef]
28. Jiang, Y.; Zhu, Z.; Yu, L.; Zhou, L.; Zhang, R.; Ma, L. Investigation of the fracture characteristics of granite and green sandstone under different thermal treatments. *Theor. Appl. Fract. Mech.* **2022**, *118*, 103217. [CrossRef]
29. *GB/T 50266-2013*; Standard for Engineering Rock Test Methods. General Administration of Quality Supervision, Inspection and Quarantine of the People's Republic of China: Beingjing, China, 2013.
30. Brown, E.T. *Rock Characterization, Testing & Monitoring, ISRM Suggested Methods*; Commission on Testing Methods, International Society for Rock Mechanics by Pergamon Press: Oxford, UK, 1981.
31. Zhang, S.; An, D.; Zhang, X.; Yu, B.; Wang, H. Research on size effect of fracture toughness of sandstone using the center-cracked circular disc samples. *Eng. Fract. Mech.* **2021**, *251*, 107777. [CrossRef]
32. Hou, C.; Jin, X.; He, J. Study on freeze-thaw damage characterization of pore structure of hard gypsum rock based on nuclear magnetic resonance technique. *J. Southwest Jiaotong Univ.* **2024**, 1–10.
33. Sun, Z. Research on Micro-Scale Pore-Throat Inhomogeneity and Fluid Storage Law in Tight Sandstone Reservoirs. Master's Thesis, Xi'an Shiyou University, Xi'an, China, 2024.
34. Li, J.; Kaunda, R.B.; Zhou, K.P. Experimental investigations on the effects of ambient freeze-thaw cycling on dynamic properties and rock pore structure deterioration of sandstone. *Cold Reg. Sci. Technol.* **2018**, *154*, 133–141. [CrossRef]
35. Zhang, J.; Deng, H.; Deng, J.; Gao, R. Fractal Analysis of Pore Structure Development of Sandstone, A Nuclear Magnetic Resonance Investigation. *IEEE Access* **2019**, *7*, 47282–47293. [CrossRef]
36. Yan, J.; Wen, D.; Li, Z.; Geng, B.; Cai, J.; Liang, Q.; Yan, Y. The quantitative evaluation method of low permeable sandstone pore structure based on nuclear magnetic resonance (NMR) logging. *Chin. J. Geophys.* **2016**, *59*, 1543–1552.
37. Li, X.; Che, X.; Li, H.; Qi, C. Macro- and fine-scale mechanical modeling of direct tensile fracture in brittle rocks with dry and wet cycles. *Geotechnics* **2024**, *07*, 1–12.
38. Han, X.; Wang, B.; Feng, J. Relationship between fractal feature and compressive strength of concrete based on MIP. *Constr. Build. Mater.* **2022**, *322*, 126504. [CrossRef]
39. Gu, X.; Liu, B.; Li, Z.; Wang, H.; Liu, J.; Nehdi, M.; Zhang, Y. Optimizing the pore structure and permeability of calcined clay limestone cement through shell powder. *J. Build. Eng.* **2024**, *91*, 2352–7102. [CrossRef]
40. Gong, S.; Zhao, Y.; Zhang, H.; Zhou, Y.; Zhao, C.; Wang, W. Experimental Study on the Anisotropy of Fracture Toughness and Fracture Process Zone of Anthracite. *J. China Coal Soc.* **2024**, *49*, 3399–3424.
41. Zhang, Z.; Weller, A. Fractal dimension of pore-space geometry of an Eocene sandstone formation. *Geophysics* **2014**, *79*, 377–387. [CrossRef]
42. Li, K. Analytical derivation of Brooks–Corey type capillary pressure models using fractal geometry and evaluation of rock heterogeneity. *J. Pet. Sci. Eng.* **2010**, *73*, 20–26. [CrossRef]
43. Hu, J.; Ren, Q.; Yang, D. Cross-scale characteristics of backfill material using NMR and fractal theory. *Trans. Nonferrous Met. Soc. China* **2020**, *30*, 1347–1363. [CrossRef]
44. Zhang, C.; Chen, Z.; Zhang, Z. Study on pore fractal structure of reservoir rock based on NMR T2 spectrum distribution. *Oil Gas Technol.* **2007**, *29*, 80–86.
45. Brooks, R.H.; Corey, A.T. Hydraulic properties of porous media. *Colo. State Univ. Hydrol. Pap.* **1964**, *3*, 7280.
46. Li, H.; Ma, L.; Yang, C.; Sima, Y.; Liu, J.; Li, Y. Experimental investigation on the damage and deterioration of sandstone subjected to cycling pore water pressure. *Bull. Eng. Geol. Environ.* **2023**, *82*, 107. [CrossRef]
47. Lin, Z.; Wang, J. Experimental study on the softening mechanism of red sandstone disintegration in water. *Roadbed Eng.* **2019**, 94–98.
48. Xie, H.; Chen, Z. Fractal geometry and fracture of rock. *Acta Mech. Sin.* **1988**, *4*, 255–264.
49. Xu, P.; Li, C.; Qiu, S.; Sasmito, A.P. A fractal network model for fractured porous media. *Fractals-Complex Geom. Patterns Scaling Nat. Soc.* **2016**, *24*, 1650018. [CrossRef]
50. Shi, J.; Chun, Q.; Feng, S.; Liu, C.; Liu, Z.; Wang, D.; Zhang, Y. Pore structure and fractal dimension analysis of ancient city wall bricks in China. *J. Build. Eng.* **2023**, *76*, 107324. [CrossRef]
51. Dang, H.L.; Wang, X.F.; Cui, P.X.; Hou, B.C.; Li, T. Characterization of seepage-absorption oil drive in low-permeability tight sandstone reservoirs based on nuclear magnetic resonance technology. *Adv. Geophys.* **2020**, *35*, 1759–1769.
52. Liu, Y.; Deng, H.; Jiang, Z. Pore structure characteristics of artificial sand aggregate mortar. *J. Build. Eng.* **2024**, *94*, 109940. [CrossRef]

53. Jiang, Z.; Cai, G.; Liu, Y. Pore structure and mechanical characteristics of CRS mortar based on NMR and fractal theory. *Constr. Build. Mater.* **2024**, *457*, 139459. [CrossRef]
54. Lai, J.; Wang, G.; Fan, Z.; Zhou, Z.; Chen, J. Fractal analysis of tight shaly sandstones using nuclear magnetic resonance measurements. *AAPG Bull.* **2018**, *102*, 175–193. [CrossRef]
55. Yu, S.; Deng, H.; Tian, G.; Deng, J. Microscopic Characteristic Analysis on Sandstone under Coupling Effect of Freeze-Thaw and Acidic Treatment: From Nuclear Magnetic Resonance Perspective. *Appl. Sci.* **2020**, *10*, 5699. [CrossRef]
56. Li, H.; Liu, J.; Lin, B.; Shi, S.; Yang, W.; Lu, Y.; Lu, J.; Liu, M. Mechanism of pore feature evolution of Wufeng-Longmaxi Formation shale under microwave-enhanced acid erosion. *J. Coal* **2024**, 1–13. [CrossRef]
57. Liu, Y.; Deng, H.; Jiang, Z.; Tian, G.; Wang, P.; Yu, S. Research on influence laws of aggregate sizes on pore structures and mechanical characteristics of cement mortar. *J. Constr. Build. Mater.* **2024**, *442*, 137606. [CrossRef]
58. Sun, W. Research on Fine Analysis Method and Rupture Mechanism of Brazilian Splitting and Fracture Toughness Test of Brittle Rock. Ph.D. Thesis, University of Science and Technology, Beijing, China, 2022.
59. Ayatollahi, M.R.; Sistaninia, M. Mode II fracture study of rocks using Brazilian disk specimens. *Int. J. Rock Mech. Min. Sci.* **2011**, *48*, 819–826. [CrossRef]
60. Chen, C.; Zhao, B.; Zhang, L. Mechanism of strength deterioration of red sandstone on reservoir bank slopes under the action of dry-wet cycles. *Sci. Rep.* **2023**, *13*, 20027. [CrossRef] [PubMed]
61. Hua, W.; Dong, S.; Peng, F.; Li, K.; Wang, Q. Experimental investigation on the effect of wetting-drying cycles on mixed mode fracture toughness of sandstone. *Int. J. Rock Mech. Min. Sci.* **2017**, *93*, 242–249. [CrossRef]
62. Qin, L.; Zhai, C.; Liu, S.; Xu, J.; Wu, S.; Dong, R. Fractal dimensions of low rank coal subjected to liquid nitrogen freeze-thaw based on nuclear magnetic resonance applied for coalbed methane recovery. *Powder Technol.* **2018**, *325*, 11–20. [CrossRef]
63. Huang, S.; Yu, S.; Ye, Y.; Ye, Z.; Cheng, A. Pore structure change and physico-mechanical properties deterioration of sandstone suffering freeze-thaw actions. *Constr. Build. Mater.* **2022**, *330*, 127200. [CrossRef]
64. Lian, C.; Zhuge, Y.; Beecham, S. The relationship between porosity and strength for porous concrete. *Constr. Build. Mater.* **2011**, *25*, 4294–4298. [CrossRef]
65. Amiri, M.; Moomivand, H. Development of a new physical modeling method to investigate the effect of porosity on the parameters of intact rock failure criteria. *Arab. J. Geosci.* **2019**, *12*, 317. [CrossRef]
66. Wang, X.; Zhang, Q. Role of surface roughness in the wettability, surface energy and flotation kinetics of calcite. *Powder Technol.* **2020**, *371*, 55–63. [CrossRef]
67. Luo, P.; Zhong, N.; Imran, K. Effects of pore structure and wettability on methane adsorption capacity of mud rock: Insights from mixture of organic matter and clay minerals. *Fuel* **2019**, *251*, 551–561. [CrossRef]
68. Zhou, C.; Tan, X.; Deng, Y.; Zhang, L.; Wang, J. Microscopic mechanism of softening of special soft rock. *Chin. J. Rock Mech. Eng.* **2005**, *24*, 394–400.
69. Yao, W.; Li, C.; Ke, Q. Multi-scale deterioration of physical and mechanical properties of argillaceous siltstone under cyclic wetting-drying of Yangtze River water. *Eng. Geol.* **2023**, *312*, 106925. [CrossRef]
70. Miao, C.; Yang, L.; Xu, Y.; Yang, K.; Sun, X.; Jiang, M.; Zhao, W. Experimental study on strength softening of sandstone based on nuclear magnetic resonance monitoring and its microscopic mechanism. *Chin. J. Rock Mech. Eng.* **2021**, *40*, 2189–2198.
71. Liu, H.; Xie, H.; Wu, F.; Li, C.; Gao, R. A novel box-counting method for quantitative fractal analysis of three-dimensional pore characteristics in sandstone. *Int. J. Min. Sci. Technol.* **2024**, *34*, 479–489. [CrossRef]

minerals

Article

Innovative Cut-and-Fill Mining Method for Controlled Surface Subsidence and Resourceful Utilization of Coal Gangue

Yongqiang Zhao, Yingming Yang, Zhiqiang Wang, Qingheng Gu, Shirong Wei and Xuejia Li et al.

1 State Key Laboratory of Water Resource Protection and Utilization in Coal Mining, Beijing 102211, China; 20039429@ceic.com (Y.Z.); yangym1988@163.com (Y.Y.); 2021165@aust.edu.cn (Q.G.)
2 School of Safety Science and Engineering, Anhui University of Science and Technology, Huainan 232000, China; 18161655022@163.com
3 School of Mining Engineering, Anhui University of Science and Technology, Huainan 232000, China
4 China Energy Shendong Coal Group Co., Ltd., Shenmu 719315, China; answer24@yeah.net (S.W.); 15610451523@163.com (X.L.)
* Correspondence: 2021072@aust.edu.cn

Abstract: Existing coal filling mining technologies face significant challenges of controlled surface subsidence, efficient utilization of waste rock in coal mines, and a shortage of adequate filling materials. This study introduces an innovative cut-and-fill mining method designed to strategically partition the goaf into cutting and filling zones. In the cutting zone, in situ filling materials are employed to construct waste rock column supports adjacent to the filling zone, thereby achieving controlled surface subsidence. This approach is integrated with long-wall mining operations and implemented using advanced, comprehensive equipment. FLAC3D simulations were conducted to investigate the patterns of stress distribution, surface deformation, and plastic zone formation within the mining field. With the implementation of the cut-and-fill mining balance, key observations include a reduction in maximum principal stress near the center of the goaf, an increasing trend in minimum principal stress, regular displacement distributions, and intact plastic zones positioned vertically away from the stope and horizontally close to the center of the stope. Compared to traditional caving methods, the cut-and-fill technique significantly reduces maximum vertical displacement, by nearly 95%, and maximum horizontal displacement, by approximately 90%. Additionally, it minimizes energy accumulation, lowers overall energy release, and prolongs the release period. Importantly, this method facilitates the resourceful utilization of approximately 800 million tons of waste rock, potentially leading to an estimated reduction of 500 million tons in CO_2 emissions. By achieving a balance of three effects—harmonizing coal extraction and filling capacity, aligning the supply and demand of filling materials, and optimizing the balance between filling costs and mining benefits—this method provides a sustainable and eco-friendly solution for the coal mining industry. The findings of this study are crucial for guiding the industry towards more environmentally responsible practices.

Keywords: cut-and-fill mining; material self-sufficiency; mining characteristic; dual reduction; triple effect

Academic Editors: Yuye Tan, Xun Chen, Yuan Li and Abbas Taheri

Received: 24 November 2024
Revised: 15 January 2025
Accepted: 24 January 2025
Published: 31 January 2025

Citation: Yongqiang Zhao, Yingming Yang, Zhiqiang Wang, Qingheng Gu, Shirong Wei and Xuejia Li et al. . Innovative Cut-and-Fill Mining Method for Controlled Surface Subsidence and Resourceful Utilization of Coal Gangue. *Minerals* **2025**, *15*, 146. https://doi.org/10.3390/min15020146

1. Introduction

Traditional approaches to control surface subsidence in mining include partial extraction [1,2], paste filling [3–5], strata separation grouting [6,7], and highly mechanized gangue filling technology behind efficient fully mechanized mining faces [8–10]. These methods are widely applied in coal mining sites, sometimes combining two or more techniques based on specific site conditions. With the development of green mining, several new coal mining methods with a high integration capability, maturity, or originality have emerged, such as the mixed mining technology of the gangue filling and caving method [11], the novel strip-style Wongawilli coal mining method [12], short-wall block-segment green mining technology [13], and the N00 mining method [14].

The novel strip-style Wongawilli coal mining method originates from the Wongawilli colliery in Australia and involves the sequential extraction of coal seams in narrow strips. By using this technique, miners can achieve higher recovery rates while minimizing surface subsidence. Short-wall block-segment green mining technology refers to the extraction of coal using a continuous miner on shorter faces compared to long-wall mining. The block-segment approach divides the mine into manageable sections, allowing for more controlled operations and reduced impact on the surrounding rock mass. The N00 mining method represents an advanced pillarless mining technique designed to improve safety and efficiency. It eliminates the need for permanent pillars, thereby maximizing coal recovery. This method also reduces ground control issues and enhances worker safety through continuous support systems.

In China, approximately 0.15 to 0.2 tons of coal gangue is generated for every ton of raw coal produced, making coal gangue the largest accumulated and annually produced industrial waste, occupying the most land in storage yards [15–17]. Comprehensive mechanized coal gangue filling mining technology has become an effective solution for special mining methods [15,18,19]. For example, to address the major dual technical challenges of recovering the lower coal mining pillars and handling underground gangue, mines can prioritize the method of using gangue to replace coal pillar extraction. Aiming to reduce carbon emissions and use carbon dioxide, the mineralization filling method has gradually become a research hotspot, and some mining areas have been subjected to industrial tests [20–22].

From the perspective of meeting filling material requirements and conserving coal resources, the traditional methods for controlling surface subsidence have inherent flaws. Partial extraction yields extraction rates of only 40% to 60%, and as coal resources decline, the extraction of residual strip coal pillars has become an inevitable choice for extending the service life of mines in the old mining areas of east-central China [23,24]. Filling materials with paste filling requires a certain solidification time, resulting in high economic and time costs. The low filling efficiency severely constrains coal mining efficiency, making widespread adoption challenging [25–27]. Strata separation grouting can avoid interference between mining and filling, almost without affecting coal mining efficiency. However, the prediction of grouting zones is difficult, accurate grouting implementation is challenging, and the subsidence reduction effect remains a matter of debate, leading to its limited current use [28]. Gangue filling suffers from limitations in pre-determined filling and restricted production scales [29–31].

Filling mining is a key technology for green mining because it reduces environmental damage by stabilizing the surface, minimizing subsidence, and reducing waste disposal. It also limits harmful gas emissions and prevents ecological damage, promoting sustainable mining practices. However, balancing the need for filling materials, reducing filling costs, and ensuring high-efficiency production are critical bottlenecks for coal mine filling mining technology. In light of these challenges and based on the advantages and disadvantages of existing mining methods, a new method called "cut-and-fill mining" in coal mining is proposed. Cut-and-fill mining designates the goaf scientifically into the cutting area and the filling area, utilizing the in situ roof of the cutting zone as filling material for the filling zone. This creates a supporting structure of gangue pillars in the goaf, achieving partial filling to control surface subsidence. Cut-and-fill mining combines the high efficiency of long-wall caving with the effective roof control of filling methods, providing a new approach to address the challenges faced by filling mining.

2. Implementation Method of Cut-and-Fill Mining

2.1. Roadway Layout

Figure 1a shows the three-dimensional schematic diagram of the cut-and-fill mining working face. The cut-and-fill mining working face includes the coal mining face before the support and the roof cutting and filling face after the support. The back roof cutting and filling surface is divided into the roof cutting area and filling area, and the roof cutting area and filling area are arranged in sequence. The cutting device in the roof cutting area fractures the immediate roof into crushed gangue, which serves as material for the adjacent filling area. As the hydraulic supports along the boundary of the filling area advance, metal mesh preinstalled on the hydraulic support on both sides automatically unfolds, providing lateral confinement to contain the crushed gangue. The fractured gangue is then automatically conveyed to the adjacent filling area in real time via a transfer mechanism on the cutting device. Once deposited within the metal mesh, the crushed gangue is compacted by a compaction device, completely filling the filling area. With this lateral confinement, a load-bearing structure comprising the broken gangue pillar and the immediate roof is established within the filling area, serving the function of partial backfilling and effectively reducing subsidence.

As shown in Figure 1b, multiple working faces are established in a new panel area to achieve mining without roadway excavation and coal pillar extraction. On one side of the entire panel area, there are downhill return air passages and downhill track passages, while on the other side, a downhill belt passage surrounds the entire panel area and connects with the downhill return air passage after encircling the area, ultimately linking to the shaft. An open-cut roadway is arranged on the first working face, with a section of the downhill belt passage serving as the upper entry groove for this working face. During the extraction process, a goaf is left near the position adjacent to the next working face, forming an upper entry groove for the initial extraction working face. The goaf area near the next coal mining face serves as the upper entry groove for the subsequent coal mining face. The upper entry groove, open-cut roadway, lower entry groove, downhill belt passage, and downhill return air passage are sequentially connected to form the ventilation system for the working face. In the mining process of the panel area, the downhill return air passage and downhill track passage remain unchanged throughout the entire mining process, constituting fixed passages. However, the downhill belt passage and open-cut roadway gradually change with the mining goafs, representing variable passages.

To facilitate roadway-free excavation and coal pillar-free mining, the cut-and-fill mining method modifies the ventilation pathways and conditions at the working face, necessitating careful design to prevent air leakage. Strategies such as installing hanging curtains to obstruct airflow and applying grouting treatments to the goaf section using gangue pillars positioned behind the support can be implemented to improve the efficiency of the ventilation system.

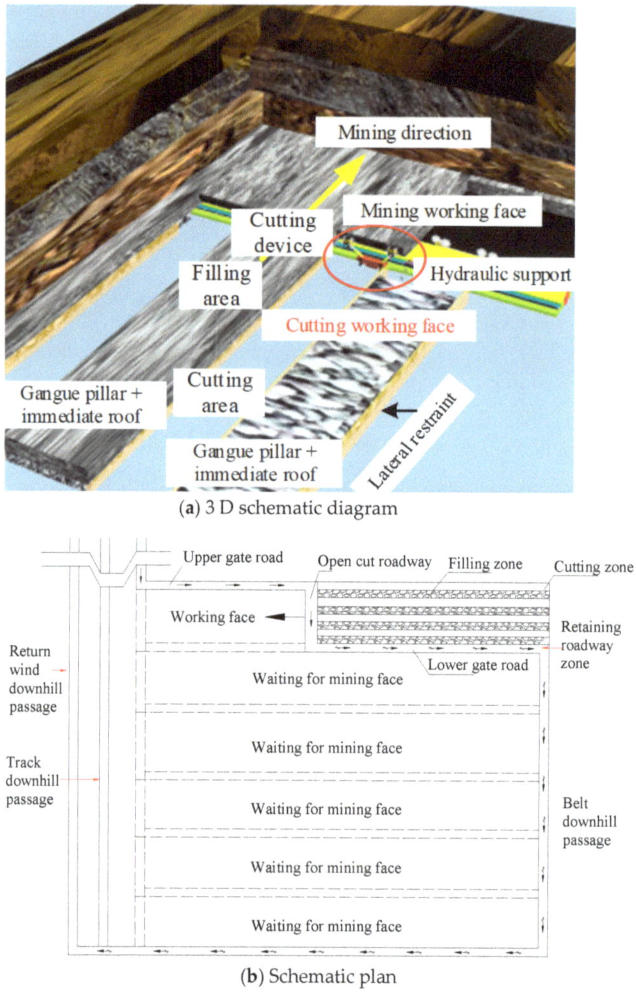

Figure 1. Layout of cut-and-fill mining [32,33].

2.2. Integrated Equipment

Figure 2 illustrates the working schematic diagram of the integrated equipment for cut-and-fill mining. In Figure 2, Area A represents the coal mining machine working zone, Area B is the pedestrian passage in the coal mining zone, Area C is the pedestrian passage in the cut-and-fill zone, and Area D is the cut-and-fill zone. The main components of the cut-and-fill integrated mining equipment include the protective support, cutting device, and compaction device.

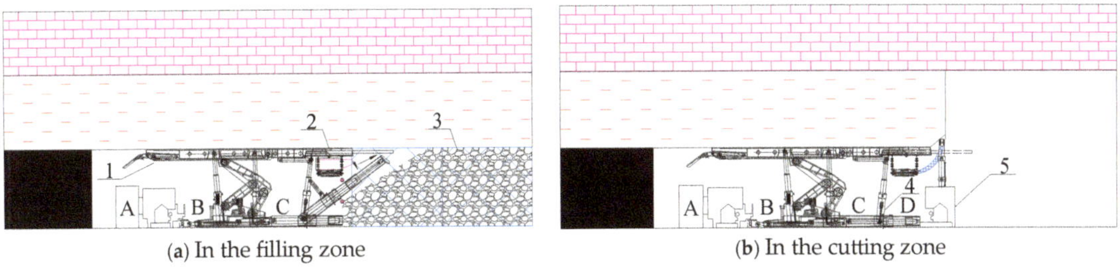

Figure 2. Schematic diagram of integrated equipment for cut-and-fill mining. 1—front support; 2—rear support; 3—lateral support metal mesh; 4—compaction device; 5—cutting device.

The protective support system features a six-column configuration, adapted from the solid backfill coal hydraulic support used in the "Mining + Separation + Filling + X" coal mining method. It consists of the bottom walking device, front support on the front side, rear support on the rear side, and compaction device. The rear support is appropriately extended to provide the space required for the cutting device. A retractable rear probing beam is set on the rear support, ensuring that the cutting and filling zone's roof is within the protective range of the support. The conveyor belt is suspended and fixed beneath the rear support, with several adjustable plow-type unloaders on the filling zone conveyor belt, controlling the discharge position of the backfilled gangue.

The front and rear supports are interconnected via an intermediate connection, while the compaction device and walking device are connected through a telescopic mechanism. This combination of the intermediate connection and telescopic device allows for independent operation of mining and filling processes. The suspension of the conveyor belt is fixed under the rear support. The conveying belt of the filling zone is equipped with multiple position-adjustable plow unloaders to control the unloading position of the filling gangue and facilitate the adjustment of the filling zone.

The cutting device of the cutting zone is derived from the scaling transformation of the continuous highwall mining system. The main body of the cutting device in the cutting zone is located within the rear support range and can travel along the inclination of the working face. When the cutting device cuts the roof of the cutting zone, the rear probing beam retracts, and the cutting part extends to cut the roof. The rear probing beam in the filling zone remains extended. The cutting device is connected to the conveyor belt through a movable belt, feeding crushed gangue onto the conveyor belt.

The lateral support metal mesh is pre-placed on the protective support on both sides of the filling zone. As the support advances forward, it automatically unfolds continuously, serving as the boundary on both sides of the filling zone. Mechanical automated anchor bolts, evenly spaced and fixed to the roof and floor, provide lateral constraint to the gangue pillars in the filling zone. Simultaneously, the compaction device applies pre-stress compaction to ensure the stability of the gangue pillars in the filling zone.

3. Simulation of Cut-and-Fill Mining

In this study, the mining effect of the cut-and-fill mining and caving method is calculated with the help of the classical numerical simulation of FLAC3D 5.0, which can obtain reliable result data and provide some data support for the validation and application of this method. Although the use of FLAC3D 5.0 for stress analysis is not new and FLAC3D 5.0 primarily models plastic failure mechanisms rather than realistic fracture formation, using indirect methods like equivalent plastic strain distributions to represent complex geological phenomena, these simulations are crucial to demonstrate the effectiveness and feasibility of our method, particularly in the context of coal mining, where ground control is a significant challenge.

3.1. Model Construction

The coal mine used in this section is in Inner Mongolia, China. The closer the rock parameters used in numerical simulation are to the actual field conditions, the better the simulated results. However, some simplifications are necessary in the actual process of model construction. Regarding the results of previous rock mechanics tests and crushed-gangue compression tests, the rock layers in the simulation model are simplified and reasonably assigned, as shown in Table 1.

Table 1. Mechanical parameters of rock formation.

Rock Type	Bulk Modulus (GPa)	Shear Modulus (GPa)	Density (kg/m^3)	Internal Friction Angle (°)	Cohesion (MPa)	Tensile Strength (MPa)
Fine Sandstone	2.270	2.050	1760	28	1.10	1.100
Mudstone	1.800	1.560	2080	32	1.50	1.200
Sandy Mudstone	2.060	1.860	2000	31	1.50	1.300
Coal	1.400	0.540	1400	32	0.50	0.500
Crushed Gangue	0.006	0.004	1280	15	0.03	0.002

As illustrated in Figure 3a, the three-dimensional computational model created for this simulation measures 300 m in length, 240 m in width, and 134 m in height. The side boundaries of the model are constrained in horizontal displacement, the bottom boundary is constrained in the vertical direction, and the upper boundary is free.

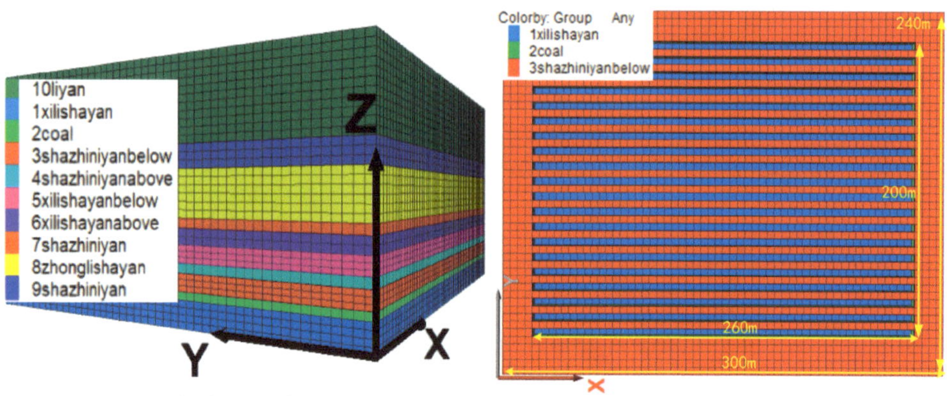

(**a**) 3 D schematic diagram (**b**) Schematic plan

Figure 3. The numerical simulation mode.

During the mining process, two monitoring lines are established: one for vertical and horizontal displacement on the surface and another for the maximum and minimum principal stresses within the rock layers. Monitoring Line 1 is positioned at the top of Layer 1 (Z = 130 m), with coordinates Y = 120. Monitoring points are placed every 10 m along the X direction to observe both the horizontal and vertical displacement of the ground. Monitoring Line 2 is located at the top of Layer 8 (Z = 78 m), and also has coordinates Y = 120. Monitoring points are set every 10 m along the X direction to measure the maximum and minimum principal stresses at that location.

As shown in Figure 3b, the mining area is divided into 20 cutting and filling strips along the strike, with each strip having a mining width of 10 m, consisting of a 5 m cutting width and a 5 m filling width. The cutting height is 3 m, and the filling height is 4 m (with a mining height of 4 m), resulting in an expansion coefficient of 1.33. A 20 m boundary is left around the perimeter of the mining area. The working face is arranged along the coal seam's dip, with a working face length of 200 m and an accumulated filling width of 100 m. The working face is mined along the coal seam's strike, with a total mining length of 260 m. The data in this paper are derived from the simulation software and edited by Origin software.

3.2. Result Analysis

3.2.1. Analysis of Stress Distribution Patterns

Based on the numerical calculation results, stress contour maps at equilibrium were obtained for Layer 8 at Z = 78 m and Y = 120 m, as shown in Figure 4. Additionally, the stress monitoring data curves at Z = 78 m are presented in Figure 5.

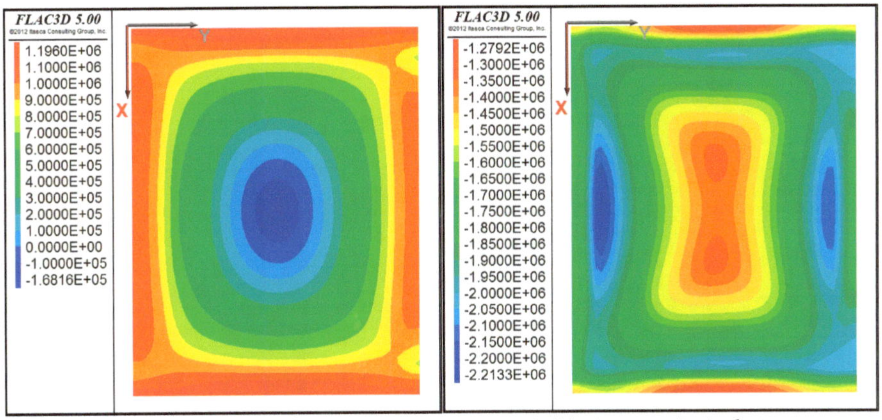

(**a**) Minimum principal stress (**b**) Maximum principal stress

Figure 4. Maximum and minimum principal stress contour at Z = 78 m when at mining equilibrium.

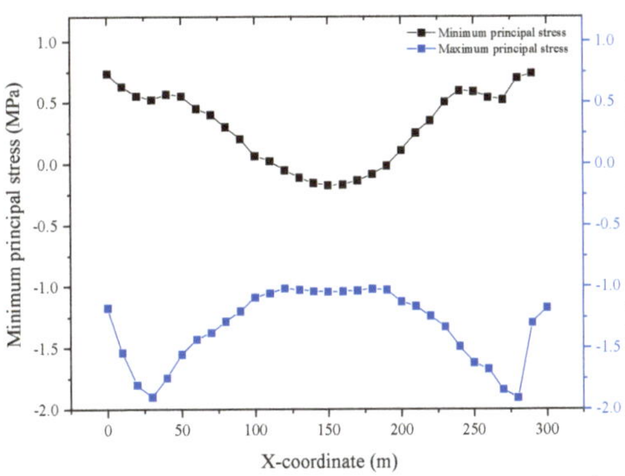

Figure 5. Principal stress curve of Y = 120 m × Z = 78 m when at mining equilibrium.

As shown in Figure 4, after cut-and-fill mining equilibrium, the distribution of the maximum principal stress generally exhibits a trend of decreasing stress near the center of the goaf. Similarly, the minimum principal stress generally shows a trend of increasing stress near the center of the goaf. That is, the σ_1-σ_3 values become smaller, indicating that in the Z = 78 m, areas closer to the center of the goaf are less prone to instability and failure. The rock layer's load-bearing capacity is stronger, reflecting the load-bearing role of the critical layer.

Figure 5 reveals that after the cut-and-fill mining equilibrium, the stress distribution along the Y = 120 m × Z = 78 m monitoring line shows a symmetrical pattern. The maximum principal stress for cut-and-fill mining is 1.93 MPa, located at X = 30 m and 280 m, while the maximum value of the minimum principal stress is 0.17 MPa, located at the center of the goaf. This pattern again suggests that the maximum principal stress generally decreases near the center of the goaf, while the minimum principal stress generally increases near the center of the goaf. This is beneficial for the rock layer to resist failure and stabilize the mining-induced stress after instability.

3.2.2. Analysis of Displacement Distribution Patterns

Based on the numerical calculation results, displacement contour maps at equilibrium were obtained for Z = 130 m and Y = 120 m, as shown in Figures 6 and 7. Additionally, displacement monitoring data at Z = 130 m × Y = 120 m are presented in Figure 8.

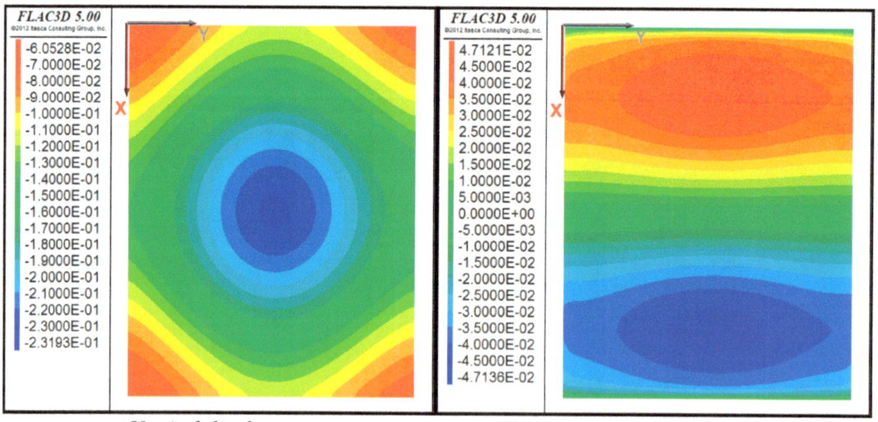

(a) Vertical displacement (b) Horizontal displacement

Figure 6. Displacement contour maps at equilibrium for Z = 130 m.

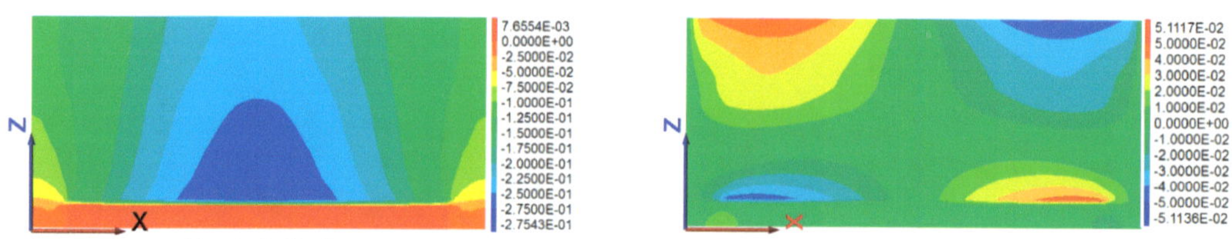

(a) Vertical displacement (b) Horizontal displacement

Figure 7. Displacement contour maps at equilibrium for Y = 120 m.

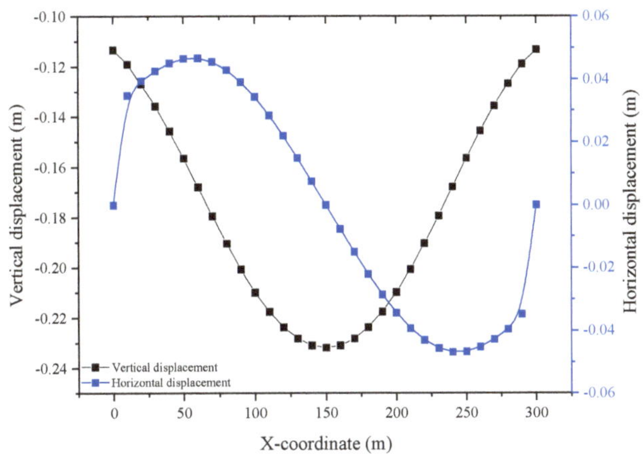

Figure 8. Displacements of Z = 130 m × Y = 120 m when at mining equilibrium.

As illustrated in Figures 6 and 7, after the cut-and-fill mining equilibrium, the vertical displacement exhibits a standard symmetric distribution, while the horizontal displacement shows a standard anti-symmetric distribution. The vertical displacement value increases with being near the mined coal seam, while the horizontal displacement value increases with being near the surface or mined coal seam.

Figure 8 demonstrates that after the cut-and-fill mining equilibrium, the maximum vertical displacement is 0.23 m, coinciding with the center of mining. The corresponding subsidence coefficient is 0.06. The horizontal displacement has two extreme centers at X = 50 m and X = 250 m, with an absolute maximum value of 0.05 m.

3.2.3. Analysis of Plastic Zone Distribution

Based on the numerical simulation results, plastic zone distribution maps at the equilibrium are shown in Figure 9. Plastic zone failure is mainly characterized by shear failure, and the distribution of plastic zones exhibits development along the edge of the mined-out area and the coal mining boundary. The overlying strata are well-preserved vertically away from the mining area and horizontally close to the central part of the mining area. In the central part of the mining area, the overlying strata exhibit mainly horizontal subsidence with small fractures. Therefore, the cut-and-fill mining method is beneficial for reducing crack formation in the overlying strata above the mining area and preserving their integrity.

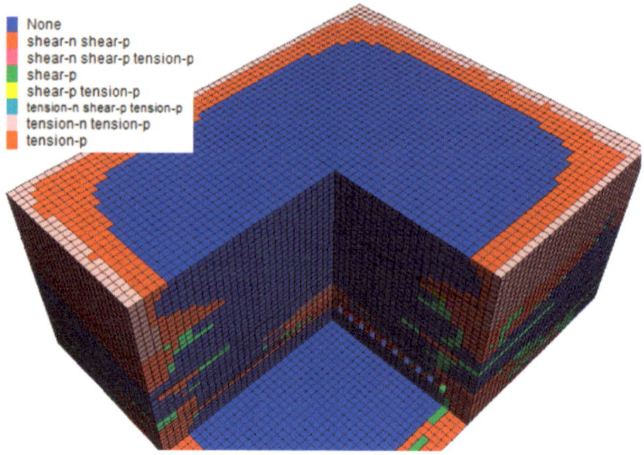

Figure 9. Distribution of plastic zones when at mining equilibrium.

4. Dual Reduction and Triple-Effect Balance

4.1. Dual Reduction Effect

4.1.1. Subsidence Reduction Effect

Continuing to use FLAC3D, we simulated the overlying strata failure characteristics during the caving method, focusing on displacement and plastic zone characteristics. The model construction is almost identical to that of the cut-and-fill mining, with the only difference being that after mining, the model parameters are halved under cut-and-fill mining conditions to simulate the damage caused to the overlying strata by coal seam mining. To prevent excessive subsidence of the roof strata, the calculation is stopped before the roof contacts the floor.

(1) Analysis of displacement distribution

According to the numerical calculation results, displacement contour maps at equilibrium for Z = 130 m and Y = 120 m are shown in Figures 10 and 11, and displacement monitoring data at Z = 130 m × Y = 120 m are presented in Figure 12. As illustrated in Figures 10–12, in the caving method, vertical displacement exhibits a standard symmetrical distribution, and horizontal displacement shows a standard anti-symmetrical distribution. As we move away from the surface, the vertical displacement values increase, but the magnitude of increase is not substantial. The maximum vertical displacement at the surface is 3.67 m, while in the goaf, it reaches 3.79 m. Compared to cut-and-fill mining, the maximum vertical displacement has increased nearly 20 times, and the subsidence coefficient is almost equal to 1. As one moves closer to the surface or the mining area, the horizontal displacement values increase. The maximum absolute value of horizontal displacement at the surface is 0.6 m, representing an almost 10-fold increase compared to cut-and-fill mining.

(2) Analysis of distribution law of plastic areas

According to the numerical calculation results, Figure 13 shows the distribution of plastic zones at the balanced state. It can be observed that the overlying strata in the mined area are almost completely damaged. Through comprehensive comparative analysis, the conclusion can be drawn that, compared to the caving method, the cut-and-fill mining significantly reduces surface deformation and the distribution of plastic zones. The stress distribution in the mined area also exhibits clear regularity, which is beneficial for better prediction and the prevention of surface subsidence.

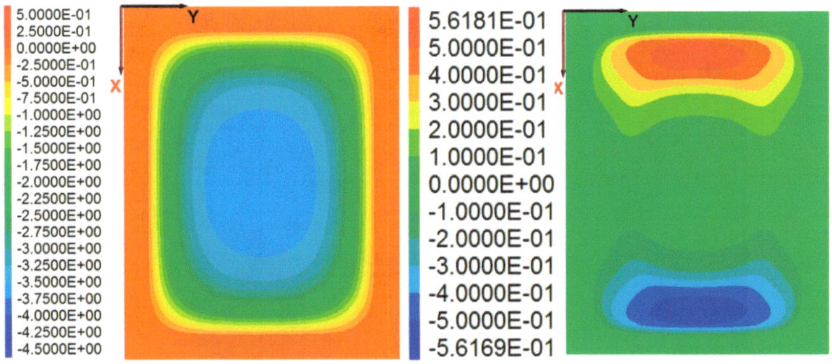

(a) Vertical displacement (b) Horizontal displacement

Figure 10. Displacement contour maps at equilibrium for Z = 130 m.

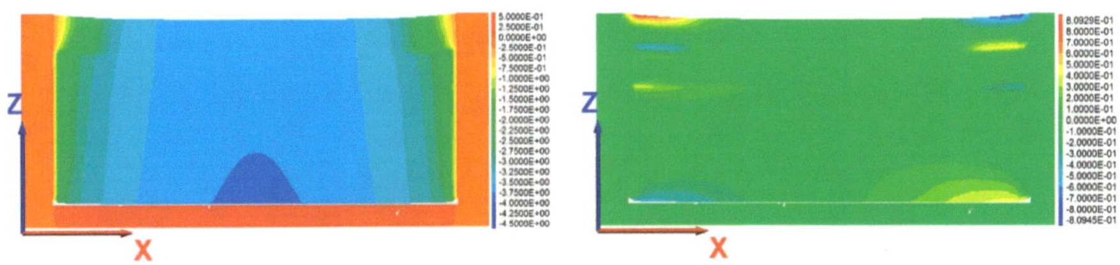

(a) Vertical displacement (b) Horizontal displacement

Figure 11. Displacement contour maps at equilibrium for Y = 120 m.

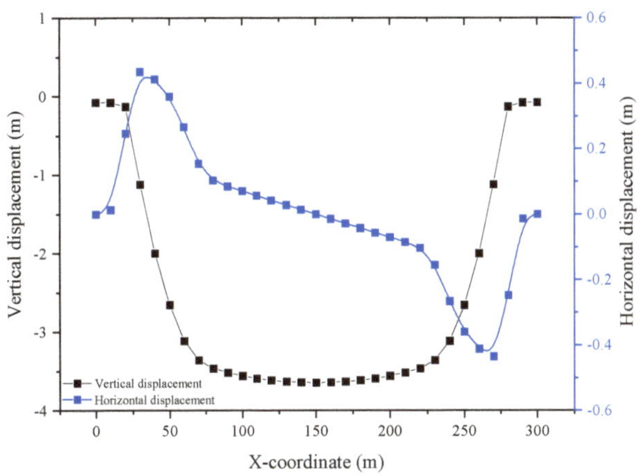

Figure 12. Displacements of Z = 130 m × Y = 120 m when at mining equilibrium.

(3) Analysis of energy change during roof failure

In the process of cut-and-fill mining, the roof has been in the process of the accumulation, dissipation, and mutual conversion of various energy from the open-off cut to the roof collapse. The energy involved includes strain energy, gravitational potential energy, kinetic energy, surface energy, and radiation energy. In the coal seam mining process, the gravitational potential energy is the source of other energy [34,35]. As shown in Figure 14, the change of roof energy in the stope can be divided into two parts: energy accumulation and energy release. Meanwhile, energy accumulation and energy release run through the whole process of roof failure.

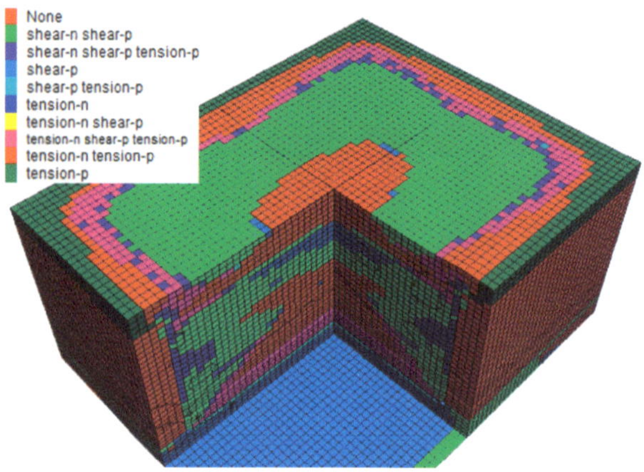

Figure 13. Distribution of plastic zones when at mining equilibrium.

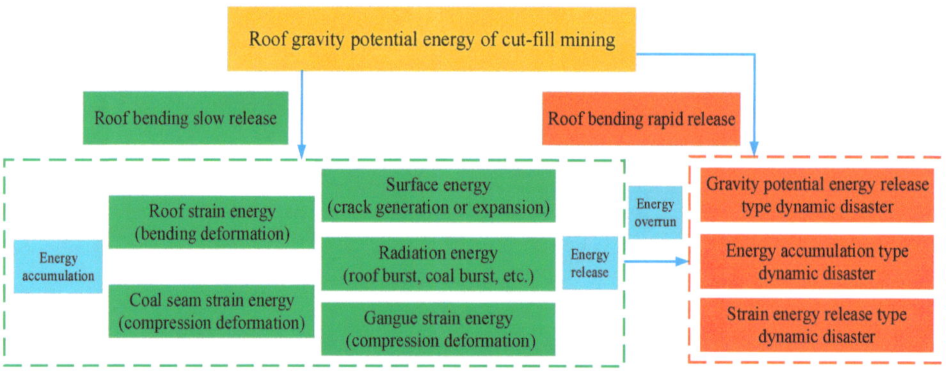

Figure 14. Energy variation in process of deformation until breaking of roof.

When the roof is broken, the energy is mainly released by surface energy, radiation energy, and kinetic energy. At this time, the energy conversion rate is almost completed in an instant, and there is a risk of dynamic disasters such as energy accumulation, strain energy release, and gravitational potential energy release. When the roof is broken, part of the gravitational potential energy and strain energy are converted into radiation energy, which is transmitted to the front roof and coal body in the form of a vibration wave. If the strain energy accumulation in the front coal rock mass is in a critical state, the disturbance of this radiation energy will cause instability and cause dynamic disasters of strain energy release.

From the simulation analysis above, it can be observed that the maximum vertical displacement in cut-and-fill mining is only 5% of that observed in caving mining, while the maximum horizontal displacement is reduced to 10% of that in caving mining. This significant reduction in vertical displacement of the roof greatly enhances safety by controlling the source of energy generation that leads to dynamic disasters. Additionally, the substantial decrease in horizontal displacement of the surrounding rock indicates that the transfer of gravitational potential energy from the roof substantially reduces the risk of local energy accumulation.

Moreover, cut-and-fill mining slows down the roof movement rate, thereby reducing the rate of energy conversion and the release rates of surface and radiation energy. This approach aligns with the strain energy change rate, promotes energy dissipation before roof fracture, and minimizes energy release during roof failure, thus mitigating the occurrence of dynamic disasters associated with strain energy release.

In conclusion, the surrounding rock deformation in cut-and-fill mining is significantly reduced, leading to a notable decrease in both the amount and rate of energy release during roof failure. This, in turn, lowers the risks associated with dynamic disasters caused by elastic energy accumulation and gravitational potential energy release. This paper analyzes the energy source responsible for dynamic disaster events, particularly the roof deformation induced by mining and the energy conversion process. The next phase of research will focus on quantifying the amount and proportion of energy conversion during specific processes in cut-and-fill mining.

4.1.2. Emission Reduction Effect

Coal gangue accounts about for 15% to 20% of China's total coal production [16,36]. Due to its low calorific value and difficulties in utilization, it is often openly piled around the mining area. As shown in Figure 15, the coal gangue production was approximately 8.13×10^8 tons in 2023. The carbon emissions caused by the traditional caving mining method mainly come from the following aspects: coal bed methane release, the spontaneous combustion of coal and coal mine waste, and machinery and equipment transportation, among others. According to some studies [37–40], the unit carbon emission of collapse coal mining is about 0.2–0.4 tons of carbon equivalent/ton of coal. Cut-and-fill mining enables the underground selection of coal gangue, facilitating on-site filling and saving energy consumption and surface disposal management costs. At the same time, it can greatly reduce the carbon equivalent/ton of coal.

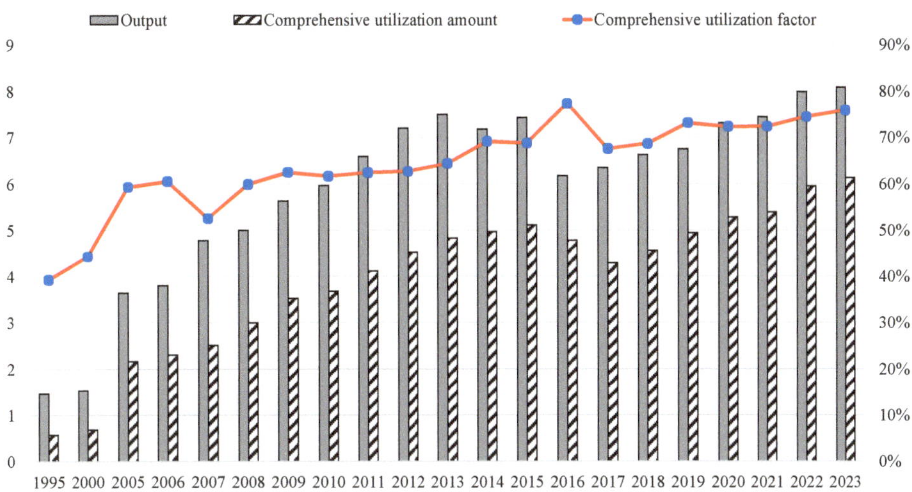

Figure 15. History of the output and utilization of coal gangue in China.

With certain combustible substances in coal gangue, spontaneous combustion can occur under certain conditions, and the toxic and harmful gasses such as nitrogen oxides, carbon oxides, sulfur dioxide, and smoke are discharged. The carbon content of coal gangue is about 25%~30% [16,17,41,42]. The ground accumulation of coal gangue easily causes slow oxidation, which generates greenhouse gasses such as CO_2, and so this is estimated to reduce the production of about 500 million tons of CO_2.

4.2. Triple-Effect Balance

4.2.1. Balancing Coal Mining Capacity with Filling Capacity

Based on practical experience from filling mining at the Daizhuang Coal Mine in China, as shown in Figure 16, the additional cost per ton of coal due to filling mining is 105 yuan. After excluding expenses such as tunneling costs, coal gangue treatment, land restoration, and policy support, the actual increase in the cost per ton of coal is approximately 24.3 yuan. The primary constraint limiting the widespread adoption of filling mining methods in China is not the increase in cost per ton of coal, but the significant mismatch between filling capacity and coal mining capacity.

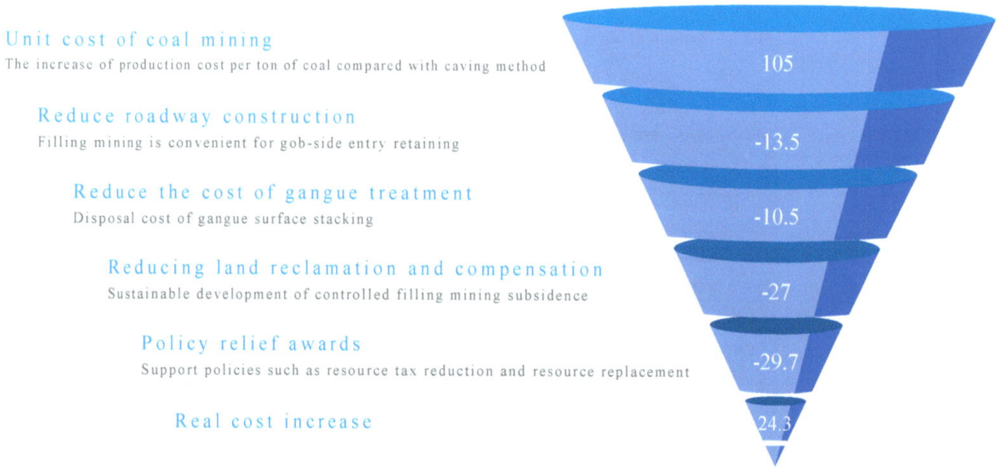

Figure 16. Increase in filling mining cost in Daizhuang Coal Mine in China.

Traditional filling mining methods severely constrain coal mining efficiency, leading to a substantial reduction in production capacity and, consequently, a significant decrease in coal mine profits. In contrast, cut-and-fill mining allows for the separation of the mining and filling processes, enabling the simultaneous operation of both. This effectively eliminates mutual interference and the time losses associated with staggered operations. Since cut-and-fill mining is a form of partial filling, it results in a shortened filling time. Additionally, the integration of long-wall mining in cut-and-fill operations ensures high coal mining efficiency.

4.2.2. Balancing Supply and Demand of Filling Materials

As shown in Figure 17, in the "Mining + Cutting + Filling" mode of cut-and-fill mining, all filling materials are sourced from the partially cut roof in the goaf. One of the fundamental principles of cut-and-fill mining is to maintain an equal volume of cutting and filling, meaning the volume of crushed gangue generated from cutting the roof should be equal to the volume of the filling material. By determining the parameters of the cutting area, the corresponding parameters of the filling area can thus be established.

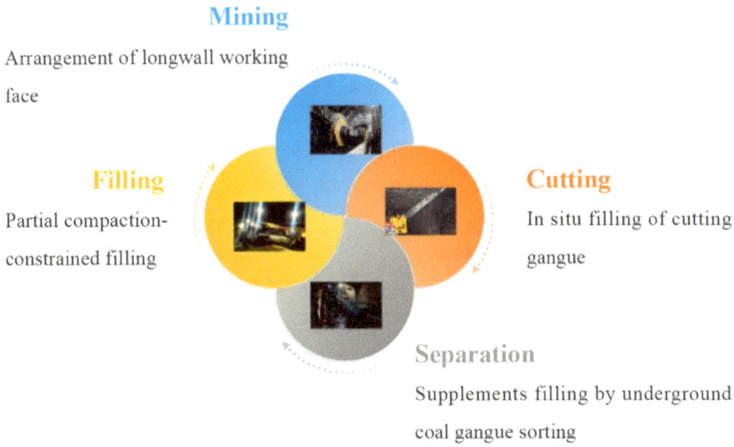

Figure 17. Mining technology and gangue source of cut-and-fill mining.

In cases where subsidence control is a critical requirement, an alternative "Mining + Cutting + Separation + Filling" mode may be adopted. In this mode, gangue selected from underground serves as auxiliary filling material for the cut-and-fill mining process. This approach enhances the filling effect without increasing the support and cutting strength of the cut-and-fill mining operation, thus minimizing the environmental impact while improving the overall effectiveness of the filling process.

4.2.3. Balancing Filling Costs and Mining Benefits

For coal mines, the primary motivation for filling is to avoid resettlement conflicts and land compensation. However, once the cost of filling exceeds that of land compensation, coal mines often choose not to backfill production under profit-driven motives, leading to irreparable environmental losses.

As shown in Figure 18, cut-and-fill mining significantly improves filling efficiency, reduces costs related to filling materials, and decreases the emissions of mining damage and coal gangue. This is beneficial for achieving a balance between filling costs and mining benefits, promoting the application of filling mining methods.

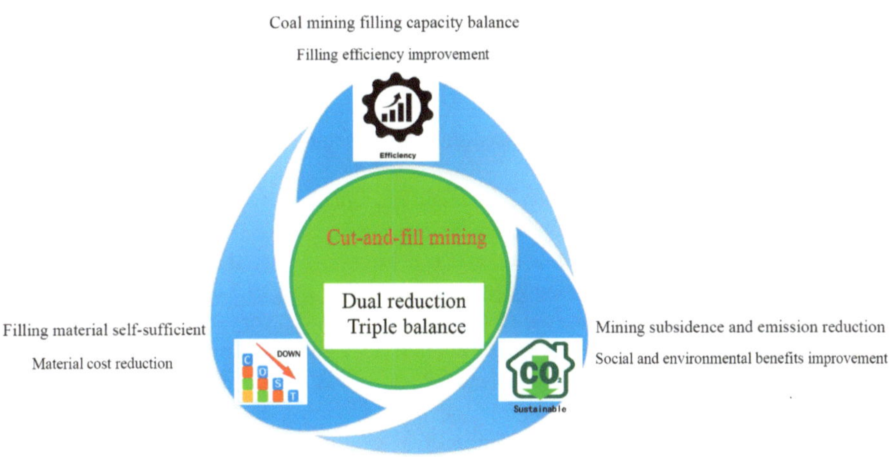

Figure 18. Double reduction and triple effect of cut-and-fill mining.

5. Discussion

5.1. Existing Technical Foundation

The key to cut-and-fill mining lies in the stability of the cutting space and the safety of the cutting operation. Achieving these objectives requires innovation based on the existing coal gangue filling technology system. After nearly 20 years of development, coal gangue filling technology has evolved through four generations, forming a new paradigm for modern green coal mining known as 'mining and filling + X'. The primary research and development goal at this stage is to minimize ecological damage at the source of mining methods. The specific process is outlined in Figure 19 [8–10].

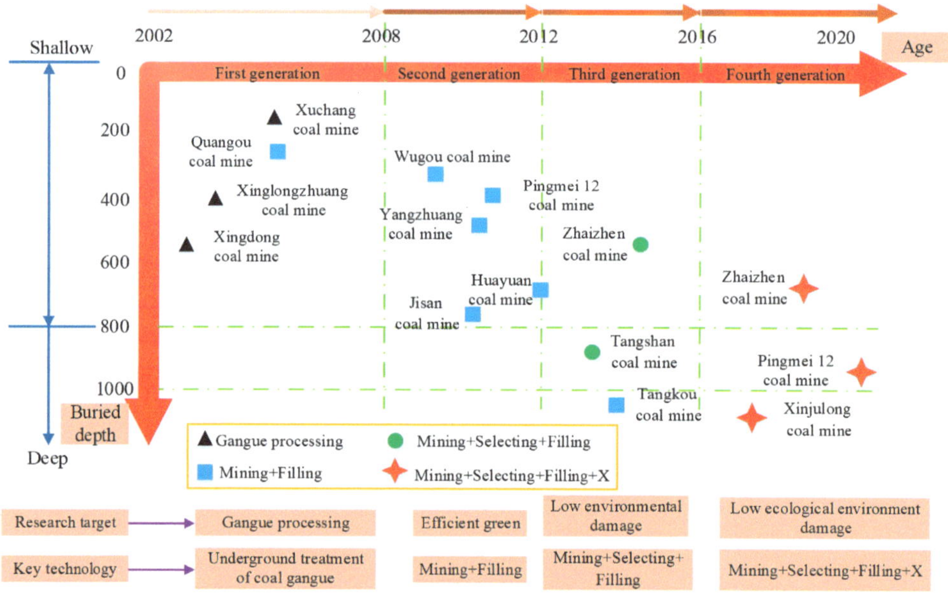

Figure 19. Development history of mining–dressing–backfilling + X.

Throughout the development of coal gangue filling technology, numerous new technologies and processes have emerged, particularly with the continuous upgrading of coal gangue filling hydraulic supports. This evolution has undergone four generations of innovation. Currently, 'mining and filling + X' coal mining has achieved the parallel operation of mining and filling, as well as the integration of mining and filling equipment. The main challenges remain the shortage of gangue materials and the complexities of filling and transportation.

The starting point of cut-and-fill mining is to ensure the self-reliance of filling materials and in situ underground filling, aiming to reduce and control subsidence. This approach provides a feasible solution to the current issues faced by 'mining and filling + X'. However, cut-and-fill mining requires the addition of a roof cutting device behind the support, necessitating the transformation and upgrading of existing coal gangue filling hydraulic supports. With its development history and the current strong manufacturing industry as a support, it is feasible to further modify existing filling supports to meet the needs of cut-and-fill mining.

The coal mining filling technology, and potential underground coal gangue separation in cut-and-fill mining, are largely consistent with the current fully mechanized solid filling method and can serve as the foundation for cut-and-fill mining. The main distinction between cut-and-fill mining and current 'mining and filling + X' coal mining is the roof cutting process, and the roof cutting process can be realized by advanced manufacturing.

5.2. Distinct Advantages

Compared to traditional caving and filling mining methods, cut-and-fill mining presents several distinct advantages:

(1) Efficiency: it allows for simultaneous mining and filling processes, minimizing interference and time losses, which maintains a high coal mining efficiency.
(2) Economic balance: It achieves a better balance between coal mining capacity and filling capacity, optimizing the supply and demand of filling materials, and providing a favorable cost–benefit ratio.
(3) Technological innovation: it incorporates advanced technology into the mining process, such as hydraulic supports equipped with cutting devices, which not only facilitates the roof cutting process but also represents an evolution from traditional coal gangue filling technology.

In summary, the cut-and-fill mining method investigated in this study provides a sustainable approach that enhances mine safety, efficiency, and ecological conservation. Its adoption could lead to significant advancements in addressing technical challenges faced by the coal industry, making it a promising practice for future mining operations.

5.3. Prospects for Further Research

According to industry statistics, the total number of coal mines in China is about 4000, and about 200 coal mines adopt the filling mining process. The proportion of filling mining in China is about 5%, and the proportion of the gangue filling process is even less. The main reason why it is difficult to popularize is the shortage of adequate filling materials; the cut-and-fill mining proposed in this paper can realize the partial filling and the partial filling materials for cut-and-fill mining come from part of the cutting roof of the goaf, which can realize self-sufficiency and in situ filling. However, there are several areas where improvements can be made.

(1) Refining the cut-and-fill mining system: developing advanced methods and equipment specifically designed for the unique requirements of cut-and-fill mining.
(2) Comparative analysis of parameters: To better understand the parameters influencing the cut-and-fill mining method, a comparative analysis was conducted between different coal mines and similar operations in other regions worldwide. This comparison revealed variations in geological conditions, equipment capabilities, and operational practices, highlighting the need for site-specific optimizations. For instance, differences in strata thickness, rock hardness, and water content can significantly impact the effectiveness of the cut-and-fill method.
(3) Influence of key parameters on simulation results: One significant advantage of numerical simulation is its ability to conveniently study the influence of key parameters on calculation results. Variables such as the depth of cut, spacing of cuts, material properties, and support systems were systematically varied to assess their impacts on mining stability and productivity. The simulations allowed us to identify optimal parameter settings that maximize efficiency while minimizing risks. For example, increasing the depth of cut may enhance productivity but could also lead to greater stress concentrations and potential instability. Therefore, finding a balance through iterative simulations is crucial.
(4) Detailed analysis of energy changes during deformation: A more detailed analysis of the change in energy during the deformation process before roof failure is essential for predicting and preventing catastrophic failures. Numerical models provided insights

into how energy accumulates within the rock mass as it deforms, eventually leading to sudden release at the point of failure. By monitoring energy levels throughout the deformation process, we can anticipate critical thresholds and implement preemptive measures to stabilize the roof structure. Understanding these energy dynamics allows for the development of early warning systems and the improved design of support structures, enhancing mine safety.

6. Conclusions

This study has focused on investigating the cut-and-fill mining method, specifically its effects on stress distribution, displacement patterns, plastic zone formation, and energy changes during the mining process. To achieve this, numerical simulations using FLAC3D were conducted to analyze the behavior of overlying strata and the extent of surface deformation under different mining conditions compared to traditional caving methods.

(1) Stress distribution and displacement patterns: The simulation results demonstrate that near the mined-out area center, the maximum principal stress decreases, while the minimum principal stress increases, after mining with cut-and-fill methods. The analysis of displacement shows a symmetric vertical displacement pattern and an anti-symmetric horizontal displacement pattern. Notably, the maximum vertical displacement at the mining center is 0.23 m, with a subsidence coefficient of 0.06, and horizontal displacement peaks at 0.05 m. These findings indicate significantly reduced surface deformation compared to caving mining.
(2) Plastic zone analysis: Plastic zone development primarily occurs along the edges of the mined-out area, while the overlying strata remain largely intact in the central region, exhibiting mainly horizontal subsidence with minimal fractures. This suggests that cut-and-fill mining can preserve the integrity of the overlying strata and minimize plastic zones, thereby reducing the risk of surface fractures.
(3) Energy release mitigation: Cut-and-fill mining notably reduces the amount and rate of energy release during roof failure, lowering the risks associated with dynamic disasters caused by elastic energy accumulation and gravitational potential energy release. By controlling the source of energy generation and slowing down the roof movement rate, this method minimizes energy release during roof failure.
(4) Environmental and economic benefits: One of the most significant advantages of cut-and-fill mining is its positive impact on the environment and economy. It decreases coal gangue emissions through on-site filling, reducing the need for surface disposal and lowering carbon emissions, potentially cutting CO_2 production by approximately 500 million tons. Moreover, it offers economic benefits by improving filling efficiency, reducing costs related to filling materials, and achieving a favorable balance between filling costs and mining benefits.

Author Contributions: All the authors contributed to publishing this paper. Writing—review and editing and funding acquisition, Y.Z.; resources, Y.Y.; data curation, Z.W.; formal analysis, Q.G.; software, S.W.; visualization, X.L.; conceptualization and writing—original draft, C.W. All authors have read and agreed to the published version of the manuscript.

Funding: The study was supported by the State Key Laboratory of Water Resource Protection and Utilization in Coal Mining (WPUKFJJ2022-15), National Natural Science Foundation of China (No.52304198).

Data Availability Statement: All the data in this paper are available.

Conflicts of Interest: Authors Shirong Wei and Xuejia Li were employed by the China Energy Shendong Coal Group Co., Ltd. The remaining authors declare that the research was conducted in the absence of any commercial or financial relationships that could be construed as a potential conflict of interest.

References

1. Xu, Y.; Ma, L.; Ngo, I.; Zhai, J. Prediction of the Height of Water-Conductive Fractured Zone under Continuous Extraction and Partial Backfill Mining Method-a Case Study. *Sustainability* **2022**, *14*, 6582. [CrossRef]
2. Lyu, K.; Jiang, N.; Yin, D.; Meng, S.Y.; Gao, Z.Y.; Lyu, T. Deterioration of Compressive Properties of Coal Rocks Under Water and Gas Coupling. *J. Cent. South Univ.* **2024**, *31*, 477–495. [CrossRef]
3. Huang, W.; Song, T.; Li, H.; Liu, Y.; Hou, T.; Gao, M.; Zheng, Y. Design of Key Parameters for Strip-Filling Structures Using Cemented Gangue in Goaf—A Case Study. *Sustainability* **2023**, *15*, 4698. [CrossRef]
4. Wang, X.; Silva, M.C.E.; Pereira, M.F.C.; Brás, H.; Paneiro, G. Strength Behavior and Deformability Characteristics of Paste Backfill with the Addition of Recycled Rubber. *Int. J. Min. Reclam. Environ.* **2023**, *37*, 713–730. [CrossRef]
5. Wu, P.; Zhao, J.; Jin, J. Similar Simulation of Overburden Movement Characteristics under Paste Filling Mining Conditions. *Sci. Rep.* **2023**, *13*, 12550. [CrossRef] [PubMed]
6. Xuan, D.; Xu, J. Longwall Surface Subsidence Control by Technology of Isolated Overburden Grout Injection. *Int. J. Min. Sci. Technol.* **2017**, *27*, 813–818. [CrossRef]
7. Shen, B.; Poulsen, B. Investigation of Overburden Behaviour for Grout Injection to Control Mine Subsidence. *Int. J. Min. Sci. Technol.* **2014**, *24*, 317–323. [CrossRef]
8. Zhang, Q.; Wang, Z.; Zhang, J.; Jiang, H.; Wang, Y.; Yang, K.; Tian, X.; Yuan, L. Integrated Green Mining Technology of Coal Mining-Gangue Washing-Backfilling-Strata Control-System Monitoring-Taking Tangshan Mine as a Case Study. *Environ. Sci. Pollut. Res.* **2022**, *29*, 5798–5811. [CrossRef] [PubMed]
9. Huang, P.; Zhang, J.; Yan, X.; Spearing, A.J.S.; Li, M.; Liu, S. Deformation Response of Roof in Solid Backfilling Coal Mining Based on Viscoelastic Properties of Waste Gangue. *Int. J. Min. Sci. Technol.* **2021**, *31*, 279–289. [CrossRef]
10. Sun, K.; Zhang, J.; He, M.; Li, M.; Guo, S. Control of Surface Deformation and Overburden Movement in Coal Mine Area by an Innovative Roadway Cemented Paste Backfilling Method Using Mining Waste. *Sci. Total Environ.* **2023**, *891*, 164693. [CrossRef] [PubMed]
11. Fang, K.; Zhang, J.X.; Zhang, Q.; Sun, Q.; Yin, W.; Zhou, F. Fully Mechanised Mixed Mining Technology Involving Solid Backfilling and Caving Methods in Longwall Workface. *Min. Technol.* **2016**, *125*, 205–211. [CrossRef]
12. Ma, L.; Jin, Z.; Liu, W.; Zhang, D.; Zhang, Y. Wongawilli Roadway Backfilling Coal Mining Method—A Case Study in Wangtaipu Coal Mine. *Int. J. Oil Gas Coal Technol.* **2019**, *20*, 342–359. [CrossRef]
13. Zhang, Y.; Liu, Y.; Lai, X.; Gao, J. Physical Modeling of the Controlled Water-Flowing Fracture Development during Short-Wall Block Backfill Mining. *Lithosphere* **2021**, *2021*, 2860087. [CrossRef]
14. Zhang, J.; He, M.; Shimada, H.; Wang, Y.; Hou, S.; Liu, B.; Yang, G.; Zhou, P.; Li, H.; Wu, X. Similar Model Study on the Principle of Balanced Mining and Overlying Strata Movement Law in Shallow and Thin Coal Seam Based on N00 Mining Method. *Eng. Fail. Anal.* **2023**, *152*, 107457. [CrossRef]
15. Bo, L.; Yang, S.; Liu, Y.; Zhang, Z.; Wang, Y.; Wang, Y. Coal Mine Solid Waste Backfill Process in China: Current Status and Challenges. *Sustainability* **2023**, *15*, 13489. [CrossRef]
16. Xu, F.; Xiao, J.; Ye, S.; Liu, W.; Yang, M.; Yao, Z.; Zhong, Q. Closed-Loop Cleaning Treatment System- Resource Recovery of Coal Gangue. *Energy Sources Part A Recovery Util. Environ. Eff.* **2024**, *46*, 5236–5253. [CrossRef]
17. Zhang, Q.; Zhang, J.; Wu, Z.; Chen, Y. Overview of Solid Backfilling Technology Based on Coal-Waste Underground Separation in China. *Sustainability* **2019**, *11*, 2118. [CrossRef]
18. Yin, S.; Guo, Z.; Wang, Q.; Zhao, Y.; Li, Y. Effects of Particle Size on Compressive Deformation Characteristics of Broken Rock Mass in Gangue Rib of Automatically Formed Gob-Side Entry Retaining. *Powder Technol.* **2023**, *430*, 118987. [CrossRef]
19. Guo, Y.; Zhang, J.; Li, M.; Wang, L.; Li, Z. Preventing Water Inrush Hazards in Coal Mines by Coal Gangue Backfilling in Gobs: Influences of the Particle Size and Stress on Seepage Characteristics. *Environ. Sci. Pollut. Res.* **2023**, *30*, 104374–104387. [CrossRef]
20. Ngo, I.; Ma, L.; Sajib, M.H.; Zhang, H.; Zhao, Z.; Yu, K.; Zhang, Z.; Peng, C. Examination of the Use of Binderless Zeolite Blend as Backfill Materials to Enhance CO_2 Adsorption in Coal Mine Working Face. *Case Stud. Constr. Mater.* **2024**, *21*, e03699. [CrossRef]
21. Ngo, I.; Ma, L.; Zhao, Z.; Zhai, J.; Yu, K.; Wu, Y. Sol–Gel-Stabilized CO_2 Foam for Enhanced In-Situ Carbonation in Foamed Fly Ash Backfill Materials. *Geomech. Geophys. Geo-Energy Geo-Resour.* **2024**, *10*, 80. [CrossRef]
22. Ngo, I.; Ma, L.; Zhai, J.; Wang, Y. Enhancing Fly Ash Utilization in Backfill Materials Treated With CO_2 Carbonation under Ambient Conditions. *Int. J. Min. Sci. Technol.* **2023**, *33*, 323–337. [CrossRef]
23. Ma, J.; Ding, Y.; Zhang, H. Application of Pedrail Powered Support on Strata Control in Short-Wall Coal Mining. *Geofluids* **2022**, *2022*, 5690659. [CrossRef]
24. Heritage, Y. Mechanics of Rib Deformation Observations and Monitoring in Australian Coal Mines. *Int. J. Min. Sci. Technol.* **2019**, *29*, 119–129. [CrossRef]

25. Cacciuttolo, C.; Marinovic, A. Experiences of Underground Mine Backfilling Using Mine Tailings Developed in the Andean Region of Peru: A Green Mining Solution to Reduce Socio-Environmental Impacts. *Sustainability* **2023**, *15*, 12912. [CrossRef]
26. Shahsavari, M.; Jafari, M.; Grabinsky, M. Simulation of Cemented Paste Backfill (CPB) Deposition through Column Experiments: Comparisons of Field Measurements, Laboratory Measurements, and Analytical Solutions. *Can. Geotech. J.* **2023**, *60*, 1505–1514. [CrossRef]
27. Naguleswaran, N.; Nagaratnam, S.; Ryan, L.V. Flow Characteristics of Cemented Paste Backfill. *Geotech. Geol. Eng.* **2018**, *36*, 2261–2272.
28. Wang, X.; Wu, W.; Wu, B. Grouting of Bed Separation Spaces to Control Sliding of the High-Located Main Key Stratum during Longwall Mining. *Q. J. Eng. Geol. Hydrogeol.* **2020**, *53*, 569–578. [CrossRef]
29. Li, M.; Zhang, J.; Deng, X.; Ju, F.; Li, B. Measurement and Numerical Analysis of Water-Conducting Fractured Zone in Solid Backfill Mining Under an Aquifer: A Case Study in China. *Q. J. Eng. Geol. Hydrogeol.* **2017**, *50*, 81–87. [CrossRef]
30. Li, M.; Zhang, J.; Huang, Y.; Zhou, N. Effects of Particle Size of Crushed Gangue Backfill Materials on Surface Subsidence and Its Application under Buildings. *Environ. Earth Sci.* **2017**, *76*, 603. [CrossRef]
31. Zhang, J.; Li, B.; Zhou, N.; Zhang, Q. Application of Solid Backfilling to Reduce Hard-Roof Caving and Longwall Coal Face Burst Potential. *Int. J. Rock Mech. Min. Sci.* **2016**, *88*, 197–205. [CrossRef]
32. Wang, Y.; He, M.; Zhang, J.; Wang, Q.; Yang, J.; Hou, S. Roof Control Mechanism and Design Methods of Gob-Side Entry Retained by N00 Coal Mining Method. *Rock Mech. Rock Eng.* **2023**, *57*, 621–638. [CrossRef]
33. Zhang, J.; He, M.; Yang, G.; Wang, Y.; Hou, S. N00 Method with Double-Sided Roof Cutting for Protecting Roadways and Surface Strata. *Rock Mech. Rock Eng.* **2023**, *57*, 1629–1651. [CrossRef]
34. Zhou, N. *Mechanism of Preventing Dynamic Hazards under Hard Roof by Solid Backfilling Technology*; China University of Mining and Technology: Xuzhou, China, 2014.
35. Ji, S.; Lai, X.; Cui, F.; Liu, Y.; Pan, R.; Karlovšek, J. The Failure of Edge-Cracked Hard Roof in Underground Mining: An Analytical Study. *Int. J. Rock Mech. Min. Sci.* **2024**, *183*, 105934. [CrossRef]
36. Dong, C.; Zhou, N.; Ferro, G.A.; Yan, H.; Xu, J.; Wang, H.; Liu, S.; Zhang, Z. Research on Proportion and Performance Optimization of Pure Gangue Backfilling Slurry Based on Multi-Objective Differential Evolution Algorithm. *Constr. Build. Mater.* **2024**, *418*, 135432. [CrossRef]
37. Azadi, M.; Northey, S.A.; Ali, S.H.; Edraki, M. Transparency on Greenhouse Gas Emissions from Mining to Enable Climate Change Mitigation. *Nat. Geosci.* **2020**, *13*, 100–104. [CrossRef]
38. Gao, J.; Guan, C.; Zhang, B. China's CH_4 Emissions from Coal Mining: A Review of Current Bottom-Up Inventories. *Sci. Total Environ.* **2020**, *725*, 138295. [CrossRef] [PubMed]
39. Jakob, M.; Steckel, J.C.; Jotzo, F.; Sovacool, B.K.; Cornelsen, L.; Chandra, R.; Edenhofer, O.; Holden, C.; Löschel, A.; Nace, T.; et al. The Future of Coal in a Carbon-Constrained Climate. *Nat. Clim. Change* **2020**, *10*, 704–707. [CrossRef]
40. Yang, B.; Bai, Z.; Zhang, J. Environmental Impact of Mining-Associated Carbon Emissions and Analysis of Cleaner Production Strategies in China. *Environ. Sci. Pollut. Res.* **2021**, *28*, 13649–13659. [CrossRef] [PubMed]
41. Valluri, S.; Claremboux, V.; Kawatra, S. Opportunities and Challenges in CO_2 Utilization. *J. Environ. Sci.* **2022**, *113*, 322–344. [CrossRef] [PubMed]
42. Li, J.; Wang, J. Comprehensive Utilization and Environmental Risks of Coal Gangue: A Review. *J. Clean. Prod.* **2019**, *239*, 117946. [CrossRef]

minerals

MDPI

Article

Quantitative Analysis of Yield Stress and Its Evolution in Fiber-Reinforced Cemented Paste Backfill

Shili Hu [1], Jingping Qiu [1], Qingsong Zhang [1], Zhenbang Guo [1] and Chen Liu [2,*]

1 School of Resource and Civil Engineering, Northeastern University, Shenyang 110819, China; 2010738@stu.neu.edu.cn (S.H.); qiujingping@mail.neu.edu.cn (J.Q.); 2101006@stu.neu.edu.cn (Q.Z.); 2010364@stu.neu.edu.cn (Z.G.)

2 Department of Materials and Environment (Microlab), Faculty of Civil Engineering and Geoscience, Delft University of Technology, 2628 CN Delft, The Netherlands

* Correspondence: c.liu-12@tudelft.nl

Abstract: Fiber-reinforced cemented paste backfill (FR-CPB) has attracted considerable attention in modern mining applications due to its superior mechanical properties and adaptability. Despite its potential, understanding its rheological behavior remains limited, largely because of the absence of quantitative methods for assessing fiber packing behavior within CPB. This study develops a rheology-based approach to determine the maximum packing fraction of polypropylene fibers in fresh CPB, revealing that shorter fibers (3 mm) achieve a maximum packing fraction of 0.661, significantly higher than longer fibers (12 mm) with 0.534. Building on these findings, a quantitative model for the static yield stress of FR-CPB was developed, showing that under a high fiber content (0.9%) and with longer fibers (12 mm), the yield stress reached 274.34 kPa, a 40% increase compared to shorter fibers. Additionally, the study modeled the time-dependent evolution of yield stress, achieving a prediction accuracy with a correlation coefficient of 0.92. These advancements enable the optimization of FR-CPB composition, which can reduce material usage, enhance pipeline transport efficiency, and improve backfill stability in underground voids. By minimizing the risk of structural failure and optimizing resource allocation, this research provides a theoretical foundation for safer and more cost-effective mining operations.

Keywords: cemented paste backfill; polypropylene fiber; maximum packing fraction; yield stress; yield stress evolution

Academic Editors: Abbas Taheri, Yuye Tan, Xun Chen and Yuan Li

Received: 16 December 2024
Revised: 4 January 2025
Accepted: 14 January 2025
Published: 16 January 2025

Citation: Hu, S.; Qiu, J.; Zhang, Q.; Guo, Z.; Liu, C. Quantitative Analysis of Yield Stress and Its Evolution in Fiber-Reinforced Cemented Paste Backfill. *Minerals* **2025**, *15*, 81. https://doi.org/10.3390/min15010081

1. Introduction

Cemented paste backfill (CPB), composed of tailings, cementitious materials, and water, is an environmentally friendly construction material extensively utilized in mine backfill operations [1–5]. This material not only facilitates the resourceful utilization of tailings but also ensures the stability of underground structures by filling voids created during mining [6–8]. However, the CPB's inherent brittleness under complex in situ loading conditions can result in catastrophic failure, posing significant safety risks to underground mines [9]. To address this limitation, fiber-reinforced CPB (FR-CPB) has emerged as a prominent research focus, offering substantial potential to enhance toughness and deformation resistance [10,11].

In recent years, research on FR-CPB has grown significantly. Studies have demonstrated that incorporating fibers such as polypropylene fibers [12–14], steel fibers [15,16], or polyacrylonitrile fibers [17,18] into CPB markedly enhances its tensile strength, toughness, and shear resistance. Different fiber types exhibit unique reinforcement characteristics, and researchers have extensively investigated the effects of fiber type, content, and distribution on the mechanical properties of FR-CPB [12,13,19,20]. Findings have revealed that fiber addition effectively suppresses crack propagation and enhances deformation resistance by forming a three-dimensional network structure [21–23]. Furthermore, fiber incorporation significantly improves the impact resistance of material, increasing its resilience under sudden stress [24]. In fact, impact resistance can be enhanced by up to 30%–40% depending on the type and quantity of the fiber incorporated [25]. Beyond mechanical properties, the rheological behavior of CPB is a critical factor influencing its engineering applications, as it directly affects pipeline transportation efficiency [26]. However, existing research predominantly focuses on the hardened properties of FR-CPB, with relatively few studies examining its rheological behavior, particularly the packing behavior of fibers in backfill materials and their significant influence on rheological performance. Understanding the mechanisms through which fibers affect the rheological properties of backfill is crucial for optimizing FR-CPB design and enhancing construction efficiency.

The maximum packing fraction (ϕ_{fm}) of fibers is a key factor influencing the rheological properties of fiber-reinforced cement-based materials [27]. Martinie et al. [28] proposed that ϕ_{fm} is an important parameter for describing the yield stress of cementitious materials containing steel fibers. This conclusion was also drawn by Guo et al. [29] in their study of cement pastes incorporating flexible fibers. The packing behavior of fibers in these materials is closely linked to their length, diameter, surface morphology and the matrix's water-to-cement ratio [30]. In CPB, the high water content and low cement dosage result in more complex fiber distribution and packing behavior compared to ordinary cement paste [31]. Consequently, existing studies [27–29] on fiber packing in cement paste cannot be directly applied to CPB. For example, the high water content in CPB promotes fiber aggregation, which negatively impacts slurry uniformity and, in turn, its rheological performance [32]. Currently, effective methods for measuring the ϕ_{fm} of fibers in CPB are lacking, posing challenges for studying and optimizing the rheological properties of FR-CPB. Therefore, systematic research on ϕ_{fm} of fibers in CPB, particularly regarding measurement methods, is essential.

Static yield stress (hereafter referred to as yield stress) is a key rheological parameter for assessing the shear resistance of CPB, significantly influencing its stability and transport performance [33]. Excessive yield stress can hinder material transportation, reducing construction efficiency, while insufficient yield stress may compromise the stability of the backfill within voids. Thus, precise control of yield stress is essential to ensure the transportability and long-term stability of CPB. Furthermore, understanding the evolution of yield stress is critical for predicting the material's behavior under prolonged loading. In practical applications, CPB must maintain long-term stability within voids, and both the magnitude and evolution of yield stress are directly linked to the backfill's reliability and safety [34]. However, quantitative studies on the yield stress and its evolution in FR-CPB are limited, constraining efforts to predict and optimize its performance in complex underground environments.

In response to the aforementioned issues, this study prepared polypropylene (PP) fibers of varying lengths and formulated FR-CPB with different fiber contents. Additionally, the packing behavior of PP fibers in CPB, as well as the yield stress and its evolution in FR-CPB, were investigated. This study addresses the aforementioned gaps by introducing a novel rheology-based method to determine ϕ_{fm} of PP fibers in CPB. This innovative method is inspired by the dynamic response of particle flocs under shear, allowing the quantitative characterization of fiber packing behavior within fresh CPB. To the best of our knowledge, this is the first systematic attempt to apply such a methodology in the context of FR-CPB, enabling a more precise optimization of fiber content and length for tailored material performance. The findings are expected to provide a robust scientific foundation and technical guidance for the practical application of FR-CPB in engineering projects.

2. Methodology

Unlike rigid fibers, flexible fibers such as PP fibers tend to deform under shear, deviating from their original conformation and orientation [35]. This is similar to the flocs' structural response when subjected to shear, i.e., deformation, deflocculation, and rearrangement (Figure 1a). The structural changes in flocs are accompanied by changes in the rheological properties of the suspension. Based on this, Tregger et al. [36] obtained ϕ_{fm} of cement particles. Therefore, flexible fibers can be regarded as a unique form of flocs, and the rheological response corresponding to the change in fiber morphology can serve to obtain ϕ_{fm} of the fibers. This is the core idea of the shear rheology-based approach proposed to measure ϕ_{fm} of flexible fibers.

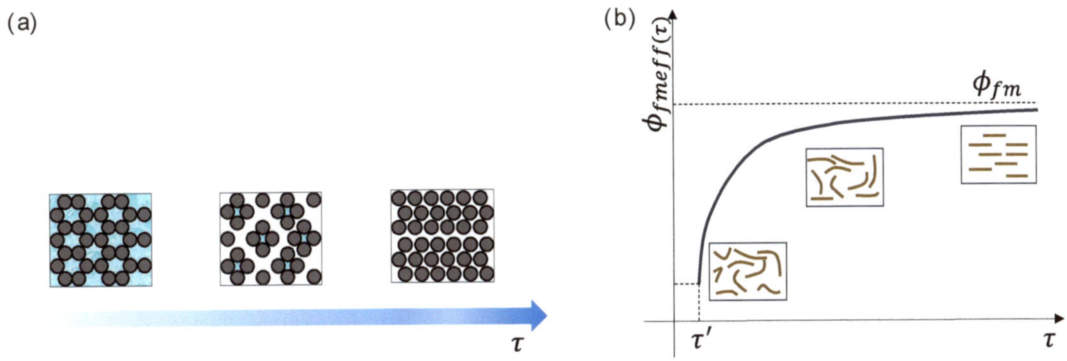

Figure 1. (**a**) Evolution of floc morphology with shear stress (adapted from [36]). (**b**) Evolution of $\phi_{fmeff}(\tau)$ with shear stress (adapted from [29]).

According to the results of Sultangaliyeva et al. [27], the viscosity of fiber-reinforced cement-based materials with respect to the volume fraction of fibers is as follows:

$$\frac{\eta'}{\eta''} = \left(1 - \frac{\phi_f}{\phi_{fm}}\right)^{-2} \tag{1}$$

where η' and η'' represent the viscosity of the suspension and the suspending medium, respectively, ϕ_f is the fiber volume fraction, and ϕ_{fm} is the maximum packing fraction of the fibers.

It should be noted that, due to the different conformations and orientations of flexible fibers under various shear conditions, ϕ_{fm} in Equation (1) more accurately represents the flexible fibers' packing fraction in a specific deformation state. Considering the dynamic nature of the packing fraction, ϕ_{fm} in Equation (1) can be replaced by the effective maximum packing fraction $\phi_{fmeff}(\tau)$. As the shearing effect increases, the flexible fibers continuously straighten and rearrange, causing $\phi_{fmeff}(\tau)$ to increase until it stabilizes (Figure 1b). In addition, to enhance the model's flexibility, an adjusting parameter "k" is introduced, referencing the results of Tregger et al. [36]. In summary, Equation (1) can be further expressed as

If $\tau > \tau'$

$$\eta(\tau) = \left[k\left(1 - \frac{\phi_f}{\phi_{fmeff}(\tau)}\right)\right]^{-2} \tag{2}$$

If $\tau \to \infty$ $(\tau \gg \tau')$

$$\eta = \left[k'\left(1 - \frac{\phi_f}{\phi_{fm}}\right)\right]^{-2} \tag{3}$$

where $\eta(\tau)$ represents the ratio of the suspension's viscosity to that of the suspending medium, with this ratio being dependent on the shear stress τ; τ' denotes the yield stress of the suspension. When $\tau \gg \tau'$, $\eta(\tau)$ becomes independent of the shear effect, corresponding to the adjusting parameter k'.

According to Equations (2) and (3), ϕ_{fm} can be directly obtained through experiments with the following steps. First, conduct stress ramp-up tests on FR-CPB samples with different fiber contents to obtain the viscosity–shear stress curves (Figure 2a). Then, plot $1 - [\eta(\tau)]^{-1/2}$ against the fiber volume fraction under different shear stresses, as shown in Figure 2b. When ϕ_f approaches $\phi_{fmeff}(\tau)$, $\eta(\tau)$ tends to infinity, and thus $1 - [\eta(\tau)]^{-1/2}$ approaches one. By fitting the data, the $\phi_{fmeff}(\tau_1)$ of flexible fibers under a specific shear stress τ_1 can be obtained. Repeat the above steps to obtain the $\phi_{fmeff}(\tau)$—τ curve; ϕ_{fm} is the volume fraction corresponding to when the curve tends to be stable (Figure 2c).

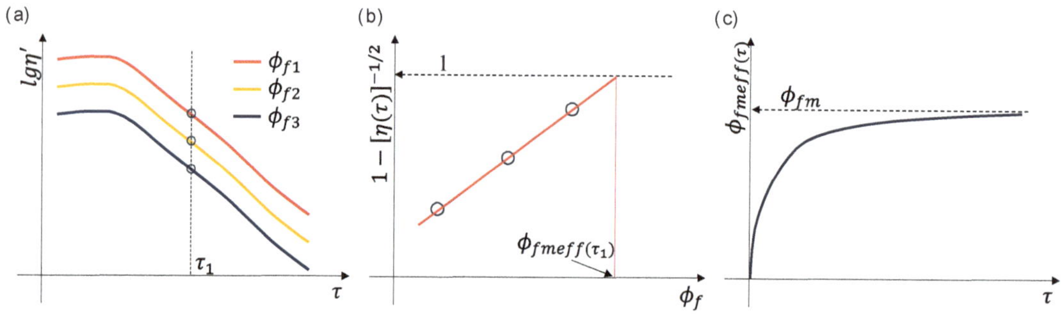

Figure 2. (**a**) Flow curves of FR-CPB varying in fiber dosages (ϕ_{f1}, ϕ_{f2}, ϕ_{f3}) under stress ramp-up tests [29]; (**b**) $\phi_{fmeff}(\tau_1)$ corresponding to the fiber configuration under τ_1 [29]; (**c**) determination of ϕ_{fm} [29].

3. Materials and Experimental Methods

3.1. Raw Materials and Sample Preparation

The materials used to prepare the FR-CPB include tailings, binder, water, and fibers. The tailings used came from a gold mine in northeastern China, with the main components being quartz, calcite, and dolomite (Figure 3a). The content of fines (<20 μm) is 20.66%, classifying it as coarse tailings. The uniformity coefficient and coefficient of curvature of the tailings are 6.81 and 1.81, respectively. The detailed particle size distribution of the tailings is shown in Figure 3b. The specific gravity of the tailings is 2.65. Commercial PO 42.5R ordinary Portland cement (Hailuo brand) was used as the binder in this study to provide the required strength for the material [37]. The particle size distribution of the cement can be seen in Figure 3b. The main chemical components of the cement, determined by X-ray fluorescence (XRF) analysis, are CaO (62.34%), SiO_2 (21.43%), Fe_2O_3 (5.06%), and Al_2O_3 (4.25%), among others [38]. Laboratory tap water was utilized as the mixing water for preparing the backfill samples. Four different lengths of PP fibers (3 mm, 6 mm, 9 mm, and 12 mm) were selected for the experiments in this study. The diameter, density, tensile strength, and Young's modulus of the fibers are 40 μm, 0.91 g/m^3, 398 MPa, and 4.2 GPa, respectively. It should be noted that PP fibers were selected for this study due to their superior mechanical properties (high tensile strength and elasticity modulus), excellent chemical stability in high-alkalinity environments, and cost effectiveness compared to other high-performance fibers [39]. Their lightweight and high toughness enable the formation of a three-dimensional network, enhancing the rheological and mechanical performance of cemented paste backfill [6]. Additionally, their widespread availability and potential applicability to other fiber types make them a practical and insightful choice for this research.

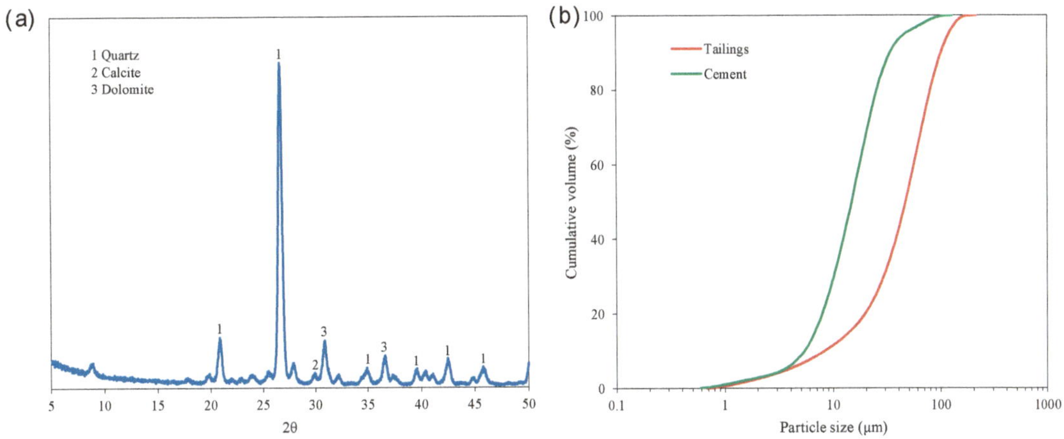

Figure 3. (**a**) XRD results of the gold tailings used. (**b**) Particle size distribution of the tailings and cement used.

For the mix proportions of the FR-CPB, the solid content and binder dosage were set to 76% and 10%, respectively. As for the fiber content, it varies in different experimental projects. The main scenarios are divided into four categories: determination of ϕ_{fm} of PP fibers, yield stress measurement, model validation, and time-dependent yield stress. The specific mix proportions are shown in Table 1. The proportions in Table 1 were selected based on a balance between commonly used industrial practices and the specific requirements for evaluating the rheological and mechanical behavior of FR-CPB. After all the raw materials were prepared, the dry materials (cement, tailings, and PP fibers) were mixed for 3 min, and then water was added and mixed for a further 5 min to obtain fresh FR-CPB.

Table 1. Summary of the mix compositions of the prepared FR-CPB samples.

Trial Mix	Solid Content (%)	Binder Dosage (%)	Fiber Content		Fiber Length (mm)
			Vol.%	Wt.%	
ϕ_{fm} determination	76	10	0	0	3, 6, 9, 12
			0.08	0.027	
			0.1	0.034	
Yield stress measurement			0	0	
			0.891	0.3	
			1.782	0.6	
			2.674	0.9	
Model validation			0.891	0.3	
			1.782	0.6	
			2.674	0.9	
Time-dependent yield stress			0	0	
			0.891	0.3	
			1.782	0.6	
			2.674	0.9	

3.2. Experimental Methods

A Brookfield rheometer (RSX-SST), equipped with a four-bladed vane rotor measuring 40 mm in length and 20 mm in diameter, was used in this experiment. Before the main test, pre-shearing was applied at a shear rate of 100 s^{-1} for 60 s followed by a 30 s resting period. This procedure was intended to eliminate the sample's shear history and ensure a uniform starting condition for testing [40,41].

3.2.1. Stress Ramp-Up Test

A stress ramp-up test was performed to generate the flow curves of FR-CPB and to further determine ϕ_{fm} of the PP fibers in CPB. During this test, the shear stress increased at a constant rate of 1 Pa/s until reaching 100 Pa. This gradual increase allowed for precise monitoring of the material's response to varying stress levels, ultimately providing a comprehensive flow curve. The choice of this rheological protocol is, on the one hand, based on [29], and on the other hand, this shear method indeed allows for a comprehensive capture of the configurational changes of the fibers in CPB.

3.2.2. Yield Stress Test

Yield stress tests were conducted to determine the yield stress of FR-CPB. During the experiment, a shear rate of 0.01 s^{-1} was applied to the samples for 100 s, generating shear stress–time curves. The peak stress on the curve was identified as the yield stress (Figure 4a).

To investigate the evolution of yield stress, tests were performed at intervals of 0, 30, 60, 90, and 120 min. It should be noted that newly prepared samples were used for testing at each time point to prevent structural disturbances or performance variations caused by repeated testing, ensuring the reliability of the results. After all tests were completed, a yield stress–time curve was constructed, and its slope was calculated to represent the yield stress evolution rate (A_τ) of the FR-CPB (Figure 4b).

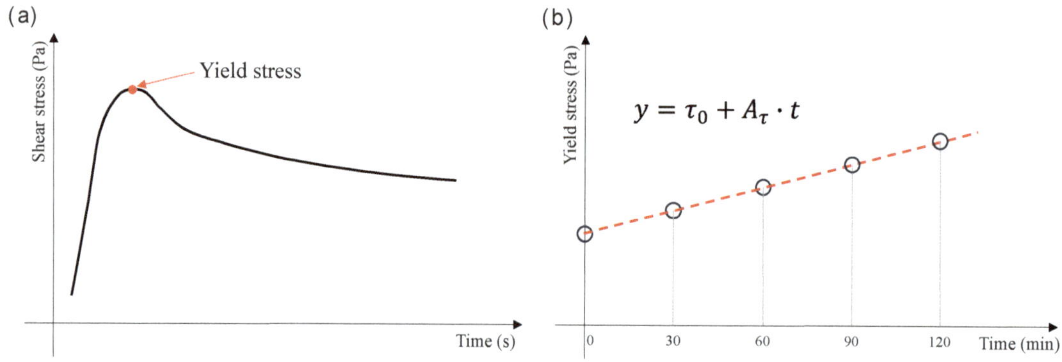

Figure 4. Acquisition of rheological data for FR-CPB: (**a**) yield stress; (**b**) yield stress evolution rate.

4. Results and Discussion

4.1. Maximum Packing Fraction

Figure 5a shows the dynamic nature of the packing fraction of the PP fibers when subjected to shear. Initially, regardless of the fiber length, the curve exhibits a plateau region when the shear stress remains below 75 Pa. This reflects the adaptability of flexible fibers to shear, which results in FR-CPB having a higher yield stress compared to traditional CPB. As the shear stress further increases, the PP fibers gradually straighten and rearrange, resulting in a continuous increase in $\phi_{fmeff}(\tau)$ until it reaches ϕ_{fm}, which corresponds to the optimal conformation of the fibers. This dynamic characteristic of the fiber packing fraction under shear also validates the correctness of the method proposed in this study. To quantify this dynamic characteristic and obtain ϕ_{fm}, the following equation was employed:

$$\phi_{fmeff}(\tau) = \phi_{fmi} + \left(\phi_{fm} - \phi_{fmi}\right)\left\{1 - exp\left[-\left(\frac{\tau}{\tau_r}\right)^{\alpha}\right]\right\} \quad (4)$$

where ϕ_{fmi} is the initial packing fraction corresponding to the fiber conformation before the test; τ_r and α are the characteristic stress and characteristic coefficient, respectively.

This equation effectively describes the dynamic characteristics of the fibers (Figure 5a). Through this equation, ϕ_{fm} of different fibers can be obtained. The ϕ_{fm} values for PP fibers of 3, 6, 9, and 12 mm are 0.661, 0.630, 0.592, and 0.534, respectively. Figure 5b illustrates the relationship between the fiber aspect ratio and ϕ_{fm}. The figure shows a significant negative linear correlation between the aspect ratio and ϕ_{fm}, with a correlation coefficient of 0.98. Thus, fibers with smaller aspect ratios exhibit a greater capacity to fill space, leading to higher packing efficiency. This finding is consistent with the observations of Nan et al., who reported that the fiber aspect ratio significantly influences ϕ_{fm}, with fibers of higher aspect ratios tending to exhibit lower ϕ_{fm} [42]. Additionally, Toll also indicated that an increase in the aspect ratio alters the orientation and distribution of fibers, thereby reducing packing efficiency [43].

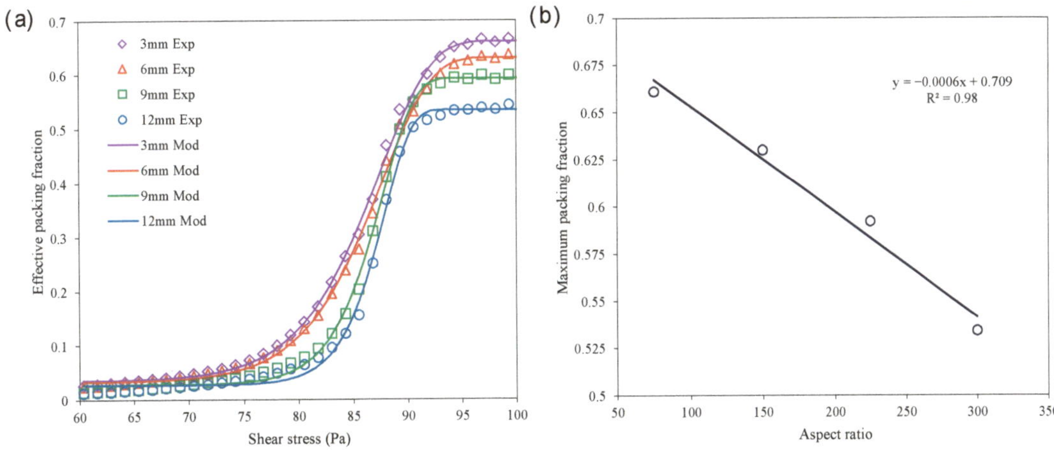

Figure 5. (**a**) Evolution of $\phi_{fmeff}(\tau)$ of different fibers with shear stress. (**b**) Relationship between fiber aspect ratio and maximum packing fraction.

4.2. Yield Stress

Figure 6a presents the yield stress results of the FR-CPB under different lengths (3 mm, 6 mm, 9 mm, and 12 mm) and contents (0%, 0.3%, 0.6%, and 0.9%) of PP fibers. Overall, the yield stress of the FR-CPB increases with the fiber length and content. Similar phenomena were observed by Zhao et al. [6]. For short fibers (3 mm and 6 mm), the increase in yield stress is relatively limited. This may be because short fibers are insufficiently long to interweave and form a stable fiber skeleton structure within the slurry, resulting in weaker constraints on particle flow. Hannant [44] also noted that fibers exhibit significant limitations in forming effective skeleton structures, thus contributing relatively less to enhancing the slurry's mechanical properties. In contrast, longer fibers (9 mm and 12 mm) effectively interweave to create a more stable skeleton structure in the slurry, substantially improving shear resistance and yielding a notable increase in yield stress. Regarding fiber content, at a low content (0.3%), the increase in yield stress is relatively slow, suggesting limited reinforcement; however, at medium to high contents (0.6% and 0.9%), the yield stress increases significantly. This indicates that a higher fiber content promotes the formation of denser fiber networks, enhancing the overall yield stress of the slurry. Notably, there is a synergistic effect between fiber length and content. Under the combination of high content (0.9%) and long fibers (12 mm), the yield stress reaches a maximum of 274.34 kPa, demonstrating a significant enhancement of the slurry's shear resistance. This finding also suggests that the packing fraction of fibers plays a crucial role in influencing yield stress, especially at higher fiber contents and longer lengths, where this trend becomes more pronounced.

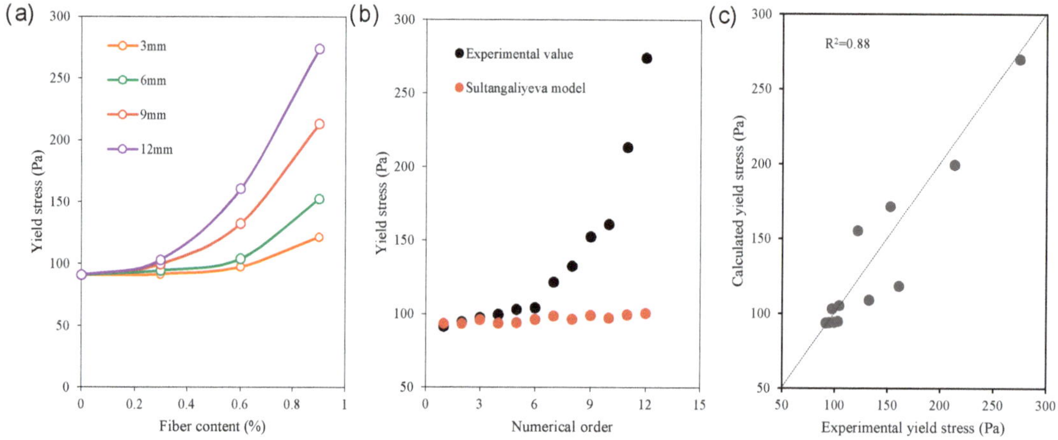

Figure 6. (**a**) Effect of fiber length and content on the yield stress of CPB. (**b**) Comparison between Sultangaliyeva model predictions and actual yield stress values. (**c**) Comparison of calculated yield stress and experimental yield stress.

The Sultangaliyeva model (Equation (5)) [27] is widely employed to predict the yield stress of fiber-reinforced cement paste, with its predictions primarily relying on the packing fraction of fibers. Here, this model was applied for calculations, and the results were compared with experimental measurements, as illustrated in Figure 6b. The comparison reveals noticeable deviations between the model's predictions and the measured values, particularly under higher fiber content and length, where the discrepancies become more pronounced. These deviations suggest that the Sultangaliyeva model has limitations when applied to CPB, potentially neglecting critical mechanisms influencing yield stress. For instance, fiber–fiber interactions and the reinforcing effects of fibers on the matrix are inadequately accounted for, which likely explains the model's significant underestimation of yield stress at high fiber contents. Consequently, refining the model is essential to more accurately capture the influence of fibers on the yield stress of CPB, thereby improving its predictive accuracy and applicability.

$$\tau_{fc} = \tau_c\left[\left(1 - \frac{\phi_f}{\phi_{fm}}\right)^{-2}\right] \tag{5}$$

where τ_{fc} and τ_c represent the yield stress of FR-CPB and the CPB matrix, respectively.

Guo et al. [29] demonstrated that fiber–fiber interactions and the reinforcing effects of fibers on the matrix can be expressed as a power function of ϕ_f/ϕ_{fm}. Consequently, Equation (5) can be modified as

$$\tau_{fc} = \tau_c \left[\left(1 - \frac{\phi_f}{\phi_{fm}} \right)^{-2} + a \left(\frac{\phi_f}{\phi_{fm}} \right)^b \right] \quad (6)$$

where a and b are coefficients related to fiber interactions, determined through experimental data.

Using Equation (6) and experimental data, the specific form of the model was obtained through fitting, as shown in Equation (7). As illustrated in Figure 6c, while the data exhibit some dispersion, likely due to the limited sample size, the overall trend closely follows the dashed line (ideal fitting line), with a goodness of fit of 0.88.

$$\tau_{fc} = \tau_c \left[\left(1 - \frac{\phi_f}{\phi_{fm}} \right)^{-2} + 9086676.13 \left(\frac{\phi_f}{\phi_{fm}} \right)^{5.14} \right] \quad (7)$$

Model Validation

To validate the proposed model, FR-CPB was prepared using various fiber systems, followed by rheological testing and result comparison. In this study, an experimental design was implemented using equi-volume pairwise mixtures of fibers with different lengths to construct composite fiber systems, labeled as 3–6, 3–9, 3–12, 6–9, 6–12, and 9–12, respectively. Here, "3–6" represents a composite system consisting of 3 mm and 6 mm fibers, with other labels following a similar pattern. These composite fibers were used to prepare CPB under fixed experimental conditions: a solid content of 76%, a cement dosage of 10%, and fiber contents of 0.3%, 0.6%, and 0.9%. The yield stress of each system was subsequently measured.

Figure 7a illustrates the ϕ_{fm} of composite fiber systems determined using the proposed method. The results reveal that when fibers of different lengths are mixed in equal volumes, the ϕ_{fm} of the composite fiber system lies between the ϕ_{fm} values of the individual fibers that constitute the system. For example, the ϕ_{fm} of the 3–6 combination is 0.655, slightly lower than that of 3 mm fibers (0.661) but higher than that of 6 mm fibers (0.630). A similar trend is observed in other combinations: the ϕ_{fm} of the 3–9 system is 0.618, exceeding that of 9 mm single fibers (0.592), while the 3–12 combination achieves a ϕ_{fm} of 0.596, surpassing the 0.534 of 12 mm single fibers. In contrast, combinations of longer fibers (e.g., 6–9, 6–12, and 9–12) exhibit a gradual decline in the ϕ_{fm}, with values of 0.598, 0.584, and 0.558, respectively. These findings suggest that in mixed fiber systems, shorter fibers play a more significant role in improving overall packing performance.

The comparison between the experimental yield stress values and the model predictions for the validation group is illustrated in Figure 7b. Most data align closely with the ideal fitting line, achieving a correlation coefficient of 0.93. These results confirm that the improved model (Equation (7)) effectively captures the yield stress characteristics of FR-CPB, demonstrating high predictive accuracy and broad applicability. However, it should be noted that a noticeable deviation occurs around the 100 Pa range where alignment is reduced. This deviation can be attributed to several factors: (1) the transition zone of fiber interaction, where intermediate fiber content and length combinations result in an underdeveloped fiber network, leading to variability in the packing fraction and fiber distribution; (2) material heterogeneity, as incomplete bridging and inconsistent fiber orientation at intermediate stress levels increase local variations in the rheological behavior; (3) limitations of the model, which assumes uniform fiber dispersion and interaction but may lose predictive accuracy under these conditions; and (4) experimental uncertainty, where slight variations in sample preparation or testing protocols might amplify discrepancies at this range. To address these issues, future studies could incorporate additional data for intermediate stress levels, refine the model to better account for local heterogeneities, and utilize imaging techniques to analyze fiber distribution and orientation within the material.

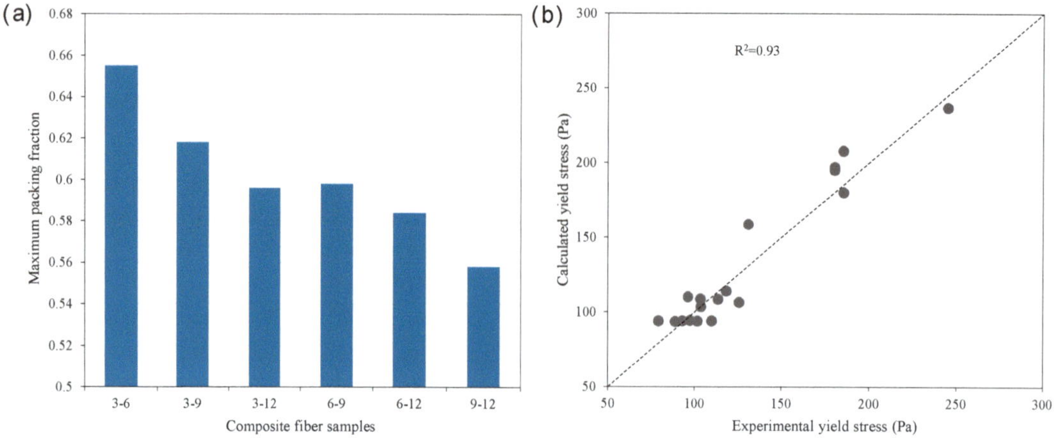

Figure 7. (**a**) Maximum packing fraction of combined fibers. (**b**) Comparison of calculated yield stress and experimental yield stress.

4.3. Time-Dependent Yield Stress

The effects of fiber length and content on the yield stress evolution rate of the FR-CPB (within 2 h) are presented in Figure 8a. The figure shows that the yield stress evolution rate increases with fiber content. This phenomenon can be attributed to the three-dimensional network structure formed by fiber-reinforced cement-based materials. Fibers bridge cement particles to form a continuous network, which enhances the system's overall stability and reduces cement particle sedimentation [45]. This uniform distribution improves the contact efficiency between cement particles and water, thereby accelerating the hydration reaction [46]. When the fiber content reaches 0.6%, particularly for fibers with a length of 12 mm, the yield stress evolution rate of the samples is significantly higher than that of shorter fibers (3 mm and 6 mm). Additionally, long fibers create a more stable and continuous three-dimensional network within the slurry, promoting the even distribution of cement particles and reducing sedimentation, which in turn enhances the cement hydration efficiency. Conversely, shorter fibers are less effective at bridging, with limited connections between fibers. This reduces contact opportunities among cement particles, leading to incomplete hydration reactions and a less pronounced reinforcement effect. Moreover, long fibers can span larger particle gaps, providing stronger reinforcement, while short fibers fail to achieve this. Thus, the synergistic interaction between fiber length and content determines the variation in the yield stress evolution rate of FR-CPB.

Figure 8b compares the relative yield stress with its evolution rate. Both parameters exhibit a similar upward trend with an increasing relative fiber volume fraction. Notably, the relative yield stress consistently remains lower than the relative yield stress evolution rate, with the gap between them widening as the relative volume fraction increases. This observation suggests that, in addition to fiber–fiber interactions and the reinforcing effects of fibers on the matrix, time-dependent factors significantly influence the yield stress evolution rate. This finding is consistent with previous analyses, indicating that a higher relative fiber volume fraction enhances cement hydration efficiency, thereby accelerating the yield stress evolution rate. Therefore, Equation (7) cannot be directly applied to describe the yield stress evolution rate of FR-CPB without fully considering the effects of cement hydration. Recognizing the critical role of exponential parameters in determining the function's outcomes, the framework of Equation (7) was retained. By combining experimental data and optimizing the exponential parameters, a yield stress evolution rate model for FR-CPB (Equation (8)) was developed. Figure 8c compares the model predictions with the observed yield stress evolution rate values. Most data align closely with the ideal fitting line, achieving a correlation coefficient of 0.92. This indicates that the model provides an accurate prediction of the yield stress evolution rate for FR-CPB.

$$R_{fc} = R_c\left[\left(1 - \frac{\phi_f}{\phi_{fm}}\right)^{-2} + 9086676.13\left(\frac{\phi_f}{\phi_{fm}}\right)^5\right] \tag{8}$$

where R_c and R_{fc} represent the yield stress evolution rates of CPB and FR-CPB, respectively.

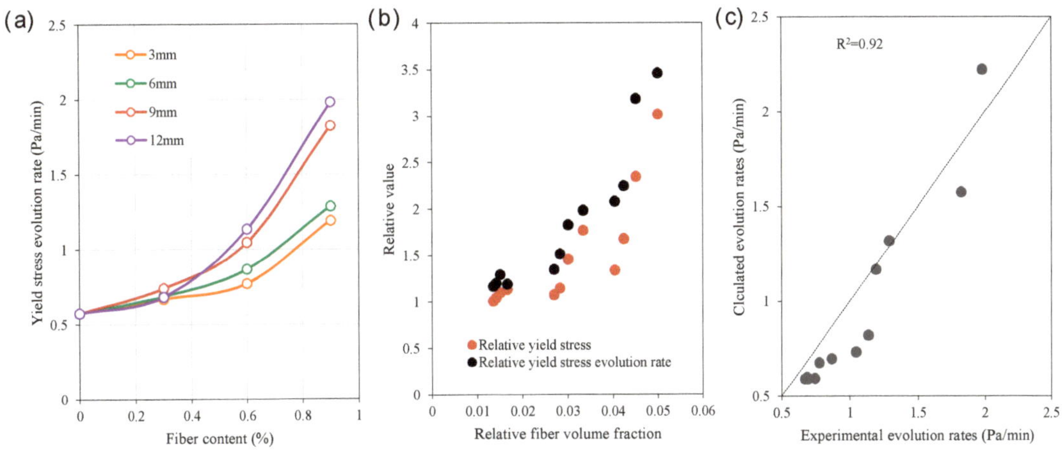

Figure 8. (**a**) Effect of fiber length and content on yield stress evolution rate of FR-CPB within 2 h. (**b**) Variation in relative yield stress and relative yield stress evolution rate with relative fiber volume fraction. (**c**) Comparison of calculated yield stress evolution rates and experimental yield stress evolution rates.

5. Conclusions

Inspired by the response of particle flocs under shear, this study proposes a method to determine the maximum packing fraction of PP fibers in CPB. Building on this, the maximum packing fraction is used to analyze and model the yield stress and yield stress evolution rate of FR-CPB. The main conclusions are as follows:

- The proposed rheology-based method effectively captures the dynamic characteristics of PP fibers in CPB during the shearing process, enabling the accurate determination of the maximum packing fraction. This approach provides a reliable tool for gaining deeper insights into fiber packing behavior within CPB.
- The maximum packing fraction decreased with increasing fiber length, ranging from 0.661 for 3 mm fibers to 0.534 for 12 mm fibers, highlighting the significant influence of fiber geometry on packing efficiency. Moreover, shorter fibers demonstrate greater advantages in space filling and packing efficiency, while the inclusion of longer fibers dilutes the overall packing performance.
- The improved yield stress model demonstrated high predictive accuracy, with a correlation coefficient of 0.93 between the predicted and experimental values. Additionally, the relative yield stress was always less than the relative yield stress evolution rate due to the influence of cement hydration efficiency. By adjusting the exponential parameters in the yield stress model, the model successfully achieves an accurate description of the yield stress evolution rate.

While the proposed models have demonstrated strong potential for optimizing FR-CPB, future research should explore their applicability to other fiber types, such as steel or natural fibers, to evaluate generalizability. Scaling up the models for large-scale industrial use, particularly in mining operations, is essential to validate their performance under practical field conditions. Additionally, further studies are needed to assess the models' reliability under varying environmental conditions, such as temperature, humidity, and chemical exposure, as well as their long-term performance considering fiber degradation. Addressing these aspects will refine the models and expand their applicability to optimize FR-CPB systems across diverse and complex conditions.

Author Contributions: Conceptualization, S.H.; formal analysis, J.Q.; data curation, J.Q., Q.Z. and Z.G.; writing—original draft preparation, S.H.; writing—review and editing, S.H., J.Q. and C.L.; supervision, C.L. All authors have read and agreed to the published version of the manuscript.

Funding: This research was funded by the National Natural Science Foundation of China (52374116).

Data Availability Statement: Data are contained within the article.

Acknowledgments: The authors would like to thank Fan Yao from Shiyanjia Lab (www.shiyanjia.com) (accessed on 13 January 2025) for technical support of tests.

Conflicts of Interest: The authors declare no conflicts of interest.

References

1. Zhang, C.; Taheri, A.; Du, C.; Xia, W.; Tan, Y. Mechanical Characteristics and Macro-Microscopic Response Mechanisms of Cemented Paste Backfill under Different Curing Temperatures. *Minerals* **2024**, *14*, 433. [CrossRef]
2. Sagade, A.; Fall, M. Study of Fresh Properties of Cemented Paste Backfill Material with Ternary Cement Blends. *Constr. Build. Mater.* **2024**, *411*, 134287. [CrossRef]
3. Wu, A.; Wang, Y.; Ruan, Z.; Xiao, B.; Wang, J.; Wang, L. Key Theory and Technology of Cemented Paste Backfill for Green Mining of Metal Mines. *Green Smart Min. Eng.* **2024**, *1*, 27–39. [CrossRef]
4. Yang, L.; Jia, H.; Wu, A.; Jiao, H.; Chen, X.; Kou, Y.; Dong, M. Particle Aggregation and Breakage Kinetics in Cemented Paste Backfill. *Int. J. Miner. Metall. Mater.* **2024**, *31*, 1965–1974. [CrossRef]
5. Wang, D.; Zhang, Q.; Liu, B.; Zhu, D.; Chen, Q. Enhanced Immobilization of Fluoride in Phosphogypsum-Based Cement-Free Paste Backfill Modified by Polyaluminum Chloride and Its Mechanism. *Constr. Build. Mater.* **2025**, *458*, 139622. [CrossRef]
6. Zhao, X.; Wang, H.; Luo, G.; Dai, K.; Hu, Q.; Jin, J.; Liu, Y.; Liu, B.; Miao, Y.; Zhu, K.; et al. Study on the Rheological and Thixotropic Properties of Fiber-Reinforced Cemented Paste Backfill Containing Blast Furnace Slag. *Minerals* **2024**, *14*, 964. [CrossRef]
7. Guner, N.U.; Yilmaz, E.; Sari, M.; Kasap, T. Cementitious Backfill with Partial Replacement of Cu-Rich Mine Tailings by Sand: Rheological, Mechanical and Microstructural Properties. *Minerals* **2023**, *13*, 437. [CrossRef]
8. Carnogursky, E.A.; Fall, M.; Haruna, S. Rheology and Setting Time of Saline Cemented Paste Backfill. *Miner. Eng.* **2023**, *202*, 108258. [CrossRef]
9. Cui, L.; McAdie, A. Experimental Study on Evolutive Fracture Behavior and Properties of Sulfate-Rich Fiber-Reinforced Cemented Paste Backfill under Pure Mode-I, Mode-II, and Mode-III Loadings. *Int. J. Rock Mech. Min. Sci.* **2023**, *169*, 105434. [CrossRef]
10. Hou, Y.; Yang, K.; Yin, S.; Yu, X.; Kou, P.; Wang, Y. Enhancing Workability, Strength, and Microstructure of Cemented Tailings Backfill through Mineral Admixtures and Fibers. *J. Build. Eng.* **2024**, *84*, 108590. [CrossRef]
11. Yang, J.; Zhao, K.; Yu, X.; Yan, Y.; He, Z.; Zhou, Y.; Lai, Y. Fracture Evolution of Fiber-Reinforced Backfill Based on Acoustic Emission Fractal Dimension and b-Value. *Cem. Concr. Compos.* **2022**, *134*, 104739. [CrossRef]
12. Chakilam, S.; Cui, L. Effect of Polypropylene Fiber Content and Fiber Length on the Saturated Hydraulic Conductivity of Hydrating Cemented Paste Backfill. *Constr. Build. Mater.* **2020**, *262*, 120854. [CrossRef]
13. Xue, G.; Yilmaz, E.; Song, W.; Cao, S. Fiber Length Effect on Strength Properties of Polypropylene Fiber Reinforced Cemented Tailings Backfill Specimens with Different Sizes. *Constr. Build. Mater.* **2020**, *241*, 118113. [CrossRef]
14. Xu, W.; Li, Q.; Zhang, Y. Influence of Temperature on Compressive Strength, Microstructure Properties and Failure Pattern of Fiber-Reinforced Cemented Tailings Backfill. *Constr. Build. Mater.* **2019**, *222*, 776–785. [CrossRef]
15. Li, X.; Cao, S.; Yilmaz, E. Microstructural Evolution and Strengthening Mechanism of Aligned Steel Fiber Cement-Based Tail Backfills Exposed to Electromagnetic Induction. *Int. J. Miner. Metall. Mater.* **2024**, *31*, 2390–2403. [CrossRef]
16. Cao, S.; Che, C.; Zhang, Y.; Shan, C.; Liu, Y.; Zhao, C.; Du, S. Mechanical Properties and Damage Evolution Characteristics of Waste Tire Steel Fiber-Modified Cemented Paste Backfill. *Int. J. Min. Sci. Technol.* **2024**, *34*, 909–924. [CrossRef]

17. Xue, G.; Yilmaz, E.; Song, W.; Yilmaz, E. Influence of Fiber Reinforcement on Mechanical Behavior and Microstructural Properties of Cemented Tailings Backfill. *Constr. Build. Mater.* **2019**, *213*, 275–285. [CrossRef]
18. Zhao, K.; Zhao, K.; Yan, Y.; Yang, J.; Wu, J.; Lai, Y.; Liu, L.; Zeng, X. Influence of Different Fibers on Compressive Toughness and Damage of Early Age Cemented Tailings Backfill. *Environ. Sci. Pollut. Res.* **2023**, *30*, 37449–37461. [CrossRef]
19. Yin, S.; Hou, Y.; Chen, X.; Zhang, M. Mechanical, Flowing and Microstructural Properties of Cemented Sulfur Tailings Backfill: Effects of Fiber Lengths and Dosage. *Constr. Build. Mater.* **2021**, *309*, 125058. [CrossRef]
20. Nematollahi, M.; Ahmadi, M.S.; Fattahi, S. Mechanical Properties of Glass and Jute Fiber Reinforced Thermoplastic Composites with Waste PET Needle-Punched Carpet Matrix. *World J. Eng.* 2024; *ahead of print*.
21. Cao, P.; Feng, D.; Zhou, C.; Zuo, W. Study on Fracture Behavior of Polypropylene Fiber Reinforced Concrete with Bending Beam Test and Digital Speckle Method. *Comput. Concr.* **2014**, *14*, 527–546. [CrossRef]
22. Adlparvar, M.R.; Esmaeili, M.; Taghavi Parsa, M.H. Strength Properties of Fiber Reinforced Concrete Including Steel Fibers. *World J. Eng.* **2024**, *21*, 194–202. [CrossRef]
23. Sathe, S.; Dandin, S.; Wagale, M.; Mali, P.R. Effective Combinations of Different Fiber Reinforcements and Super-Plasticizers for Optimal Performance of Geo-Polymer Concrete. *World J. Eng.* 2024; *ahead of print*.
24. Xue, G.; Yilmaz, E.; Feng, G.; Cao, S.; Sun, L. Reinforcement Effect of Polypropylene Fiber on Dynamic Properties of Cemented Tailings Backfill under SHPB Impact Loading. *Constr. Build. Mater.* **2021**, *279*, 122417. [CrossRef]
25. Pan, J.; Ping, P.; Ding, B.; Zhu, B.; Lin, Y.; Ukrainczyk, N.; Zhang, H.; Cai, J. Impact Behaviour of 3D Printed Fiber Reinforced Cementitious Composite Beams. *Compos. Part A Appl. Sci. Manuf.* **2024**, *182*, 108175. [CrossRef]
26. Cheng, H.; Wu, S.; Li, H.; Zhang, X. Influence of Time and Temperature on Rheology and Flow Performance of Cemented Paste Backfill. *Constr. Build. Mater.* **2020**, *231*, 117117. [CrossRef]
27. Sultangaliyeva, F.; Carré, H.; La Borderie, C.; Zuo, W.; Keita, E.; Roussel, N. Influence of Flexible Fibers on the Yield Stress of Fresh Cement Pastes and Mortars. *Cem. Concr. Res.* **2020**, *138*, 106221. [CrossRef]
28. Martinie, L.; Rossi, P.; Roussel, N. Rheology of Fiber Reinforced Cementitious Materials: Classification and Prediction. *Cem. Concr. Res.* **2010**, *40*, 226–234. [CrossRef]
29. Guo, Z.; Qiu, J.; Huang, D.; Liu, K.; Kirichek, A.; Liu, C.; Chen, B.; Zhao, Y.; Qu, Z. Rheology of Flexible Fiber-Reinforced Cement Pastes: Maximum Packing Fraction Determination and Structural Build-up Analysis. *Compos. Struct.* **2025**, *352*, 118662. [CrossRef]
30. Li, Q.; Bai, Y. A Novel Approach to Modified Polyethylene Fibres with Size Effect in Cement Mortar. *Case Stud. Constr. Mater.* **2024**, *21*, e03916. [CrossRef]
31. Liu, C.; Li, Z.; Nie, S.; Skibsted, J.; Ye, G. Structural Evolution of Calcium Sodium Aluminosilicate Hydrate (C-(N-) ASH) Gels Induced by Water Exposure: The Impact of Na Leaching. *Cem. Concr. Res.* **2024**, *178*, 107432. [CrossRef]
32. Chen, F.; Bai, S.; Guan, X.; Qiao, J.; Gou, H. Influence of Type and Particle Size of Superabsorbent Polymer on Early Water Distribution and Internal Curing Zone Properties of Cement Paste. *Cem. Concr. Compos.* **2024**, *150*, 105526. [CrossRef]
33. Wang, X.; Wang, H.; Wu, A.; Jiang, H.; Peng, Q.; Zhang, X. Evaluation of Time-Dependent Rheological Properties of Cemented Paste Backfill Incorporating Superplasticizer with Special Focus on Thixotropy and Static Yield Stress. *J. Cent. South Univ.* **2022**, *29*, 1239–1249. [CrossRef]
34. Guo, Z.; Qiu, J.; Pel, L.; Zhao, Y.; Zhu, Q.; Kwek, J.W.; Zhang, L.; Jiang, H.; Yang, J.; Qu, Z. A Contribution to Understanding the Rheological Measurement, Yielding Mechanism and Structural Evolution of Fresh Cemented Paste Backfill. *Cem. Concr. Compos.* **2023**, *143*, 105221. [CrossRef]
35. Poodipeddi, S.K.K.; Singampalli, A.; Rayala, L.S.M.; Ravula, S.S.N. Structural and Fatigue Analysis of Car Wheel Rims with Carbon Fibre Composites. *World J. Eng.* **2024**, *21*, 503–509. [CrossRef]
36. Tregger, N.A.; Pakula, M.E.; Shah, S.P. Influence of Clays on the Rheology of Cement Pastes. *Cem. Concr. Res.* **2010**, *40*, 384–391. [CrossRef]
37. Sathe, S.; Patil, S.; Bhosale, Y.N. Investigation of Strength, Durability, and Microstructure Properties of Concrete with Waste Marble Powder as a Partial Replacement of Cement. *World J. Eng.* 2024; *ahead of print*.
38. Guo, Z.; Qiu, J.; Kirichek, A.; Zhou, H.; Liu, C.; Yang, L. Recycling Waste Tyre Polymer for Production of Fibre Reinforced Cemented Tailings Backfill in Green Mining. *Sci. Total Environ.* **2024**, *908*, 168320. [CrossRef] [PubMed]
39. Silva, E.R.M.; Silva, J.C.R.; Fernandes, E.K.; dos Santos Leite, J.; da Costa, A.M.P.; Tapajós, L.S.; dos Santos Leite, J. Experimental Study on the Behavior of Polypropylene Fiber-Reinforced Concrete Beams. *Cad. Pedagógico* **2024**, *21*, e8020. [CrossRef]
40. DVSK, C.; Satish Kumar Ch, N. A Review on Concrete's Rheological Properties. *World J. Eng.* **2024**, *21*, 53–70.
41. Qu, Z.; Yu, Q.; Ong, G.P.; Cardinaels, R.; Ke, L.; Long, Y.; Geng, G. 3D Printing Concrete Containing Thermal Responsive Gelatin: Towards Cold Environment Applications. *Cem. Concr. Compos.* **2023**, *140*, 105029. [CrossRef]
42. Nan, W.; Wang, Y.; Ge, Y.; Wang, J. Effect of Shape Parameters of Fiber on the Packing Structure. *Powder Technol.* **2014**, *261*, 210–218. [CrossRef]
43. Toll, S. Packing Mechanics of Fiber Reinforcements. *Polym. Eng. Sci.* **1998**, *38*, 1337–1350. [CrossRef]
44. Hannant, P.J. *Fibre Cements and Fibre Concretes*; Wiley John & Sons, Limited: Chichester, UK, 1978.

45. Hidayat, B.A.; Sabdono, P.; Hung, C.-C.; Indriyantho, B.R. Compressive and Flexural Behavior of Fiber-Reinforced Mortar: An Experimental Study. In Proceedings of the AIP Conference Proceedings, Malang, Indonesia, 12 August 2021; Volume 2447.
46. Niu, H.M.; Xing, Y.M.; Zhao, Y.R. A Review on the Methods of Improving Fiber Distribution of Engineered Cementitious Composites (ECC). *Adv. Mater. Res.* **2013**, *683*, 46–50. [CrossRef]

minerals

Review

Pipeline Transport Performance of Paste Backfill Slurry in Long-Distance Underground Backfilling: A Review

Wei Wang [1], Bin Yu [2,3], Wenyuan Xu [2,3], Kai Yang [2,3], Yanying Yin [2,3] and Mengyuan Li [2,3,*]

1 School of Resources and Environmental Engineering, Wuhan University of Technology, Wuhan 430070, China; 64810@whut.edu.cn
2 BGRIMM Technology Group, Beijing 100160, China; yu_b@bgrimm.com (B.Y.); xuwenyuan@bgrimm.com (W.X.); yangkai@bgrimm.com (K.Y.); yanyingyin2000@163.com (Y.Y.)
3 National Centre for International Research on Green Metal Mining, Beijing 102628, China
* Correspondence: limengyuan@bgrimm.com

Abstract: This paper reviews recent advancements in the pipeline transport performance of paste backfill slurry in long-distance underground backfilling operations, with a primary focus on applications in metal mines. Key aspects, including flow performance, energy consumption during transport, and operational stability, are discussed in detail. Slurry concentration and rheological properties, including viscosity, yield stress, and flow behavior, as well as particle size distribution, are examined for their effects on transport efficiency. The relationship between these characteristics and pipeline resistance is also examined. Factors like pipeline orientation, configuration, diameter, length, elbow design, and elevation gradients are explored, demonstrating that careful design can optimize flow performance, reduce energy consumption, and minimize the risk of blockages and bursts. Additionally, the roles of commonly used additives, such as water reducers, foaming agents, antifreeze agents, and thickeners, are discussed in terms of their impact on slurry flowability, stability, and resistance losses. Optimal slurry regulation, strategic pipeline design, and effective additive utilization improve flow efficiency, extend service life, and reduce maintenance costs, thereby ensuring reliable backfill operations. Future research should focus on innovative pipeline designs, such as improving material selection and configuration to optimize flow stability and reduce energy consumption. Advanced additives, including thickeners and water reducers, could further enhance slurry flowability, reduce pipeline resistance, and improve system reliability.

Keywords: paste backfill slurry; pipeline transport; rheological properties; pipeline design; additives

Citation: Wang, W.; Yu, B.; Xu, W.; Yang, K.; Yin, Y.; Li, M. Pipeline Transport Performance of Paste Backfill Slurry in Long-Distance Underground Backfilling: A Review. *Minerals* **2024**, *14*, 1238. https://doi.org/10.3390/min14121238

Academic Editors: Abbas Taheri, Yuye Tan, Xun Chen and Yuan Li

Received: 27 October 2024
Revised: 19 November 2024
Accepted: 3 December 2024
Published: 5 December 2024

1. Introduction

Rapid industrialization has led to a significant increase in demand for non-ferrous metal minerals, driving the need for more efficient and sustainable transportation solutions [1–3]. Pipeline transportation plays a crucial role in addressing these demands, offering a cost-effective and environmentally friendly method for conveying materials in mining operations [4–6].

The backfill mining method, widely adopted in the industry, significantly reduces solid waste discharge and mitigates land subsidence, thereby enhancing ecological sustainability [7–9]. Moreover, it addresses underground voids, improving stope stability and increasing mineral recovery rates [10–13]. As a key part of this process, pipeline transportation has evolved from hydraulic backfill systems to the current paste backfill technologies [14–16]. These advancements ensure the safe and efficient transport of ore waste and backfill materials, playing a critical role in the sustainable development of mining practices [17–19], as shown in Figure 1 [20].

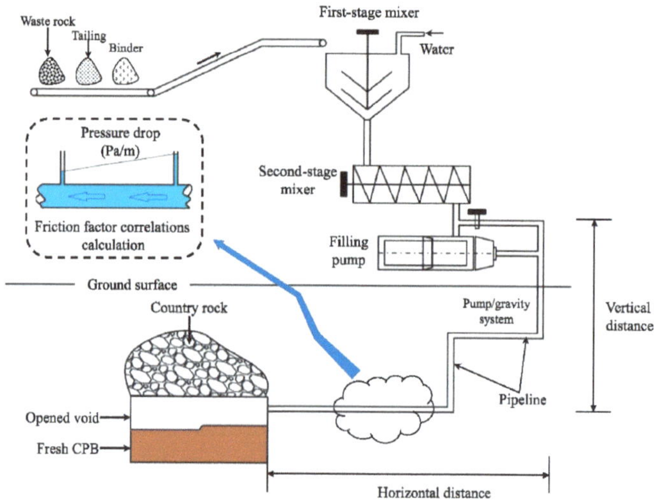

Figure 1. Backfill pipeline transportation technology. Adapted from Ref. [20].

However, pipeline transportation faces several technical and economic challenges. The design and material selection for pipelines must align with the specific properties of the transported materials to maintain system stability [21–23]. Fluid dynamics issues, such as blockages and pipe wall erosion, require extensive research and solutions [20,24,25]. Additionally, pipeline construction and operational costs remain a significant concern, requiring optimization to ensure cost-effectiveness under resource and economic constraints [2,4,26].

This paper reviews the key factors influencing pipeline transportation performance, including slurry characteristics, pipeline layout, and the use of additives. By examining these factors, we aim to advance solutions to challenges related to system stability, reliability, and cost, while contributing to the efficient and sustainable operation of pipeline transportation in mining.

2. Impact of Slurry Characteristics on Pipeline Transport Performance

The characteristics of the backfill slurry serve as the principal determinants of pipeline transportation performance, exerting a direct influence on the flowability, stability, and ultimate backfill efficacy during transport. Therefore, a precise understanding and control of slurry characteristics is essential. The characteristics of backfill slurry encompass, but are not limited to, particle size, distribution uniformity, packing density, density, and rheological properties. Packing density, in particular, is closely related to particle size and plays a significant role in determining the flowability and resistance of the slurry during pipeline transport. The influence of various slurry characteristics, including packing density, on pipeline transportation performance is both diverse and complex. This section concentrates on synthesizing and analyzing recent research findings concerning the impact of slurry characteristics on pipeline transportation performance.

Figure 2 depicts the influence of feed velocity on the sedimentation behavior of backfill slurry particles within the pipeline [27]. Figure 2a demonstrates that, when particle size and backfill slurry concentration are kept constant, the particle settling velocity diminishes progressively with increasing feed velocity. Figure 2b clearly illustrates the variation in backfill slurry concentration at various heights within the pipeline cross-section. As feed velocity increases, the concentration disparity between the bottom and top of the slurry cross-section consistently diminishes. Figure 2c,d respectively depict the impact of feed velocity on particle sedimentation in the pipeline from two distinct directions, revealing that higher feed velocities impede particle separation from the fluid flow. The particle settling velocity is minimized when the feed velocity is 3 m/s.

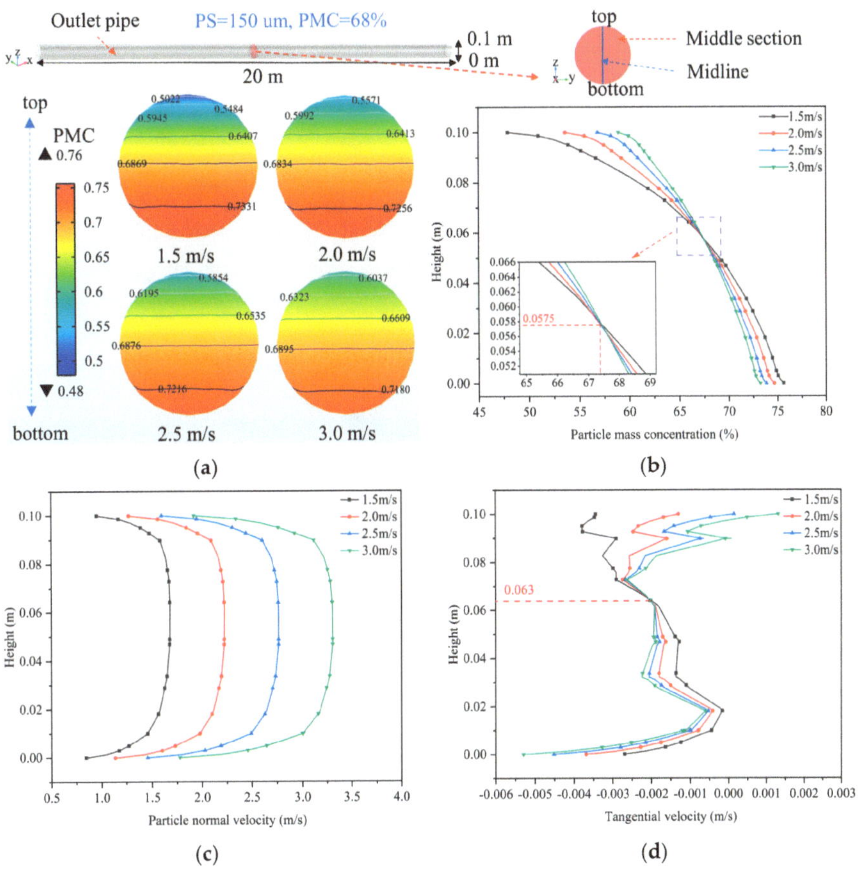

Figure 2. Particle concentration distribution of backfilled slurry in the middle section of the outlet pipe at different inlet velocities (example with particle size (PS) = 150 um and particle mass concentration (PMC) = 68%). Adapted from Ref. [27]. (**a**) PMC distribution, (**b**) Changes in PMC on the middle cutoff, (**c**) Particle normal velocity (x-axis direction positive), (**d**) Particle tangential velocity (z-axis direction positive).

Figures 3 and 4 depict the flow characteristics of fill slurry at varying concentrations during backfill transportation [28]. Figure 3 demonstrates the flow and dispersion characteristics at the nozzle in gravity flow tests for fill slurry across various concentrations, whereas Figure 4 illustrates the distribution of pipeline transport resistance under differing slurry concentrations. Investigations into flow characteristics under varying conditions reveal that increased fill slurry concentrations lead to heightened pipeline transport resistance [29–31]. The concentration of fill slurry is a pivotal factor influencing slurry rheological parameters. As the concentration of fill slurry increases, both the yield stress and viscosity of the slurry escalate significantly, as depicted in Figure 5 [32]. Consequently, to minimize energy consumption while satisfying the backfill strength requirements, lower concentration fill slurry is generally preferred for pipeline transportation.

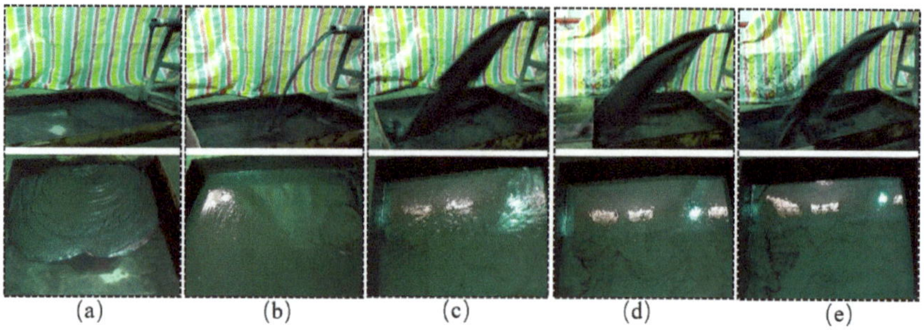

Figure 3. Flowing and diffusing states at the nozzle in the gravity flow test of different concentrations of slurry: (**a**) 80%, (**b**) 78%, (**c**) 76%, (**d**) 74%, (**e**) 72%. Adapted from Ref. [28].

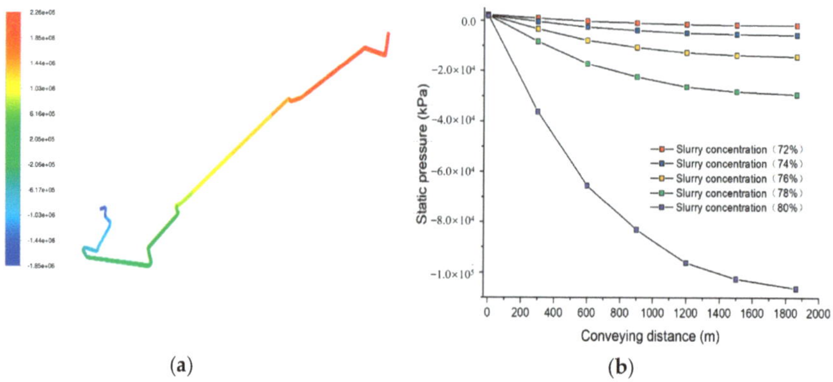

Figure 4. Analysis of flow field characteristics under different transporting concentrations. Adapted from Ref. [28]. (**a**) Distribution diagram of the pipeline transportation resistance, (**b**) Distribution curve of pipeline transportation resistance.

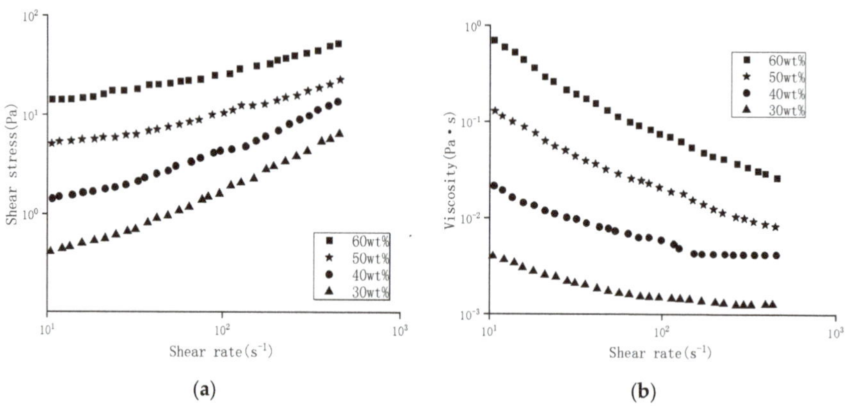

Figure 5. The effect of slurry concentration on the rheological properties of slurry. Adapted from Ref. [32]. (**a**) Relationship between shear stress and shear rate, (**b**) Relationship between viscosity and shear rate.

The thixotropy of fill slurry is another critical factor influencing the transportation process in slurry backfilling [33–36]. Thixotropy describes the property of a fluid where its viscosity changes under shear stress, gradually returning to its original state once the stress is removed. Investigating the thixotropy of fill slurry offers critical insights for optimizing long-distance pipeline transportation [37–40]. Figure 6 illustrates the thixotropic behavior of rheological parameters. The relative slip between slurry flow layers is directly governed by the magnitude of the slurry's yield stress. In essence, the magnitude of the fill slurry's yield stress directly dictates the radius of the plug flow and the thickness of the boundary layer. Figure 6a reveals that yield stress decreases with increasing shear time, following a negative exponential trend. Figure 6b indicates that plastic viscosity generally declines over time but surges sharply at the onset of shear (shear transition zone). Viscosity rapidly increases with shear time, initially reaching a peak, then gradually decreasing before stabilizing [41,42]. The mixing time for preparing the fill slurry should generally exceed the duration of the shear transition zone.

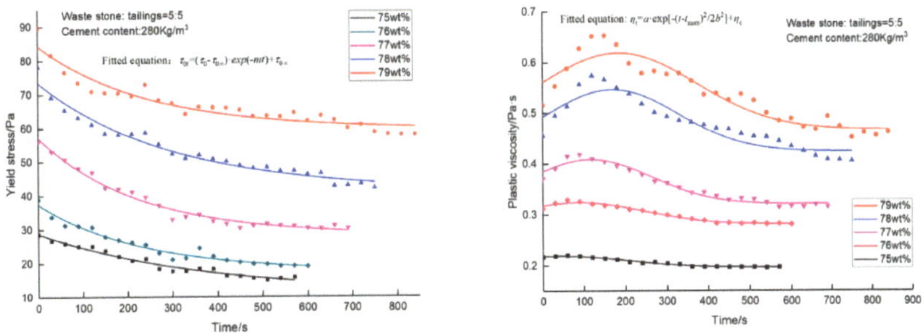

(a) (b)

Figure 6. Thixotropic model for rheological parameters. Adapted from Ref. [41]. (**a**) The fitted curves of yield stress over shear time, (**b**) The fitted curves of plastic viscosity over shear time.

Additionally, particle size distribution significantly influences the transport performance of fill slurry [43,44]. Figure 7 demonstrates the impact of incorporating fine particles on the particle concentration distribution within the fill slurry [45,46]. In Figure 7, panels A, B, and C depict the particle volume concentration distributions under conditions of low flow rate with low concentration, low flow rate with high concentration, and high flow rate with high concentration, respectively. The addition of 10% fine particles results in minimal changes to the coarse particle concentration distribution; however, introducing 20% fine particles leads to a gradual alteration in the concentration distribution of coarse particles. This suggests that a sufficient quantity of fine particles is required to modify the flow behavior of the fill slurry. Incorporating an optimal proportion of fine particles into the fill slurry can significantly enhance the suspension of coarse particles, reduce the frictional resistance between coarse particles and the pipe wall, and consequently improve the overall transport performance of the slurry. Concurrently, as the flow rate increases, the impact of fine particles on the coarse particle concentration distribution progressively diminishes. This may be attributed to the fact that, as the flow rate rises, the collision forces among coarse particles intensify, overshadowing the buoyancy effect imparted by the fine particle carrier phase.

In addition to particle size, another important characteristic influencing pipeline transport performance is the packing density (or compactness) of the slurry [47–49]. The packing density refers to how tightly the particles are arranged in the slurry, which affects the overall flowability and resistance to movement within the pipeline. It is influenced by several factors, including the particle size distribution, particle shape, and slurry concentration. The packing density of the slurry is closely related to the particle size distribution, as finer particles tend to occupy the spaces between coarser particles, leading to a higher packing density. This increased packing density can result in higher frictional resistance during transport, as the slurry becomes more viscous and more difficult to flow. On the other hand, lower packing densities (often associated with larger particles or a more uniform particle size distribution) may reduce flow resistance but could also affect the stability of the slurry during transport, leading to potential sedimentation or separation issues. Research has shown that optimizing the packing density can improve the transport performance by balancing these opposing effects [50,51]. For example, a well-balanced slurry with an optimal packing density can minimize pipeline resistance while maintaining the stability of the slurry and preventing excessive settling of particles. Thus, understanding the relationship between packing density and particle size is essential for optimizing slurry formulations for long-distance pipeline transport.

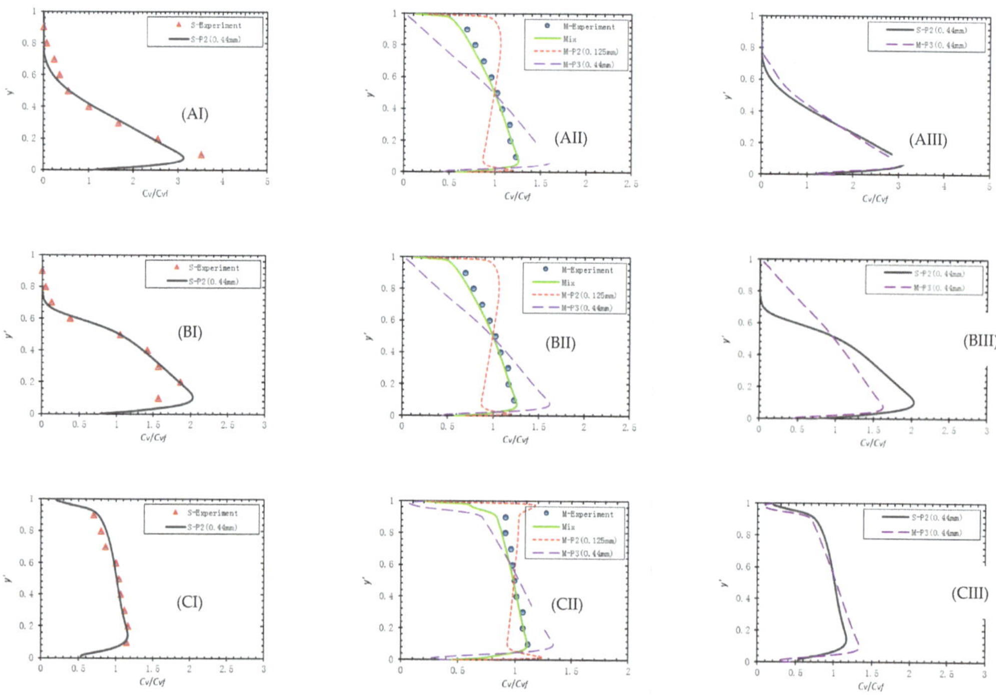

(**a**) (**b**) (**c**)

Figure 7. Effect of particle size distribution on particle concentration distribution. Adapted from Ref. [45]. (**a**) Coarse particle concentration distribution determined by simulations and experiments, (**b**) After adding fine particles, concentration distribution of fine particles and mixtures, (**c**) Comparison of coarse particle concentration and single size coarse particle concentration after adding fine particles.

Figure 8 illustrates the influence of feed velocity on the transport characteristics of fill slurry. Figure 8a depicts the effect of feed velocity on the transport resistance loss of fill slurry. Under constant conditions, the resistance loss within the pipeline escalates with an increase in feed velocity. Figure 8b illustrates the effect of feed velocity on the particle settling behavior of the fill slurry. Observations reveal that particles are uniformly distributed at the pipeline inlet (X = 0 m); however, as the distance from the inlet increases, the particle distribution across all slurry velocities gradually becomes uneven. The distribution of the fill slurry can be categorized into three distinct vertical zones: the aggregate volume reduction zone, the steady flow zone, and the coarse aggregate volume increase zone. Significant variations are observed in the extent of these zones at different locations for varying flow rates. Within the flow rate range of 1.0–2.5 m/s, the height of both the aggregate volume reduction zone and the increase zone progressively increases with transport distance, attributed to the continuous deposition of particles. At a constant position, as the flow rate increases, the height of the two zones progressively decreases, likely due to the increased fluctuating kinetic energy. At a flow rate of 3.0 m/s, the aggregate volume reduction zone demonstrates a similar variation pattern as under other conditions, with increased transport distance and the absence of an aggregate volume increase zone in the lower part of the pipeline, indicating the highest uniformity of the fill slurry [52].

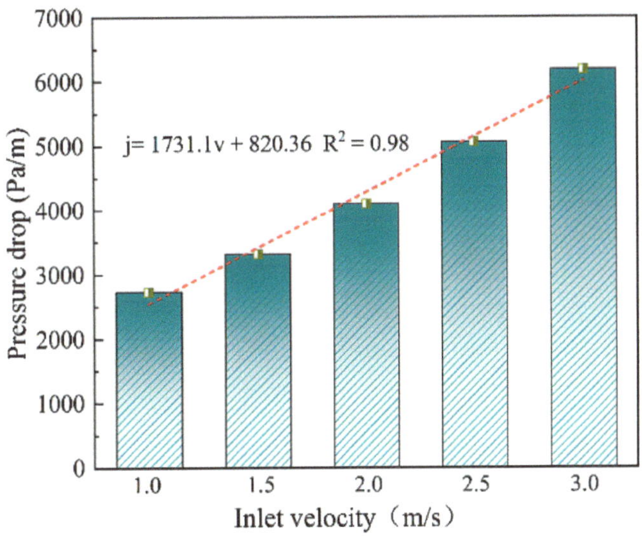

(a)

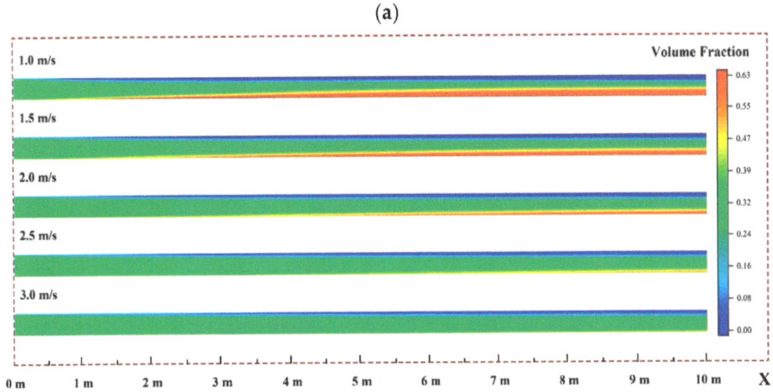

(b)

Figure 8. Effect of feeding speed on slurry conveying characteristics. Adapted from Ref. [52]. (**a**) Pressure drop at different inlet velocity, (**b**) The volume fraction distribution on the central vertical axial plane.

In practical applications, for established and operational backfill systems, the preparation and pumping equipment for the fill slurry are typically fixed. Consequently, adjusting the characteristics of the fill slurry becomes the most cost-effective and efficient approach for enhancing pipeline transport performance. It is crucial to emphasize that in the domain of mining backfill, the quality and evaluation of the backfill body are of paramount importance. Enhancing the characteristics of the fill slurry must be aligned with meeting the performance criteria of the backfill body.

3. Impact of Pipeline Layout on Pipeline Transport Performance

Pipeline layout involves multiple dimensions, including pipe orientation and configuration, pipe diameter and length, elbow and bend designs, branch and connection point arrangements, and elevation changes and gradients [53–55]. An optimal pipeline layout is crucial for guaranteeing the efficient and stable operation of the transport system. A well-engineered and meticulously planned layout not only optimizes transport efficiency and minimizes energy consumption but also mitigates underfilled flow issues, reduces the likelihood of blockages and bursts during slurry transport, and prolongs the system's operational lifespan. This section reviews recent research insights into how pipeline layout influences the performance of fill slurry transport systems.

Figure 9a–c illustrate the effects of fill gradient, elbow curvature radius, and pipe diameter on longitudinal resistance loss, as revealed by recent studies [56].

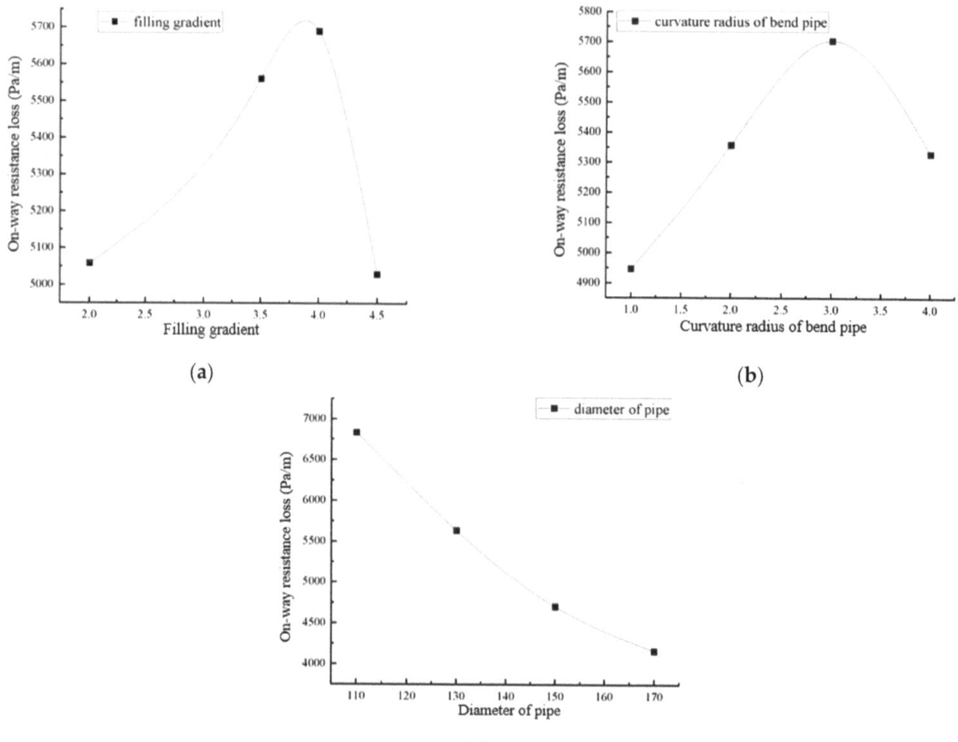

Figure 9. The impact of pipeline parameters on resistance loss: (**a**) resistance loss at different backfill gradients; (**b**) resistance loss at different elbow curvature radii; (**c**) resistance loss at different pipe inner diameters. Adapted from Ref. [56].

Figure 9a reveals that the relationship between fill gradient and resistance loss is neither purely linear nor inversely proportional. Within the studied parameters, resistance loss exhibits an exponential variation with increasing fill gradient, with a notable inflection point occurring at a fill gradient of 4. Specifically, resistance loss escalates as the fill gradient increases from 2 to 4. However, as the fill gradient rises from 4 to 4.5, resistance loss decreases. This phenomenon can be attributed to the pipeline resistance exceeding the gravitational force of the slurry when the fill gradient surpasses 4, hindering the slurry's ability to transport via self-flow. Consequently, a pumped fill method becomes necessary when the fill gradient exceeds 4.

The complex underground mining environment necessitates the use of elbows to connect fill pipelines, making their design crucial for minimizing resistance loss. Figure 9b illustrates that, within the studied parameters, resistance loss at elbows varies exponentially with the curvature radius. Specifically, when the curvature radius ranges from 1 m to 3 m, resistance loss increases as the curvature radius enlarges, which correlates with the increased circumference of the elbow and consequently greater resistance along the pipeline. However, when the curvature radius extends from 3 m to 4 m, resistance loss diminishes. This reduction occurs because a larger curvature radius leads to a decreased slurry velocity gradient, thereby reducing pipeline resistance.

Figure 9c indicates a negative correlation between pipe diameter and resistance loss, revealing that as the pipe diameter increases, resistance loss decreases. This phenomenon occurs because, for a constant flow rate, a reduction in pipe diameter elevates the velocity of the fill slurry, leading to a proportional increase in frictional resistance along the pipeline.

Figure 10 illustrates the research findings on the application of auxiliary pipelines to enhance resistance and mitigate pressure [57]. The figure demonstrates that auxiliary pipelines are installed within vertical pipe sections, with their placement adjusted based on the underground pipeline configuration and pressure distribution. The primary purpose of these auxiliary pipelines is to augment frictional loss between the fill slurry and the pipe wall, thereby increasing resistance and reducing pressure. In practical applications, the auxiliary pipelines can be connected to the main pipeline using electric valves. When the fill gradient is low, the auxiliary pipelines are activated to increase transport resistance and decrease the velocity and pressure of the slurry reaching the pipeline's base. Conversely, when the fill gradient is high, the auxiliary pipelines remain inactive to avoid pipeline blockage.

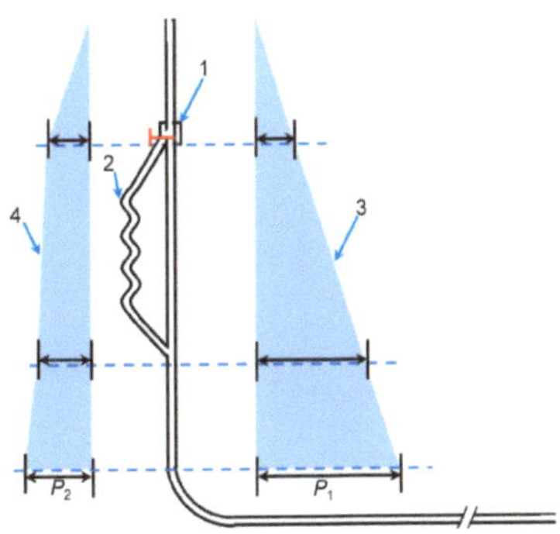

Figure 10. Increasing resistance and reducing pressure schematic of auxiliary pipe: (**1**) program-controlled electric valve; (**2**) auxiliary pipe; (**3**) pressure change of vertical pipe of L-shaped pipe; (**4**) pressure change of vertical pipe after using auxiliary pipe. Adapted from Ref. [57].

A scientifically and rationally designed pipeline layout is crucial for ensuring the reliability and long-term stability of the backfill system. Future research should concentrate on exploring advanced pipeline layout technologies, integrating cutting-edge engineering innovations and technological advancements. This includes, but is not limited to, investigations into intelligent pipeline systems, advanced fluid dynamics simulations, and adaptive pipeline designs. Through comprehensive research in these areas, professionals can gain a deeper understanding of the mechanisms by which pipeline design influences backfill system performance, leading to the development of more efficient engineering practices and management strategies.

4. Impact of Additives on Pipeline Transport Performance

An additional effective strategy for enhancing the transport performance of fill slurry is the incorporation of additives. Additives encompass a variety of types; in mining backfill applications, frequently utilized additives include water reducers, foaming agents, antifreeze agents, and thickeners. The introduction of these additives is intended to refine the physical and chemical properties of the fill slurry, thereby further optimizing pipeline transport performance and ultimately enhancing production efficiency and resource utilization. This section elucidates recent research findings regarding the impact of various additives on fill slurry performance.

Figures 11 and 12 illustrate the effects of foaming agents on both the rheological properties and pipeline transport performance of fill slurry [58]. Foaming agents, functioning as surfactants, organize their bilayer electronic structure at the liquid film surface, thereby effectively reducing the liquid's surface tension and facilitating the formation of bubbles with the surrounding air. With an increasing bubble ratio, both the apparent viscosity and shear stress of the fill slurry progressively decrease. Consequently, foaming agents can markedly enhance the flowability of the fill slurry. Initially, the bubbles are large and irregular upon entering the slurry. As transportation of the fill slurry commences, the large bubbles ascend uniformly to the pipe wall surface and the surrounding space. Under the persistent shear force exerted by the pipe wall, large bubbles fragment into smaller bubbles, creating a new turbulent viscous layer on the pipe wall surface that replaces the existing viscous layer of the slurry. Hence, the incorporation of foaming agents can diminish resistance loss during the transportation process and enhance pipeline transport efficiency.

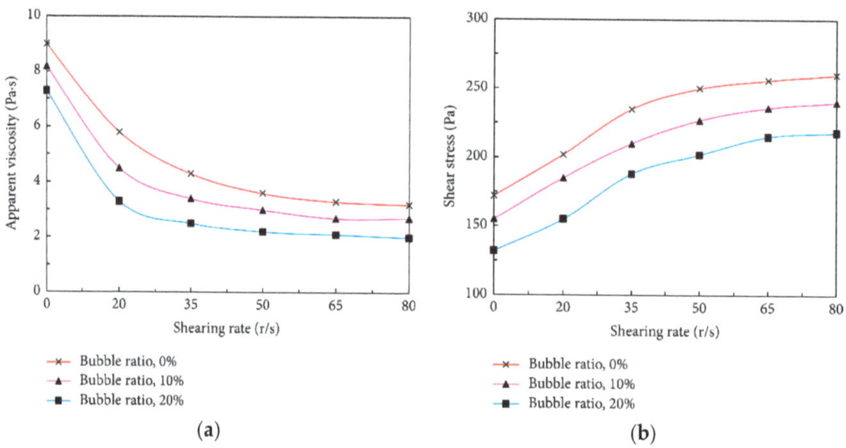

Figure 11. The impact of foaming agent on rheological parameters. Adapted from Ref. [58]. (**a**) Effect of bubble ratio on viscosity, (**b**) Effect of bubble ratio on shear stress.

Figure 13 illustrates the effect of high-efficiency water reducers on the rheological properties of fill slurry [59]. Water reducers, primarily anionic surfactants, function to disperse cement particles, minimize water usage per unit, enhance the flowability of the mixture, and improve the transport performance of the fill slurry [60,61]. As the concentration of the tailings slurry increases, both its yield stress and plastic viscosity rise accordingly [62,63]. The yield stress and plastic viscosity of fill slurry containing high-efficiency water reducers are markedly lower compared to those of tailings slurry devoid of such reducers. This effect is attributed to the molecular brush structure of high-efficiency water reducers, which features numerous cationic and anionic functional groups along the main chain [64]. These groups adsorb onto the tailings surface, thereby enhancing electrostatic repulsion, while the side chains create steric hindrance to improve flowability [65,66]. It is noteworthy that high-efficiency water reducers should be applied judiciously based on the concentration of the tailings slurry. At lower concentrations, they may induce stratification, whereas at higher concentrations, they may result in a loss of flowability. Conversely, when employed at elevated tailings slurry concentrations, they can bolster fill body strength and reduce the consumption of cementitious materials. Consequently, the economic value of incorporating high-efficiency water reducers is typically greater at higher concentrations of tailings slurry.

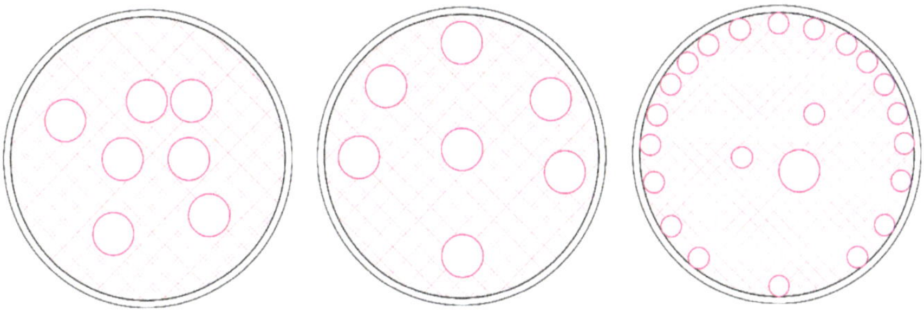

Figure 12. Change process of bubbles in pipeline transportation. Adapted from Ref. [58]. (**a**) Big and unordered bubble before transportation, (**b**) Big and ordered bubble when transportation, (**c**) Small and ordered bubble in transportation.

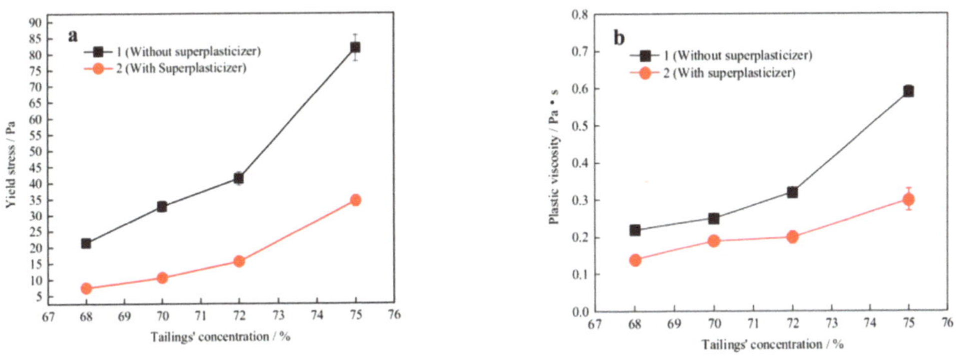

Figure 13. Effect of a superplasticizer on the rheological behavior of slurry. Adapted from Ref. [59]. (**a**) Yield stress, (**b**) Plastic viscosity.

In backfill operations, thickeners are employed to improve the water retention capacity of the fill slurry and to mitigate issues related to segregation and blockage during pipeline transport. Figure 14 illustrates the mechanism of action of a thickening agent [67]. The molecules of the thickening agent possess a multi-segment structure. Upon adsorption onto the particle surfaces, the segments of the thickening agent molecules undergo rotational movement, resulting in the aggregation of particles into larger floccules. These floccules form interconnections within the paste slurry, creating a network-like flocculation structure. Figure 15 depicts the effects of the thickening agent on the flowability, rheological properties, and compressive strength of the fill slurry. Figure 15a demonstrates that the thickening agent effectively decreases the flowability of the fill slurry, leading to particle aggregation and enhancing the slurry's resistance to bleeding and segregation. Figure 15b,c illustrate that, at a constant shear rate, a higher content of thickening agent results in an increase in the apparent viscosity and shear stress of the fill slurry. Figure 15d reveals that the compressive strength of hardened fill body specimens increases substantially with the amount of thickening agent.

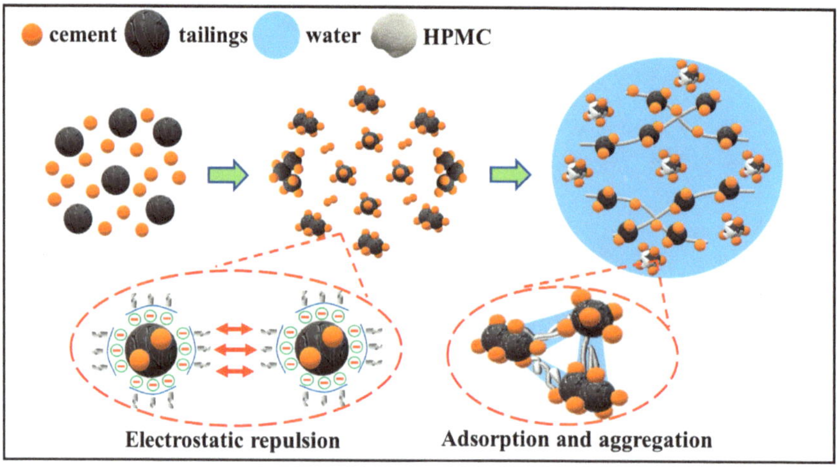

Figure 14. Action mechanism of a thickener. Adapted from Ref. [67].

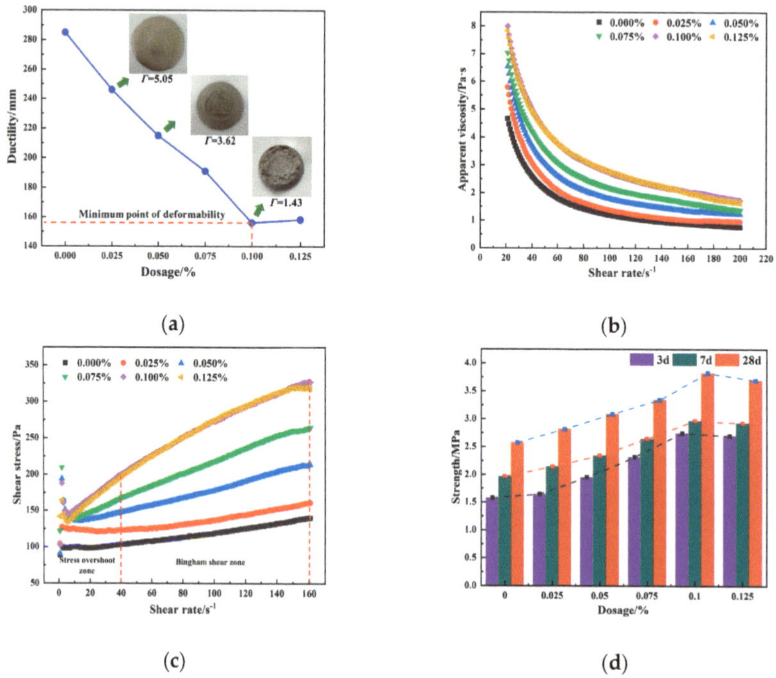

Figure 15. The influence of a thickening agent on slurry properties. Adapted from Ref. [67]. (**a**) Effect on ductility, (**b**) Effect on apparent viscosity, (**c**) Effect on the relationship between shear stress and shear rate, (**d**) Effect on strength.

Figure 16 illustrates the effect of non-chloride antifreeze on the rheological properties of fill slurry [68]. To mitigate the risk of freezing during the transportation of fill slurry in sub-zero environments, antifreeze agents are frequently employed. Research indicates that chloride-containing antifreeze additives not only lower the freezing point of fill slurry and other cement-based materials but also enhance their rheological properties in sub-zero conditions [69,70]. However, the use of chloride-containing antifreeze additives in fill slurry accelerates pipeline corrosion, markedly shortening the service life of the pipelines. To mitigate corrosion caused by chlorides in fill pipelines, non-chloride antifreeze mixtures should be incorporated into the fill slurry whenever feasible. Figure 16a demonstrates that as the antifreeze concentration increases, the yield stress of each sample decreases to varying degrees. Figure 16b reveals that the plastic viscosity of the fill slurry increases within the antifreeze concentration range of 0–15 g/L and subsequently decreases within the concentration range of 15–35 g/L.

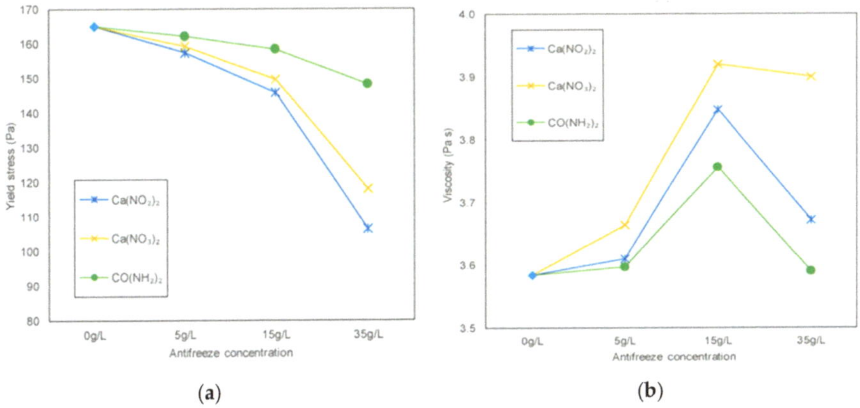

Figure 16. The impact of chloride-free antifreeze on the rheological properties of slurry. Adapted from Ref. [68]. (**a**) Yield stress, (**b**) Viscosity.

By employing scientific and precise additive formulations, the rheological parameters of fill slurry can be optimized, pipeline resistance minimized, and flowability enhanced. This approach not only enhances delivery efficiency but also optimizes the performance of the fill slurry to adapt to varying operational conditions, thereby ensuring the continuity and stability of the transportation process.

Future research should focus on the development of novel additives with multifunctional properties, such as simultaneously enhancing flowability, reducing resistance, and improving the hardened strength of the backfill material. Additionally, exploring the synergistic effects of additives with various fillers, like alternative tailings or fly ash, could optimize formulations for specific operational conditions. These efforts could address existing challenges and further enhance the efficiency and reliability of pipeline transport systems in backfill operations.

As societal focus on resource sustainability intensifies, pipeline transport, a pivotal component of the backfill process, encounters a range of challenges. At the technical level, the design and materials of pipelines must be meticulously aligned with the specific properties of different media to ensure system stability and reliability. Furthermore, challenges related to fluid dynamics within pipelines, such as blockages and wall wear, represent critical technical issues that necessitate comprehensive research and solutions.

Industry experts have undertaken extensive research to address these challenges, encompassing not only investigations into enhancing pipeline transport performance but also comprehensive analyses of backfill pipeline failures. Despite notable research achievements, unresolved issues and challenges persist in this field [71–73]. These challenges may include more intricate fluid dynamics phenomena, more precise alignment of pipelines and backfill materials, and continued exploration of environmentally sustainable backfill technologies.

5. Summary

This paper provides a comprehensive review of key factors affecting the long-distance pipeline transport performance of paste backfill slurry in deep mining, aiming to improve the efficiency and stability of backfill systems. The study explores the influence of slurry properties, pipeline layout, and additives on transport performance, summarizes the main challenges and solutions in existing research, and offers suggestions for future studies. The principal findings are summarized as follows:

(1) Slurry characteristics are critical to pipeline transport performance. Increasing the feed rate mitigates particle settling and improves concentration distribution. A lower slurry concentration reduces transport resistance and energy consumption, while optimal thixotropy and shear conditions further enhance transport efficiency. The addition of fine particles improves the suspension of coarse particles and reduces frictional resistance. Maintaining an appropriate flow rate is essential to ensure slurry homogeneity.

(2) Pipeline layout has a significant impact on the transport performance of fill slurry. As the backfill gradient increases, resistance loss initially rises and later declines. In scenarios with higher backfill gradients, pumping methods are generally more effective. The curvature radius of elbows plays a major role in resistance loss, with smaller radii increasing resistance and larger radii reducing it. Increasing the pipeline diameter lowers resistance loss by reducing flow velocity. Auxiliary pipelines enhance resistance and manage pressure through friction loss, depending on the backfill gradient.

(3) Additive usage enhances slurry transport performance. Foaming agents improve flowability by reducing the slurry's apparent viscosity and shear stress, which boosts transport efficiency. High-efficiency water reducers lower yield stress and plastic viscosity, optimizing rheological properties and reducing the need for cementitious materials, particularly in high-concentration slurries. Thickeners increase resistance to bleeding and segregation, thus strengthening the mechanical properties of the slurry. Chloride-free antifreeze agents effectively reduce yield stress and plastic viscosity in low-temperature environments while avoiding the corrosion issues associated with chloride-based agents.

Future research should focus on optimizing pipeline design to reduce transport resistance and developing more adaptable additives to improve slurry flow and system stability. These targeted efforts could significantly enhance the performance and sustainability of pipeline transport systems in backfill mining.

Author Contributions: Conceptualization, W.W. and B.Y.; methodology, W.X. and K.Y.; software, K.Y.; validation, M.L., W.W. and Y.Y.; formal analysis, M.L. and K.Y.; investigation, W.W., K.Y., Y.Y. and M.L.; resources, W.W., B.Y., K.Y. and M.L.; data curation, M.L. and K.Y.; writing—original draft preparation, M.L. and K.Y.; writing—review and editing, M.L. and W.W.; visualization, K.Y. and Y.Y.; supervision, M.L., B.Y. and W.X.; project administration, B.Y., K.Y. and M.L.; funding acquisition, B.Y., K.Y. and M.L. All authors have read and agreed to the published version of the manuscript.

Funding: This work was financially supported by the National Key R&D Program of China (No. 2021YFC2902102), the National Science Foundation of China (No. 52274122), the Research Fund of BGRIMM Technology Group (No. 02-2136) and the Youth Innovation Fund of BGRIMM Technology Group (No. 04-2306).

Data Availability Statement: The data supporting the findings of this study are available from the corresponding author upon request.

Acknowledgments: Thanks very much to the academic editors and reviewers for their earnest and valuable comments, which improved the quality of this manuscript.

Conflicts of Interest: Bin Yu, Wenyuan Xu, Kai Yang, Yanying Yin and Mengyuan Li are employees of BGRIMM Technology Group, National Centre for International Research on Green Metal Mining. The paper reflects the views of the scientists and not the company.

References

1. Guo, Q.; Xi, X.; Yang, S.; Cai, M. Technology strategies to achieve carbon peak and carbon neutrality for China's metal mines. *Int. J. Miner. Metall. Mater.* **2022**, *29*, 626–634. [CrossRef]
2. Li, P.; Cai, M. Challenges and new insights for exploitation of deep underground metal mineral resources. *Trans. Nonferrous Met. Soc. China* **2021**, *31*, 3478–3505. [CrossRef]

3. Hou, Y.; Zhao, Y.; Lu, J.; Wei, Q.; Zang, L.; Zhao, X. Environmental contamination and health risk assessment of potentially toxic trace metal elements in soils near gold mines—A global meta-analysis. *Environ. Pollut.* **2023**, *330*, 121803. [CrossRef] [PubMed]
4. Wang, L.; Cheng, L.; Yin, S.; Yan, Z.; Zhang, X. Multiphase slurry flow regimes and its pipeline transportation of underground backfill in metal mine: Mini review. *Constr. Build. Mater.* **2023**, *402*, 133014. [CrossRef]
5. Yu, H.; Li, S.; Wang, X. The Recent Progress China Has Made in the Backfill Mining Method, Part I: The Theory and Equipment of Backfill Pipeline Transportation. *Minerals* **2021**, *11*, 1274. [CrossRef]
6. Liu, W.; Chen, M.; Wang, F.; Ou, R.-Z.; Liu, Q. Paste pipeline transportation of pumping backfill technology with long distance and high stowing gradient in cold and high-altitude areas. *Adv. Civ. Eng.* **2021**, *2021*, 5287023. [CrossRef]
7. Xue, G.; Yilmaz, E.; Wang, Y. Progress and prospects of mining with backfill in metal mines in China. *Int. J. Miner. Metall. Mater.* **2023**, *30*, 1455–1473. [CrossRef]
8. Sun, K.; Zhang, J.; He, M.; Li, M.; Guo, S. Control of surface deformation and overburden movement in coal mine area by an innovative roadway cemented paste backfilling method using mining waste. *Sci. Total Environ.* **2023**, *891*, 164693. [CrossRef]
9. Lu, K.; Sun, W.; Gao, T.; Li, Z.; Zhao, J.; Cheng, H. Preparation of new copper smelting slag-based mine backfill material and investigation of its mechanical properties. *Constr. Build. Mater.* **2023**, *382*, 131228. [CrossRef]
10. Li, S.; Zhao, Z.; Yu, H.; Wang, X. The Recent Progress China Has Made in the Backfill Mining Method, Part II: The Composition and Typical Examples of Backfill Systems. *Minerals* **2021**, *11*, 1362. [CrossRef]
11. Xia, K.; Chen, C.; Liu, X.; Liu, X.; Yuan, J.; Dang, S. Assessing the stability of high-level pillars in deeply-buried metal mines stabilized using cemented backfill. *Int. J. Rock Mech. Min.* **2023**, *170*, 105489. [CrossRef]
12. Yu, J.; Liu, S.; Zhou, Z.; Jiao, H.; Wang, J.; Zhou, L. Influence of filling mining on ground stability in a metal mine. *Arab. J. Geosci.* **2024**, *17*, 18. [CrossRef]
13. Fan, J.; Yuan, Q.; Chen, J.; Ren, Y.; Zhang, D.; Yao, H.; Hu, B.; Qu, Y. Investigation of surrounding rock stability during proximal coal seams mining process and feasibility of ground control technology. *Process Saf. Environ. Prot.* **2024**, *186*, 1447–1459. [CrossRef]
14. Cai, M.; Li, P.; Tan, W.; Ren, F. Key engineering technologies to achieve green, intelligent, and sustainable development of deep metal mines in China. *Engineering* **2021**, *7*, 1513–1517. [CrossRef]
15. Li, S.; Zou, P.; Yu, H.; Hu, B.; Wang, X. Advantages of Backfill Mining Method for Small and Medium-Sized Mines in China: Safe, Eco-Friendly, and Efficient Mining. *Appl. Sci.* **2023**, *13*, 7280. [CrossRef]
16. Xia, K.; Chen, C.; Liu, X.; Wang, Y.; Liu, X.; Yuan, J. Estimating shear strength of high-level pillars supported with cemented backfilling using the Hoek–Brown strength criterion. *J. Rock Mech. Geotech. Eng.* **2024**, *16*, 454–469. [CrossRef]
17. Xie, G.; Liu, L.; Suo, Y.; Zhu, M.; Yang, P.; Sun, W. Study on the green disposal of industrial high salt water and its performance as activator to prepare magnesium—Coal based solid waste backfill material for mine. *J. Clean. Prod.* **2024**, *452*, 141933. [CrossRef]
18. Zhang, S.; Zhao, T.; Li, Y.; Li, Z.; Li, H.; Zhang, B.; Li, J.; Li, Y.; Ni, W. The effects and solidification characteristics of municipal solid waste incineration fly ash-slag-tailing based backfill blocks in underground mine condition, Constr. *Build. Mater.* **2024**, *420*, 135508. [CrossRef]
19. Ai, C.; Yu, A.; Liu, C.; Li, T. Analysis of physical characteristics and mechanism of retarder to stratified cemented backfill. *Sci. Rep.* **2024**, *14*, 13711. [CrossRef]
20. Yan, Z.; Yin, S.; Chen, X.; Wang, L. Rheological properties and wall-slip behavior of cemented tailing-waste rock backfill (CTWB) paste. *Constr. Build. Mater.* **2022**, *324*, 126723. [CrossRef]
21. Yang, K.; Shi, C.; Li, M. Research of the Transportation of Tailings Cemented Backfill in Partially Filled Pipe Flow. *J. Pipeline Syst. Eng.* **2024**, *15*, 04024014. [CrossRef]
22. Ahmed, H.M.; Bharathan, B.; Kermani, M.; Hassani, F.; Hefni, M.A.; Ahmed, H.A.M.; Hassan, G.S.A.; Moustafa, E.B.; Saleem, H.A.; Sasmito, A.P. Evaluation of Rheology Measurements Techniques for Pressure Loss in Mine Paste Backfill Transportation. *Minerals* **2022**, *12*, 678. [CrossRef]
23. Qiu, J.; Tian, M.; Zhu, D.; Xiao, C.; Wen, B.; Bin, F.; Chen, H.; Wang, D. Numerical Study of Resistance Loss and Erosive Wear during Pipe Transport of Paste Slurry. *Sustainability* **2023**, *15*, 11890. [CrossRef]
24. Dong, H.; Aziz, N.A.; Shafri, H.Z.M.; Ahmad, K.A.B. Computational fluid dynamics study on cemented paste backfill slurry: Review. *Constr. Build. Mater.* **2023**, *369*, 130558. [CrossRef]
25. Wu, A.; Ruan, Z.; Wang, J. Rheological behavior of paste in metal mines. *Int. J. Miner. Metall. Mater.* **2022**, *29*, 717–726. [CrossRef]
26. Yang, X.; Xiao, B.; Gao, Q.; He, J. Determining the pressure drop of cemented Gobi sand and tailings paste backfill in a pipe flow. *Constr. Build. Mater.* **2020**, *255*, 119371. [CrossRef]
27. Chen, Q.; Zhou, H.; Wang, Y.; Li, X.; Zhang, Q.; Feng, Y.; Qi, C. Resistance loss in cemented paste backfill pipelines: Effect of inlet velocity, particle mass concentration, and particle size. *Materials* **2022**, *15*, 3339. [CrossRef]
28. Gao, R.; Zhou, K.; Zhou, Y.; Yang, C. Research on the fluid characteristics of cemented backfill pipeline transportation of mineral processing tailings. *Alexandria Eng. J.* **2020**, *59*, 4409–4426. [CrossRef]
29. Zhang, X.; Zhao, M.; Liu, L.; Xia, X.; Zhang, J.; Zhao, C.; Bu, B. Phase-change heat storage backfill: Experimental study on rheological properties of backfill slurry with paraffin added. *Constr. Build. Mater.* **2020**, *262*, 120736. [CrossRef]
30. Ren, W.; Gao, R.; Zhang, Y.; Hou, M. Rheological properties of ultra-fine tailings cemented paste backfill under ultrasonic wave action. *Minerals* **2021**, *11*, 718. [CrossRef]
31. Li, W.; Guo, L.; Liu, G.; Pan, A.; Zhang, T. Analytical and experimental investigation of the relationship between spread and yield stress in the mini-cone test for cemented tailings backfill. *Constr. Build. Mater.* **2020**, *260*, 119770. [CrossRef]

32. Prasad, V.; Thareja, P.; Mehrotra, S.P. Influence of Rheology on the Flow Characteristics of Bauxite Slurries During Hydraulic Conveying Through the Pipeline. *Trans. Indian Inst. Met.* **2023**, *77*, 4133–4140. [CrossRef]
33. Wang, Y.; Huang, Y.; Hao, Y. Experimental study and application of rheological properties of coal gangue-fly ash backfill slurry. *Processes* **2020**, *8*, 284. [CrossRef]
34. Wei, H.; Xiao, B.; Gao, Q. Flow properties analysis and identification of a fly ash-waste rock mixed backfilling slurry. *Minerals* **2021**, *11*, 576. [CrossRef]
35. Li, Y.; Fu, J.; Yang, J.; Wang, J. Research on slurry flowability and mechanical properties of cemented paste backfill: Effects of cement-to-tailings mass ratio and mass concentration. *Materials* **2024**, *17*, 2222. [CrossRef]
36. An, S.; Liu, J.; Cheng, L.; Guo, L.; Zhou, D. Rheological and mechanical properties of full-tailings backfill material prepared by ultrafine-iron-tailings-powder-based consolidation agent. *Constr. Build. Mater.* **2024**, *417*, 135286. [CrossRef]
37. Haruna, S.; Fall, M. Time- and temperature-dependent rheological properties of cemented paste backfill that contains superplasticizer. *Powder Technol.* **2020**, *360*, 731–740. [CrossRef]
38. Neelakantan, R.; Vaezi, G.F.; Sanders, R.S. Effect of shear on the yield stress and aggregate structure of flocculant-dosed, concentrated kaolinite suspensions. *Miner. Eng.* **2018**, *123*, 95–103. [CrossRef]
39. Li, X.; Grasley, Z.C.; Garboczi, E.J.; Bullard, J.W. Simulation of the influence of intrinsic CSH aging on time-dependent relaxation of hydrating cement paste. *Constr. Build. Mater.* **2017**, *157*, 1024–1031. [CrossRef]
40. Cheng, H.; Wu, S.; Zhang, X.; Wu, A. Effect of particle gradation characteristics on yield stress of cemented paste backfill. *Int. J. Miner. Metall. Mater.* **2020**, *27*, 10–17. [CrossRef]
41. Yang, T.; Wang, J.; Qiao, D.; Li, G.; Cheng, H.; Zhang, X.; Shi, R. The thixotropic characteristics of rheological parameters of waste stone-tailings cemented paste backfill and the time effect of pipeline resistance. *Constr. Build. Mater.* **2024**, *423*, 135858. [CrossRef]
42. Yang, L.; Gao, Y.; Chen, H.; Jiao, H.; Dong, M.; Bier, T.A.; Kim, M. Three-dimensional concrete printing technology from a rheology perspective: A review. *Adv. Cem. Res.* **2024**, *126*, 72–86. [CrossRef]
43. Yang, L.; Jia, H.; Wu, A.; Jiao, H.; Chen, X.; Kou, Y.; Dong, M. Particle aggregation and breakage kinetics in cemented paste backfill. *Int. J. Miner. Metall. Mater.* **2024**, *31*, 1965–1974. [CrossRef]
44. Li, T.; Chen, X.; Liu, Z.; Jiang, W.; Min, H. Study on the suspending mechanism of gangue particle in coal-based solid waste slurry. *Alexandria Eng. J.* **2024**, *107*, 583–590. [CrossRef]
45. Li, M.; He, Y.; Liu, W.; Liu, Y.; Huang, C.; Jiang, R. Effect of adding finer particles on the transport characteristics of coarse-particle slurries in pipelines. *Ocean Eng.* **2020**, *218*, 108160. [CrossRef]
46. Hu, Y.; Li, K.; Zhang, B.; Han, B. Development of cemented paste backfill with superfine tailings: Fluidity, mechanical properties, and microstructure characteristics. *Materials* **2023**, *16*, 1951. [CrossRef]
47. Xing, K.; Hou, H.; Min, Z.; Yan, W.; Xian, Z. Effect of particle gradation on properties of fresh and hardened cemented paste backfill. *Constr. Build. Mater.* **2015**, *96*, 378–382. [CrossRef]
48. Qiu, J.; Guo, Z.; Yang, L.; Jiang, H.; Zhao, Y. Effects of packing density and water film thickness on the fluidity behavior of cemented paste backfill. *Powder Technol.* **2020**, *359*, 27–35. [CrossRef]
49. Li, C.; Huang, Z.; Li, X. Temperature effect analysis of paste backfill rheological properties based on inter-particle potential and water evolution. *Constr. Build. Mater.* **2023**, *404*, 132045. [CrossRef]
50. Zhu, J.; Zhang, H.; Chen, X.; Ji, Z. Modifying the linear packing model for predicting the packing density of GMZ bentonite pellet mixtures. *Constr. Build. Mater.* **2024**, *419*, 135505. [CrossRef]
51. Wang, X.; Guo, J.; Wu, A.; Wang, H.; Jiang, H.; Li, Z.; Qiu, Z. Wear characteristics of the pipeline transporting cemented paste backfill containing coarse aggregate. *Constr. Build. Mater.* **2024**, *410*, 134170. [CrossRef]
52. Feng, G.; Wang, Z.; Qi, T.; Du, X.; Guo, J.; Wang, H.; Shi, X.; Wen, X. Effect of velocity on flow properties and electrical resistivity of cemented coal gangue-fly ash backfill (CGFB) slurry in the pipeline. *Powder Technol.* **2022**, *396*, 191–209. [CrossRef]
53. Wang, J.; Wu, A.; Wang, M.; Ruan, Z. Experimental investigation on flow behavior of paste slurry transported by gravity in vertical pipes. *Processes* **2022**, *10*, 1696. [CrossRef]
54. Xue, Z.; Sun, H.; Gan, D.; Yan, Z.; Liu, Z. Wall slip behavior of cemented paste backfill slurry during pipeline based on noncontact experimental detection. *Int. J. Miner. Metall. Mater.* **2023**, *30*, 1515–1523. [CrossRef]
55. Hao, Y.; Song, X.; Wang, C.; Fan, B.; Yang, K. Dynamic characterization of water hammer in gangue fly ash slurry pipelines during valve closure. *Sci. Rep.* **2024**, *14*, 11470. [CrossRef]
56. Wang, X.; Wan, W.; Liu, Y.; Gao, R.; Lu, Z.; Tang, X. Analysis of Factors Influencing the Flow Characteristics of Paste Backfill in Pipeline Transportation. *Sustainability* **2023**, *15*, 6904. [CrossRef]
57. Wang, W.; Mu, H.; Mei, G.; Guo, L.; Lu, X.; Wang, A.; Sun, R. The Characteristics of Spiral Pipe Increasing Resistance and Reducing Pressure and the Amendment Equation of Stowing Gradient. *Metals* **2022**, *12*, 1105. [CrossRef]
58. Zhao, J.; Wang, X.; Peng, K.; Li, S. Utilization of Foaming Technology in Cemented Paste Backfill of High-Mud Superfine Unclassified Tailings. *Adv. Mater. Sci. Eng.* **2017**, *2017*, 6157869. [CrossRef]
59. Zhou, Z.; Liu, J.; Zhou, Q.; Wu, A.; Wang, H. A New High Charge Density Superplasticizer Regulates the Rheological Properties of Tailings Slurry and Cemented Paste Backfill Materials. *Min. Metall. Explor.* **2022**, *39*, 749–759. [CrossRef]
60. Li, M.; Zhao, J.; Jin, H.; Li, G.; Wang, Y. Review: The Effect Factors of Compatibility of Cement and Superplasticizer. *Adv. Mater. Res.* **2012**, *450–451*, 193–196. [CrossRef]

61. Matsuzawa, K.; Atarashi, D.; Miyauchi, M.; Sakai, E. Working mechanism of superplasticizer in cement paste with fluoride ion. *J. Adv. Concr. Technol.* **2015**, *13*, 305–310. [CrossRef]
62. Matsuzawa, K.; Shimazaki, D.; Kawakami, H.; Sakai, E. Effect of non-adsorbed superplasticizer molecules on fluidity of cement paste at low water-powder ratio. *Cem. Concr. Compos.* **2019**, *97*, 218–225. [CrossRef]
63. Zhang, G.; Li, G.; Li, Y. Effects of superplasticizers and retarders on the fluidity and strength of sulphoaluminate cement. *Constr. Build. Mater.* **2016**, *126*, 44–54. [CrossRef]
64. Šoukal, F.; Bocian, L.; Novotný, R.; Dlabajová, L.; Šuleková, N.; Hajzler, J.; Koutný, O.; Drdlová, M. The Effects of Silica Fume and Superplasticizer Type on the Properties and Microstructure of Reactive Powder Concrete. *Materials* **2023**, *16*, 6670. [CrossRef] [PubMed]
65. Jiang, Z.; Fang, Y.; Guo, Y.; Lin, T. Study on Synthesis and Properties of Vinyl Polyoxyethylene Ether Type Polycarboxylate Superplasticizer. *IOP Conf. Ser. Earth Environ. Sci.* **2020**, *571*, 012135. [CrossRef]
66. Li, H.; Yang, K.; Guan, X. Properties of sulfoaluminate cement-based grouting materials modified with LiAl-layered double hydroxides in the presence of PCE superplasticizer. *Constr. Build. Mater.* **2019**, *226*, 399–405. [CrossRef]
67. Yang, L.; Jia, H.; Jiao, H.; Dong, M.; Yang, T. The Mechanism of Viscosity-Enhancing Admixture in Backfill Slurry and the Evolution of Its Rheological Properties. *Minerals* **2023**, *13*, 1045. [CrossRef]
68. Zhou, Y.; Fall, M.; Haruna, S. Flow ability of cemented paste backfill with chloride-free antifreeze additives in sub-zero environments. *Cem. Concr. Compos.* **2022**, *126*, 104359. [CrossRef]
69. Traetteberg, A.; Ramachandran, V.S.; Grattan-Bellew, P.E. A study of the microstructure and hydration characteristics of tricalcium silicate in the presence of calcium chloride. *Cem. Concr. Res.* **1974**, *4*, 203–221. [CrossRef]
70. Jiang, H.; Fall, M. Yield stress and strength of saline cemented tailings in sub-zero environments: Portland cement paste backfill. *Int. J. Miner. Process.* **2017**, *160*, 68–75. [CrossRef]
71. Cunliffe, C.J.; Dadds, J.M.; Dennis, D.J.C. Flow correlations and transport behaviour of turbulent slurries in partially filled pipes. *Chem. Eng. Sci.* **2021**, *235*, 116465. [CrossRef]
72. Dong, H.; Aziz, N.A.; Shafri, H.Z.M.; Ahmad, K.A.B. Numerical Study on Transportation of Cemented Paste Backfill Slurry in Bend Pipe. *Processes* **2022**, *10*, 1454. [CrossRef]
73. Xie, M.; Zhao, J.; Zuo, M.J.; Tian, Z.; Liu, L.; Wu, J. Multi-objective maintenance decision-making of corroded parallel pipeline systems. *Appl. Energy* **2023**, *351*, 121822. [CrossRef]

minerals

MDPI

Article

Curing Agent for High-Concentration Unclassified Tailings Stockpiling: A Case Study of Tailings from a Gold Mine

Weixiang Wang [1,2], Kun Li [1,2,*], Lijie Guo [1,2,*], Sha Wang [1,2], Yifan Chu [1,2] and Yao Lu [1,2]

1 BGRIMM Technology Group, Beijing 102628, China; wangweixiang@bgrimm.com (W.W.); wangsha@bgrimm.com (S.W.); chuyifan@bgrimm.com (Y.C.); luyao@bgrimm.com (Y.L.)
2 National Centre for International Research on Green Metal Mining, Beijing 102628, China
* Correspondence: likun@bgrimm.com (K.L.); guolijie@bgrimm.com (L.G.)

Abstract: The disposal of tailings has always been a focal point in the mining industry. Semi-dry tailings stockpiling, specifically high-concentration tailings stockpiling, has emerged as a potential solution. To enhance the stability of tailings stockpiling and minimize its costs, the incorporation of a low-cost curing agent into high-concentration tailings is essential. Therefore, this study focuses on the development of a curing agent for high-concentration unclassified tailings stockpiling. The composition of a low-cost curing agent system is determined based on theoretical analysis, and the curing reaction mechanisms of each composition are researched. Subsequently, an orthogonal experiment is designed, and the strength of the modified unclassified tailings solidified samples at different curing ages is measured. Furthermore, the rheological properties of the modified unclassified tailings slurries are tested, and the feasibility of industrial transportation of the unclassified tailings slurries modified with the optimized curing agent is analyzed. Lastly, the microscopic morphologies of each material and the modified unclassified tailings solidified samples are characterized, their chemical compositions are tested, and the action mechanism of the curing agent in the curing system is further investigated. The results show that the optimal proportions of each material in the curing agent are as follows: slag, 58%; quicklime, 15%; cement, 8%; gypsum, 9%; and bentonite, 10%. The dominance of industrial waste slag exceeding 50% reflects the low-cost nature of the curing agent. Under this condition, the modified unclassified tailings slurry with a mass concentration of 75% exhibited a yield stress of 43.62 Pa and a viscosity coefficient of 0.67 Pa·s, which is suitable for pipeline transportation. These findings lay a foundation for subsequent decisions regarding stockpiling processes and equipment selection.

Keywords: unclassified tailings; stockpiling; curing agent; uniaxial compressive strength; rheological characteristics

check for updates

Citation: Wang, W.; Li, K.; Guo, L.; Wang, S.; Chu, Y.; Lu, Y. Curing Agent for High-Concentration Unclassified Tailings Stockpiling: A Case Study of Tailings from a Gold Mine. *Minerals* **2024**, *14*, 884. https://doi.org/10.3390/min14090884

Academic Editors: Yuye Tan, Xun Chen, Yuan Li and Carlito Tabelin

Received: 26 June 2024
Revised: 20 August 2024
Accepted: 26 August 2024
Published: 29 August 2024

1. Introduction

Tailings are waste materials left over from mineral processing [1]. The disposal of tailings has always been a focal point in the mining industry [2]. In recent years, incomplete statistics have indicated that China annually produces over a billion tons of tailings. As mineral resources are being extracted from deeper levels and ore grades decrease, the amount of tailings generated is increasing, making their disposal an increasingly challenging task [3]. Currently, there are three main methods for tailings disposal: wet tailings stockpiling, dry tailings stockpiling, and semi-dry tailings stockpiling, also known as high-concentration tailings stockpiling [4]. Wet tailings stockpiling involves storing tailings within a dam that contains a significant amount of water. While water is a factor in tailings dam disasters, it also plays a crucial role in exacerbating the risk of dam breaches [5,6], posing a serious threat to the safety of wet tailings stockpiling and potentially leading to pollution issues in soil, water bodies, and the atmosphere [7]. Dry tailings stockpiling requires dewatering the tailings slurry to form tailings cakes before disposal [8,9], but this method has drawbacks, such as high costs, a low production capacity, complex procedures, and discontinuous operations. Semi-dry tailings stockpiling involves concentrating low-concentration tailings to high-concentration levels (65%–75%), and then adding a curing agent to the mixture before discharging it through pipes for stockpiling [10]. This method overcomes the disadvantages of both wet and dry tailings stockpiling and holds promising application prospects [11–13]. Currently, in this field of research, the development of low-cost curing agents is a shared goal among scholars and industry players. Furthermore, utilizing unclassified tailings for stockpiling can reduce the cost of tailings classification, simplify processes, and lower overall stockpiling expenses. Therefore, this study focuses on the development of a low-cost curing agent for high-concentration unclassified tailings from a gold mine.

Research on curing agents in the field of unclassified tailings stockpiling is limited. A comparative analysis revealed that this study can draw insights from the research on cementitious materials in the context of mine tailings backfilling. However, there are notable differences between the two. In the mine tailings backfilling domain, there is a higher demand for the curing effectiveness of cemented tailings backfill, including requirements for high strength, rapid solidification, and good integration of the tunnel roof [14–16], whereas in unclassified tailings stockpiling, the primary focus is on achieving low costs and appropriate strength. Furthermore, the proportion of curing agent to tailings (referred to as PCGT) is a crucial factor in cost control. Typically, when a certain level of solidification strength is achieved, lower PCGTs result in reduced stockpiling costs. This study explores a low-cost curing agent under the condition of a low PCGT of 1:20.

In the initial stages of mine tailings backfilling, traditional curing agents, such as ordinary Portland cement, were commonly used. However, these agents were associated with high costs and significant environmental pollution due to the substantial CO_2 emissions generated during cement production [17]. To reduce the cost of mine tailings backfilling, researchers have been actively seeking inexpensive alternatives to cement. This has led to the development of new types of cementitious materials, which essentially involve using alkalis, salts, or composite activators to stimulate potential reactive volcanic ash materials like slag and fly ash [18,19]. The goal is to transform these materials into water-hardening substances. These innovative cementitious materials include red mud-based binders [20], slag-based binders [21], and composite cementitious materials [22], among others. Analysis reveals that slag, with its relatively low cost, is chosen more extensively among these options. Slag is considered a high-activity and high-quality industrial waste resource. Its high reactivity and easy grindability make it an excellent supplementary material, which has already been widely applied in cement-based construction materials and concrete engineering. In the context of backfilling materials, slag can effectively delay the heat of hydration, increase the thickness of the water film around particles, and improve the flowability of the backfilling material. Moreover, slag possesses inherent hydration characteristics that can be activated using cement, sodium hydroxide, calcium hydroxide, sodium silicate, or gypsum [23]. Telesca et al. [24] studied and obtained prefabricated building materials composed of 40% natural or flue gas desulfurization gypsum, 35% calcium hydroxide, and 25% fly ash. Yum et al. [25] used $Ca(OH)_2$- and $Ba(OH)_2$- to activate slag systems for eco-friendly concrete brick production, and limestone fines was studied as a supplementary material in activated slags. Li et al. [26] researched the possibility of using the residues after iron recovery from iron ore tailings as raw materials for the preparation of cementitious material, which was prepared by blending a mixture of 30% residues, 34% blast-furnace slag, 30% clinker, and 6% gypsum. Yao et al. [27] found that the pozzolanic activity of iron ore tailings is improved after mechanical activation, and the hydration products of CaO-activated tailings are C-S-H gels, as well as ettringite crystals in the presence of anhydrite. Barati et al. [28] investigated the stabilization of iron ore tailings with cement and bentonite. It was found that the introduction of cement/bentonite to iron ore tailings requires higher water content for optimum compaction and results in lower dry density. The addition of bentonite greatly improves the strength of tailings–cement mixtures, especially after a period of moist curing. Fall et al. [29] studied the feasibility of using bentonite–paste tailings as a barrier (liner, cover) material for mine waste containment facilities. Opiso et al. [30,31] found that tailings are simultaneously stabilized and can be repurposed a construction material via geopolymerization using coal fly ash, palm oil fuel ash, and a powder-based alkali activator.

Based on the factors mentioned above, we developed a curing agent to be used in high-concentration unclassified tailings stockpiling from a gold mine. The composition of this agent includes materials such as slag, quicklime, cement, gypsum, and bentonite. By analyzing the optimal proportions of each material, we aimed to create a solidified system that exhibits strong uniaxial compressive strength when solidified with the tailings. This is crucial to ensure the stability of tailings stockpiling. In addition, the rheological properties of the high-concentration unclassified tailings slurry after modification with the curing agent are vital parameters in the design of the pipeline transport and discharge processes [32–34]. These properties directly influence the sedimentation characteristics and slope of the tailings stockpiling surface [35,36], thereby affecting the safe operation and management of the tailings dam [13]. Therefore, while researching the low-cost aspects of the curing agent and ensuring it meets the strength requirements, it is also important to consider the rheological characteristics of the slurry after modification with the curing agent.

This study is based on the composition of a curing agent using the following research steps: The curing reaction mechanisms of each material in the curing agent were analyzed theoretically. Then, an orthogonal experiment was designed, the strength of the modified unclassified tailings solidified samples (MUTSSs) at different curing ages was measured, and the variation characteristics of the MUTSSs' strength and the influence of various materials at different levels on the strength were analyzed. Next, experiments were conducted to assess the rheological properties of the slurry, and the optimal proportions of each material in the curing agent were determined. Concurrently, the microscopic surface morphology of each material and MUTSSs with the optimal curing agent were characterized using scanning electron microscopy (SEM). Finally, the chemical compositions of the MUTSSs with the optimal curing agent were measured using an X-ray diffraction (XRD) test, and the action mechanism of the curing agent in the curing system was further investigated.

2. Curing Reaction Mechanisms

2.1. Reaction Mechanism of Cement

The primary mineral compositions of cement include tricalcium silicate (chemical formula $3CaO{\cdot}SiO_2$, referred to as C_3S), dicalcium silicate (chemical formula $2CaO{\cdot}SiO_2$, referred to as C_2S), tricalcium aluminate (chemical formula $3CaO{\cdot}Al_2O_3$, referred to as C_3A), and tetracalcium aluminoferrite (chemical formula $4CaO{\cdot}Al_2O_3{\cdot}Fe_2O_3$, referred to as C_4AF). The reactions of these four compositions with water are as follows [37,38]:

(1) Hydration of C_3S

At normal temperatures, tricalcium silicate reacts with water to form hydrated calcium silicate gel (C-S-H) and $Ca(OH)_2$.

$$2(3CaO{\cdot}SiO_2) + 6H_2O = 3CaO{\cdot}2SiO_2{\cdot}3H_2O + Ca(OH)_2 \quad (1)$$

(2) Hydration of C_2S

The product of dicalcium silicate reacting with water at normal temperatures is the same as that of tricalcium silicate, but the hydration rate is slower.

$$2(2CaO{\cdot}SiO_2) + 4H_2O = 3CaO{\cdot}2SiO_2{\cdot}3H_2O + Ca(OH)_2 \quad (2)$$

(3) Hydration of C_3A

The product of tricalcium aluminate's hydration is hydrated calcium aluminate (unstable), and the reaction rate is very fast.

$$3CaO{\cdot}Al_2O_3 + 6H_2O = 3CaO{\cdot}Al_2O_3{\cdot}6H_2O \tag{3}$$

(4) Hydration of C_4AF

The hydration product of tetracalcium aluminoferrite includes hydrated calcium aluminate and hydrated calcium iron oxide. The reaction rate is slower than that of tricalcium aluminate.

$$4CaO{\cdot}Al_2O_3{\cdot}Fe_2O_3 + 7H_2O = 3CaO{\cdot}Al_2O_3{\cdot}6H_2O + CaO{\cdot}Fe_2O_3{\cdot}H_2O \tag{4}$$

Tricalcium aluminate and tetracalcium aluminoferrite hydrate early in the hydration process. Their hydration products combine with gypsum to form ettringite (AFt) or monosulfate hydrate (AFm), which mainly act as binding agents between particles, enhancing the fluidity of the slurry and contributing to good strength after setting. Tricalcium silicate and dicalcium silicate generally hydrate in the middle-to-later stages of the hydration process. Their hydration products mainly consist of calcium hydroxide crystals (C-H) and C-S-H gel. Calcium silicate gel is a primary source of cement strength and durability, while C-H provides an alkaline environment for hydration reactions and can continue to react with active SiO_2 to generate additional calcium silicate gel, contributing to increased strength.

2.2. Reaction Mechanism of Quicklime

Quicklime primarily consists of CaO and readily undergoes chemical reactions upon contact with water. When introduced into the fulling tailings slurry, quicklime reacts with water to generate $Ca(OH)_2$. $Ca(OH)_2$ combines with the oxides of silicon, aluminum, and small amounts of aluminum and magnesium present in the tailings and slag, forming stable hydrated calcium silicates. These hydrated silicates possess higher gel strength than $Ca(OH)_2$ crystals. As a result, the modified fulling tailings slurry exhibits a denser structure upon curing. The reaction equations for this entire process are depicted in Equations (5)–(7) [39,40].

$$CaO + H_2O = Ca(OH)_2 \tag{5}$$

$$2SiO_2 + 3Ca(OH)_2 = 3CaO{\cdot}2SiO_2{\cdot}3H_2O \text{ (C-S-H)} \tag{6}$$

$$Al_2O_3 + 3Ca(OH)_2 + 3H_2O = 3CaO{\cdot}Al_2O_3{\cdot}6H_2O \text{ (C-A-H)} \tag{7}$$

2.3. Reaction Mechanism of Slag

The hydration process of slag begins with the volcanic ash reaction between active SiO_2, Al_2O_3 in the slag, and $Ca(OH)_2$. This reaction follows the patterns outlined in Equations (6) and (7). Continuous moisture supply in the slurry system contributes to the formation of an alkaline film on the surface of the slag. Simultaneously, the water-hydrated products infiltrate through gaps on the surface, causing further corrosion of the slag and the continuation of volcanic ash reactions until complete hydration of the slag. Additionally, gypsum present in the curing agent acts as a sulfate activator, the mechanism of which is described in Section 2.4.

2.4. Mechanism of Gypsum Hydration

The primary chemical composition of gypsum is $CaSO_4$, and its hydration product is calcium sulfate dihydrate [41].

$$CaSO_4 + 2H_2O = CaSO_4{\cdot}2H_2O \tag{8}$$

Hydrated calcium aluminate (C-A-H), formed by the reaction between active Al_2O_3 in the slag and $Ca(OH)_2$, reacts with calcium sulfate dihydrate to produce calcium sulfoaluminate, commonly known as ettringite (AFt) [42].

$$3CaO{\cdot}Al_2O_3{\cdot}6H_2O + 3(CaSO_4{\cdot}2H_2O) + 20H_2O = 3CaO{\cdot}Al_2O_3{\cdot}3CaSO_4{\cdot}32H_2O \quad (9)$$

The formation of ettringite consumes the main hydration product of hydrated calcium aluminate from slag, thus accelerating the hydration process. Simultaneously, the newly generated hydrated calcium sulfoaluminate increases the compactness of the structure itself. This product contains a high water content, significantly reducing free water, resulting in a denser structure and higher compressive strength in the solidified samples.

Moreover, the primary chemical reaction between iron phase in the slag and gypsum is as follows [43,44]:

$$\begin{aligned}4CaO{\cdot}Al_2O_3{\cdot}Fe_2O_3 + 3CaSO_4 + 36H_2O = {} & 3CaO{\cdot}[xAl_2O_3, (1-x)\,Fe_2O_3]{\cdot}3CaSO_4{\cdot}32H_2O + \\ & Ca(OH)_2 + 2xFe(OH)_3(gel) + (2-2x)Al(OH)_3(gel)\end{aligned} \quad (10)$$

If the amount of gypsum is relatively low, ettringite can react with tricalcium aluminate to convert it into monosulfate hydrate (AFm) [45,46].

$$3CaO{\cdot}Al_2O_3{\cdot}3CaSO_4{\cdot}32H_2O + 2(3CaO{\cdot}Al_2O_3) + 4H_2O = 3(3CaO{\cdot}Al_2O_3{\cdot}3CaSO_4{\cdot}12H_2O) \quad (11)$$

2.5. Reaction Mechanism of Bentonite

Bentonite's main composition is montmorillonite, accounting for about 85-90%. Montmorillonite crystals primarily consist of layered aluminum silicates with water molecules. Its structure comprises two layers of silico-oxygen tetrahedra sandwiched between a layer of aluminum–oxygen octahedra, forming a 2:1 layer structure [47,48]. In the crystal structure of montmorillonite, Al^{3+} in the octahedral layer can be substituted by ions like Mg^{2+} and Fe^{3+}, while Si^{4+} in the tetrahedral layer is often partially replaced by Al^{3+}. These lattice substitutions lead to an imbalance of charges between the layers, resulting in a negatively charged interlayer of montmorillonite. To maintain charge balance, large-radius cations, such as K^+, Na^+, and Ca^{2+}, are adsorbed between the structural units of montmorillonite through electrostatic forces [49].

In the presence of water, the interlayer ions of montmorillonite undergo hydration, causing the montmorillonite crystals to expand. When water is added to a cementitious system, the minerals on the montmorillonite surface immediately react with water in a process known as hydrolysis or hydration. This process gradually forms hexagonal plate-like hydration products: $Ca(OH)_2$, C-S-H, and AFt. This process involves a significant ion exchange, such as Ca^{2+}, Al^{3+}, Fe^{3+}, Si^{4+}, H^+, etc., further promoting the hydration reactions in the solidification system. This exchange enhances the structural stability of the solidification system. Moreover, the expanded bentonite supports the solidification system after expansion, thereby improving the compressive strength of the system.

In conclusion, in addition to their inherent cementitious activity, the excitation agents such as cement, quicklime, and gypsum in the curing agent also serve as alkaline excitation agents for the volcanic ash reactions of slag and tailings. The synergistic effects of multiple compositions cause the active compositions in slag and tailings to dissolve and react continuously. On the one hand, the alkaline compositions provide conditions for the formation of C-S-H and C-A-H, facilitating the disintegration of the slag glass network by stimulating the release of Ca^{2+} with lower bond energies in the presence of OH- from the slag glass body to the solution. The generated $Ca(OH)_2$, along with active Al_2O_3 and SiO_2 in the slag glass network and tailings, reacts to form C-S-H and C-A-H, promoting the further disintegration of the slag glass network and the dissolution of tailings. On the other hand, gypsum reacts with active Al_2O_3 in slag and tailings to generate AFt or AFm. With the continuous generation of AFt, an amount of $Ca(OH)_2$ is consumed, leading to its reduced concentration. Additionally, an amount of Al^{3+} in the solution is consumed, promoting the dissolution of Ca^{2+} in the slag glass body and providing the driving force for ongoing chemical reactions. The mutual reinforcement of these two aspects fully stimulates the potential activity of slag. AFt/AFm and C-S-H are generated during the hydration process of slag cementitious materials, playing a crucial role in the later-stage strength and stability of the hardened body. As an additive, bentonite enhances the strength of the solidified sample through two aspects: the ion exchange among various compositions, strengthening the interactions between them, and volume expansion, reinforcing the support of the solidification system.

3. Experimental Materials and Equipment

3.1. Experimental Materials

(1) Unclassified tailings

In this study, the unclassified tailings from a gold mine in Yantai City, Shandong Province, China, were selected as a case study. The basic physical properties are described in Table 1, and the size distribution curve of the unclassified tailings particles is displayed in Figure 1. It can be observed that the cumulative content of particles with a diameter smaller than 10.0 μm is approximately 24%, and the cumulative content of particles with a diameter smaller than 20.0 μm exceeds 37%. The main chemical compositions in the unclassified tailings are shown in Figure 2. As shown, the content of SiO_2 is 58.3%, which is the main component, indicating that the tailings have good inertness and are suitable for stockpiling. The contents of Al_2O_3, K_2O, and CaO are 13.35%, 5.36%, and 3.1%, respectively, which can promote the solidification of the slurry and improve the later strength of the solidified modified tailings. The content of S is 1.12%, which is relatively low and can enhance the strength of solidified modified tailings to a certain extent [50].

(2) Slag

S95-grade slag was selected, with a content of inorganic substances, like reactive calcium, silicon, and aluminum exceeding 30%, and a fineness of 325 mesh.

(3) Quicklime

Quicklime powder was chosen, with a CaO content of 86.3%, and a fineness of 325 mesh.

(4) Cement

Composite silicate cement 42.5 was selected, with the following parameters: a clinker + gypsum ratio greater than 50% and less than 80%, a physical and chemical blast furnace + volcanic ash + fly ash + limestone content ratio greater than 20% and less than 80%.

(5) Gypsum

The content of $CaSO_4 \cdot 2H_2O$ was 95%, with a fineness of 150 mesh.

(6) Bentonite

Ca-bentonite was selected, with a montmorillonite content of 85% and a fineness of 200 mesh.

Table 1. Basic physical properties of tailings.

Specific Gravity	Bulk Density (t/m)	Porosity (%)	Angle of Repose (°)
2.863	1.694	36.134	40.538

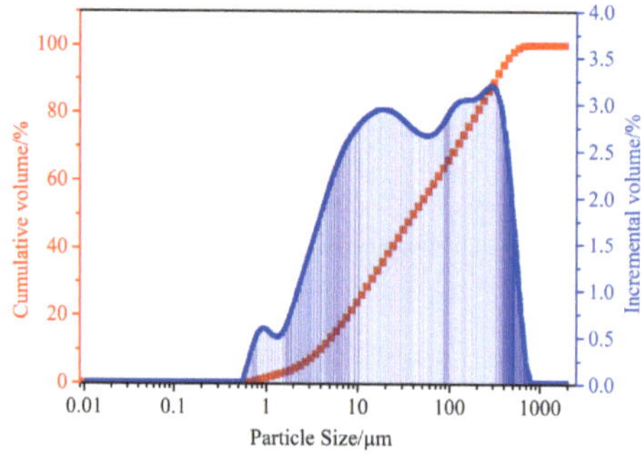

Figure 1. Unclassified tailings' particle size distribution curve.

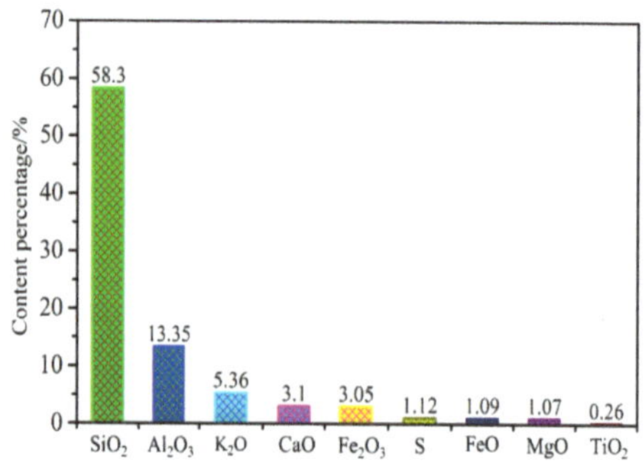

Figure 2. Unclassified tailings' chemical composition.

3.2. Experimental Equipment

(1) Load Testing Machine

An American Humboldt HM-5030 load testing machine (Humboldt Mfg. Co., Elgin, IL, USA, Figure 3a) was used. Its main technical specifications include a maximum test force of 50 kN, a maximum travel of 100 mm, and a loading rate range from 0.00001 mm/min to 89.99999 mm/min.

(2) Rheometer

An American Brookfield RST-SST Rheometer (AMETEK Brookfield, Middleboro, MA, USA, Figure 3b) was operated, with a wide torque range of 0.05–50 mNm and a variable speed range of 0.01–1000 rpm.

(3) SEM

A German Zeiss SIGMA SEM (Zeiss, Oberkochen, Germany, Figure 3c) was used, with a secondary electron image resolution of up to 1.3 nm (20 kV) and a magnification range of 20× to 1,000,000×.

(4) XRD

A Japanese X-ray diffractometer (Figure 3d), model Smartlab 9Kw (Rigaku Corporation, Tokyo, Japan), was employed. The X-ray tube focal spot size was $\leq 0.3 \times 8$ mm^2; an all-purpose matrix detector: effective detector area $\geq$300 mm^2; goniometer: radius $\geq$300 mm; rotation range: $-10°$ to $168°$.

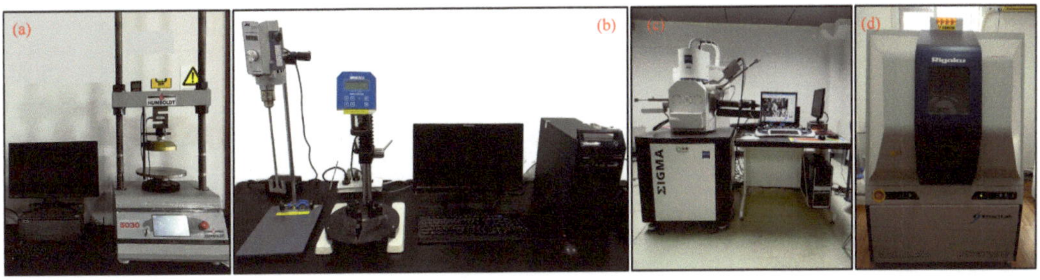

Figure 3. Experimental equipment. (**a**) Humboldt HM-5030 load testing machine; (**b**) Brookfield RST-SST Rheometer; (**c**) Zeiss SIGMA SEM; and (**d**) X-ray diffractometer.

4. Compressive Strength Testing Experiment Scheme

4.1. Orthogonal Experimental Design

During the experiment, the concentrations of the unclassified tailings slurry were set at 69%, 71%, 73%, and 75%. This study aimed to establish a solidification and stockpiling system for unclassified tailings with a relatively low PCGT. Therefore, this study focuses on the solidification effect under a PCGT of 1:20. As slag is the main composition of the curing agent and has a relatively low unit price, it is necessary to increase the proportion of slag in the curing agent to control costs, with the goal of keeping it above 50%. Therefore, when designing the proportions of the various compositions of the curing agent, this aspect should be taken into account. In addition, according to theoretical analysis, the reaction of the solidification system is faster and more complete in alkaline environments, so the proportion of quicklime should be higher. The ranges and equal divisions of the proportions of each material in the curing agent designed are shown in Table 2 [51].

Table 2. Ranges and equal divisions of proportions of each material.

	Upper Limit/%	Lower Limit/%	Equal Division/%
B (Quicklime)	30	15	5
C (Cement)	16	4	4
D (Gypsum)	12	3	3
E (Bentonite) *	10	1	3

* The remaining composition is A (slag).

Based on this, an orthogonal experiment was designed to analyze the influence of different materials on the compressive strength of the unclassified tailings solidified samples. The orthogonal experiment was designed with 5 factors and 4 levels; namely, L16(4^5), requiring at least 16 experimental trials. The orthogonal experimental design is shown in Table 3. The laboratory sample preparation presented in Table 3 was carried out for the experiments.

Table 3. Orthogonal experimental design table.

No.	O (Unclassified Tailings Slurry Concentration) *	B (Quicklime)	C (Cement)	D (Gypsum)	E (Bentonite)	A (Slag)
1	69%	30%	16%	12%	10%	32%
2	69%	25%	12%	9%	7%	47%
3	69%	20%	8%	6%	4%	62%
4	69%	15%	4%	3%	1%	77%

Table 3. *Cont.*

No.	O (Unclassified Tailings Slurry Concentration) *	B (Quicklime)	C (Cement)	D (Gypsum)	E (Bentonite)	A (Slag)
5	71%	30%	12%	6%	1%	51%
6	71%	25%	16%	3%	4%	52%
7	71%	20%	4%	12%	7%	57%
8	71%	15%	8%	9%	10%	58%
9	73%	30%	8%	3%	7%	52%
10	73%	25%	4%	6%	10%	55%
11	73%	20%	16%	9%	1%	54%
12	73%	15%	12%	12%	4%	57%
13	75%	30%	4%	9%	4%	53%
14	75%	25%	8%	12%	1%	54%
15	75%	20%	12%	3%	10%	55%
16	75%	15%	16%	6%	7%	56%

* Slurry concentration = (curing agent + tailing)/(curing agent + tailing + water).

4.2. Sample Preparation

According to the number of samples and the total mass required for the experiment, combined with the information in Table 3, the mass of each material was calculated. We used the JS30-1 electronic balance (with a range of 3 kg and accuracy of 0.1 g), Shanghai Puchun Measure Instrument Co., Ltd., Shanghai, China, to measure the corresponding masses of each material. We mixed the materials in a UJZ-15 planetary mixer for 3 min (rotation speed of 120 r/min) and then poured the mixed slurry into the three-part mold (each part size was 7.07 cm × 7.07 cm × 7.07 cm). After a certain period, the samples were demolded and placed in a standard curing chamber. The temperature was set at 20 ± 5 °C and the relative humidity was above 95%. The samples were left to cure for 7 d, 14 d, and 28 d (including the time before demolding), and then their compressive strength was tested. The process is illustrated in Figure 4. In this study, we produced 192 samples.

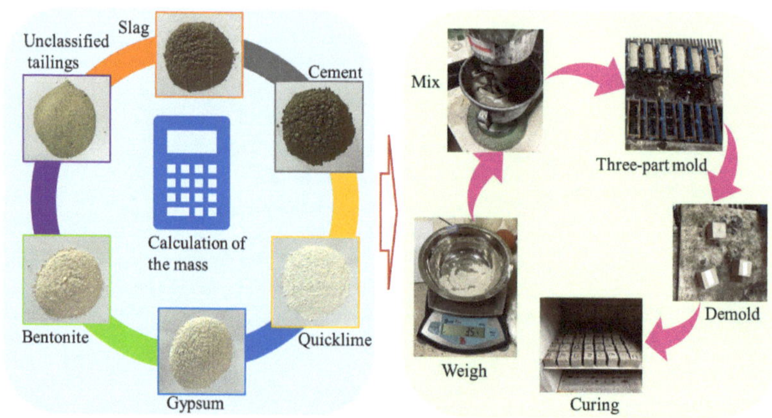

Figure 4. Schematic diagram of sample preparation process.

4.3. Compression Strength Testing

The loading rate for the test was set at 3 mm/min. The experiment was stopped when the sample was fractured and the compression displacement of the sample exceeded 18 mm. The uniaxial compression strength of the unclassified tailings solidified sample was calculated using Equation (12):

$$R_n = \frac{F}{S} \tag{12}$$

where R_n is the compression strength of the sample after n days of curing, in MPa. F is the maximum load, in N. S is the cross-sectional area of the sample perpendicular to the loading direction, in m^2.

When calculating the average compression strength for each group of samples, the arithmetic mean was calculated using Equation (13):

$$R_{\overline{n}} = \sum_{i=1}^{k} \frac{R_{ni}}{k} \quad (13)$$

where $R_{\overline{n}}$ is the average compression strength after n days of curing, in MPa. R_{ni} is the experimental value of the compression strength for the i-th sample after n days of curing, in MPa. k is the number of samples in the group.

4.4. Rheological Characteristics Testing Experiment Plan

Based on the results of the uniaxial compression tests and the determined proportions of each material in the curing agent, the rheological parameters of the slurries with curing agents were tested to analyze the flow properties of each slurry and determine the optimal proportions of each material in the curing agent. During the experiment, the corresponding weights of the unclassified tailings, curing agent, and water were weighed. The modified unclassified tailings slurries were prepared following the procedure outlined in Section 4.2. Once the slurry was uniformly mixed, it was poured into a 500 mL beaker until the liquid level reached the 500 mL mark. The beaker was then placed on the rheometer sample stage and securely fixed. The height of the rotor was adjusted, and the test was initiated. As temperature can influence the viscosity coefficient of the slurry, the experiment maintained a controlled temperature of 20 ± 2 °C.

5. Results

5.1. Compression Strength Test Results

The average uniaxial compression strength values of the samples after curing periods of 7 d, 14 d, and 28 d are summarized in Figure 5. As can be seen, at the ages of 7 d, 14 d, and 28 d, the strength of samples from No. 1 to No. 16 shows a trend of fluctuation and increase. The strength values of the same samples after 7 d are less than those after 14 d, which are, in turn, less than those after 28 d. This indicates that, within 28 d, the greater the age, the more complete the reaction of the solidified material, the better the cementation effect, and the stronger the sample strength [26]. It was also found that the greater the age, the greater the range of the sample's strength and the more obvious the difference, reflecting the different reaction degrees of the same compositions in different periods.

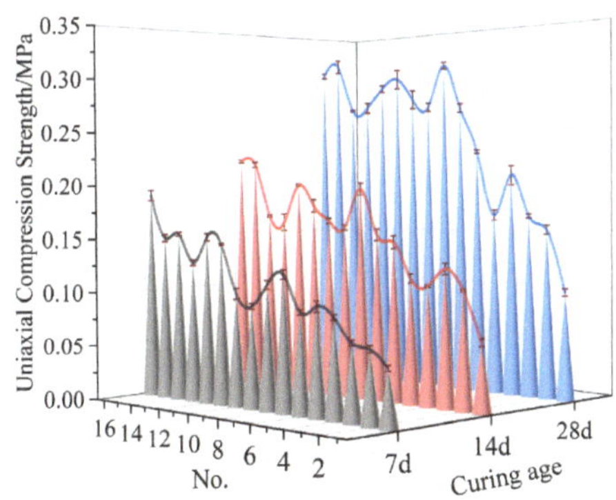

Figure 5. Average uniaxial compression strength of samples at different curing ages.

In order to intuitively show the influence of various factors at different levels on the compressive strength, the average uniaxial compressive strength values of the samples for each factor at each level were plotted, as shown in Figure 6, where *k* represents the average uniaxial compressive strength for each factor at each level. It can be seen that, with the changes at each level of factors O and B, the strength of the samples at 7 d, 14 d, and 28 d gradually increased, and the influence was obvious. The influence of each level of factors C, D, and E on the strength was inconsistent and not obvious.

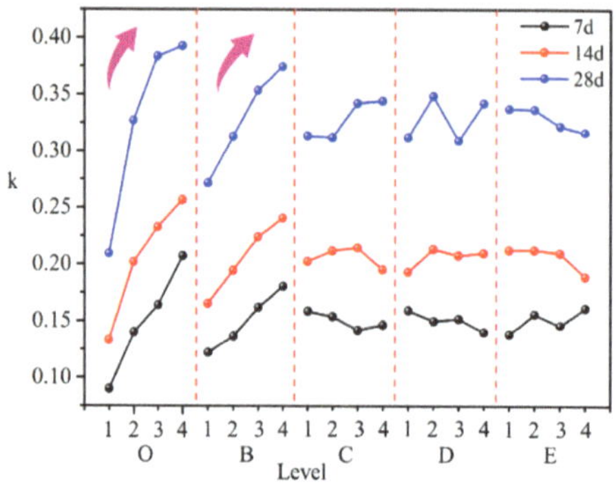

Figure 6. Influence of various factors at different levels on compressive strength.

Table 4 shows the range analysis results of the orthogonal experiments, which were found according to the data in Table 3 and Figure 5 It can be observed that the influence of various factors on the compressive strength of the samples at 7 d and 14 d follows the order of O > B > E > D > C, and the influence of various factors on the compressive strength of samples at 28 d follows the order of O > B > D > C > E, indicating that the compressive strength was primarily affected by the slurry concentration and the proportion of quicklime. Based on the test results, the optimal combination proportions of the materials in the curing agent at different curing ages are shown in Table 5.

Table 4. Results of range analysis of orthogonal experiments.

	Curing Age	O	B	C	D	E
Range R	7 d	0.117	0.059	0.017	0.019	0.023
	14 d	0.123	0.075	0.019	0.020	0.023
	28 d	0.183	0.103	0.032	0.039	0.021
Main-Sub Sequence	7 d	O > B > E > D > C				
	14 d	O > B > E > D > C				
	28 d	O > B > D > C > E				
Optimal Level	7 d	O4	B4	C1	D1	E4
	14 d	O4	B4	C3	D2	E1
	28 d	O4	B4	C4	D2	E1

Table 5. Optimal mass proportions of various materials at different curing ages.

Curing Age	Slurry Mass Concentration	Mass Proportion				
		Slag	Quicklime	Cement	Gypsum	Bentonite
7d	75%	56%	15%	16%	12%	1%
14d	75%	58%	15%	8%	9%	10%
28d	75%	62%	15%	4%	9%	10%

5.2. Rheological Characteristics Test Results

The rheological parameters of the modified tailings slurries were tested based on the three curing agent proportions above. In this study, the curing agent proportions corresponding to the best flowability were identified as the optimal proportions. The rheological characteristics of the three modified tailings slurries are illustrated in Figure 7. As shown, the regression curves of the flow curves for the modified tailings slurries I (7 d), II (14 d), and III (28 d) were represented by blue, green, and orange dashed lines, respectively. The regression equations for the flow curves of the modified tailings slurries I (7 d), II (14 d), and III (28 d) were, respectively, $\tau_{I} = 26.07 + 0.89\gamma$, $\tau_{II} = 43.62 + 0.67\gamma$, and $\tau_{III} = 30.59 + 0.71\gamma$, and all R^2 values were greater than 0.90, indicating that all three modified tailings slurries adhered to the Bingham fluid model. The yield stresses of the slurries were 26.07 Pa, 43.62 Pa, and 30.59 Pa, and the viscosity coefficients of the slurries were 0.89 Pa·s, 0.67 Pa·s, and 0.71 Pa·s, respectively.

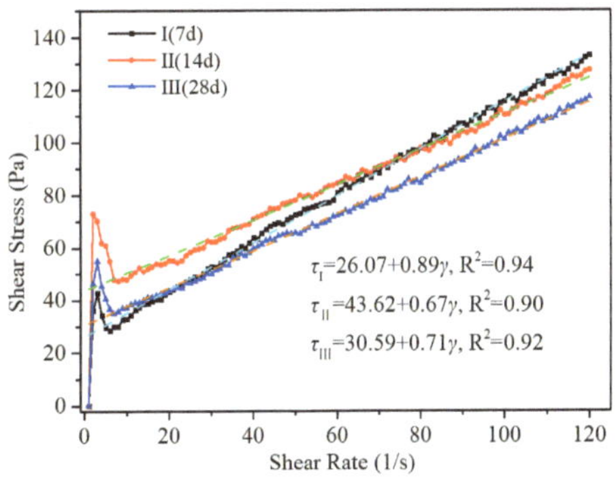

Figure 7. Rheological curves of the three modified tailings slurries.

Grabinsky et al. believed that the yield stress of high-concentration tailings is generally greater than 20 Pa but less than 100 Pa, and high-concentration tailings can be efficiently transported using centrifugal pumps [52]. Despite being high-concentration tailings, they still retain some paste-like characteristics. Their rheological behavior falls within the non-Newtonian realm, yet their yield stress is noticeably lower than that of conventional pastes. This property results in lower pipeline transportation resistance and reduced self-flowing gradients. The outcome in this study is consistent with the findings of Grabinsky et al. The yield stress indicates the threshold at which flow starts in response to shear stress exceeding the yield stress. This is typical of Bingham fluids, which only begin to flow when the applied shear stress surpasses the yield stress. The viscosity coefficient represents internal resistance to flow, and the smaller the value, the better the slurry flow [53]. Therefore, the optimal condition occurs when the yield stress is 43.62 Pa and the viscosity coefficient is 0.67 Pa·s. This corresponds to a tailings slurry concentration of 75%, with the curing agent consisting of slag (58%), quicklime (15%), cement (8%), gypsum (9%), and bentonite (10%).

Gao et al. researched the fluid characteristics of the cemented backfill pipeline transportation of mineral processing tailings [54]. It was concluded that the viscosity coefficient of the slurry concentration reached 24.42607 Pa·s at 80% and 3.73863 Pa·s at 78%. When the concentration was 76%, the viscosity coefficient decreased to 1.21629 Pa·s, and at 74% and 72%, it decreased to 0.58072 Pa·s and 0.31388 Pa·s, respectively. If the viscosity coefficient is too large, it results in excessive resistance during the transportation of slurry when only a high-pressure piston pump is feasible. Thus, the yield stress of 43.62 Pa and the viscosity coefficient of 0.67 Pa·s obtained in this study renders the slurry suitable for pipeline transportation [34,36], laying a foundation for subsequent decisions regarding stockpiling processes and equipment selection.

5.3. SEM Experimental Results

As shown in Figure 8a, slag particles primarily exhibit cubic, blocky, and irregular shapes, with distinct edges and a dense texture. The particle sizes range from approximately 1 to 10 μm. Nanoscale particles are attached to the surface of larger particles, which appear relatively smooth. Figure 8b reveals that the particles of quicklime are mainly clustered, forming uniform-sized crystal branches. The size of each branch is smaller than 1 μm, and they aggregate to form structures ranging from 1 to 10 μm, characterized by a rough surface with occasional protrusions. The aggregated structures formed by the branches appear loose and porous. Figure 8c illustrates that cement particles exist in various forms, including blocks, rods, spheres, cubes, and irregular shapes, displaying substantial shape variation. Nanoscale particles are attached to the surfaces of larger particles (size > 2 μm), leading to rough surfaces on the larger particles. Rod-like particles have dimensions of approximately 2 μm × 10 μm, and their surface roughness lies between that of the larger and spherical particles. In Figure 8d, gypsum is primarily present in flocculent, small-particle, large-block, and irregular forms. Flocculent small particles exhibit multiple protrusions and a rough surface. Figure 8e shows that most bentonite particles exhibit circular, elliptical, or similar circular shapes, with rough surfaces and sizes ranging from a few micrometers to tens of micrometers. Figure 8f demonstrates the varied sizes of the unclassified tailings particles, with the largest particle in the image measuring about 200 μm × 300 μm. The majority of particles have sizes below a few micrometers, some even in the nanometer range. In addition, most of the tailings appear as flakes, and it can be seen that the surface of the tailings is rough.

Figure 8. Microscopic morphology of various materials: (**a**) slag, (**b**) quicklime, (**c**) cement, (**d**) gypsum, (**e**) bentonite, and (**f**) unclassified tailings.

Figure 9 illustrates the cross-sectional morphology of the modified tailings solidified samples with the optimal curing agent. It can be observed that after the reaction, the various materials were fused together as a whole, and the sharp edges of the material particles disappeared. Additionally, the disappearance of the quicklime morphology indicates that the quicklime underwent a full reaction. The surfaces of some tailings particles were involved in the hydration reaction, and the rest is wrapped by the hydration products. The tailings particles that were not fully reacted, as shown by the brown arrow, acted as a skeleton, and supported the strength of the solidified samples. The hydration products were mainly AFt (prismatic ettringite) and C-S-H [51,55]. The appearance of a flocculent gel system can be observed; some gel systems exhibited clear spatial structures, as indicated by the red dashed line. The green arrows indicate the pore structure. These characteristics enhanced the strength of the solidified sample. Comprehensive analysis revealed that effective reactions occurred between the curing agent and the unclassified tailings, resulting in gel systems and spatial structures, thereby enhancing the overall strength of the solidified system and achieving effective solidification of the unclassified tailings.

Figure 9. Microscopic morphology of a modified unclassified tailings solidified sample. (AFt mainly presented as needle shaped, columnar, prismatic, etc. C-S-H mainly presented as cotton-like, network-like, and floc-like structures).

6. Discussion

Further experimentation revealed that the average uniaxial compression strength values of the modified unclassified tailings solidified samples with the optimal curing agent were 0.20 MPa, 0.25 MPa, and 0.34 MPa at 7 d, 14 d, and 28 d, respectively. We conducted XRD tests on these MUTSSs at different curing ages to further analyze the chemical reaction mechanism of each material in the curing agent. Figure 10 shows the XRD patterns of the MUTSSs at 7 d, 14 d, and 28 d. The diffraction peaks corresponding to C-S-H, AFt, Ca $(OH)_2$, $CaCO_3$, and SiO_2 can be seen in the figure [51,55]. Figure 11 illustrates the qualitative relationships among the main hydration products. As shown, with the extension of the hydration time, the contents of AFt and Ca $(OH)_2$ decreased, while the content of C-S-H increased. This phenomenon is consistent with the conclusions reported by Jha et al. [56], Telesca et al. [24], Yum et al. [25], and Li et al. [26].

When the curing time was 7 days, the AFt content was the highest, followed by C-S-H and $Ca(OH)_2$. At the initial stage of hydration, more AFt crystals formed, and C-S-H gel grew on the cement particles. Due to the growth of C-S-H and the formation of AFt crystals, strong and weak contact points were generated. At the same time, the hydration product particles connected to form a network, giving the solidified samples early strength. As the number of contact points increased, the size of the network structure increased, and consequently, the strength of the solidified samples increased. The combination of the hydration products, tailings, and unhydrated clinker is the reason for the observed early macroscopic strength. After 28 d of curing, the content of C-S-H was the highest, followed by AFt and $Ca(OH)_2$. At this time, the strength of the solidified samples was the highest. This observation can be explained as follows: After 28d of hydration, a significant amount of C-S-H was generated, enhancing the long-term strength and durability of the solidified samples. Due to the abundant presence of C-S-H and the complete formation of AFt and $Ca(OH)_2$ crystals, hydration products filled the space previously occupied by water. The spatial distribution of the hydration products was uniform and gradually interwoven, ultimately forming a dense multi-layered network structure, which increased the overall compressive strength. The above phenomena and analysis are consistent with the research presented in Section 2, and are a supplement to the theoretical analysis in Section 2, reflecting the phased nature of the curing reaction process.

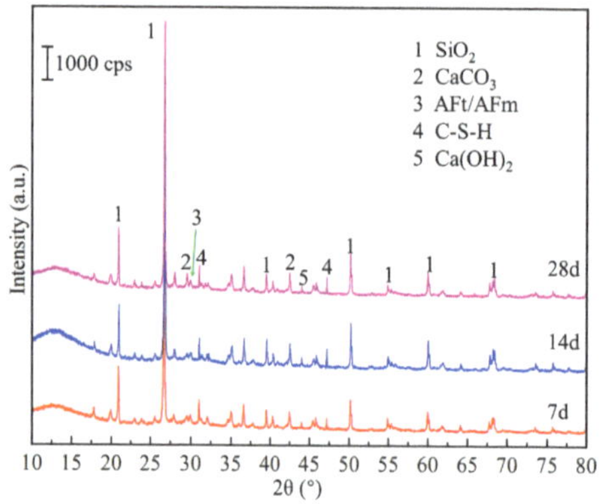

Figure 10. XRD patterns of MUTSSs.

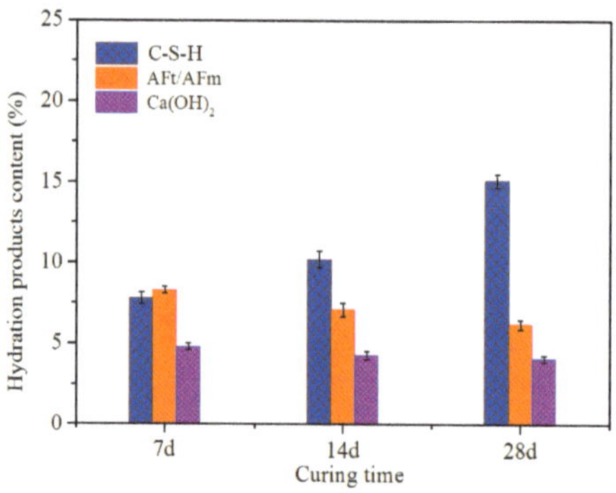

Figure 11. Qualitative relationships among the main hydration products of MUTSSs.

7. Conclusions

In this research, we conducted a theoretical analysis of chemical reactions; designed a laboratory orthogonal experiment; carried out compressive strength testing, rheological properties testing, SEM experiments, XRD experiments, etc.; and obtained low-cost curing agents suitable for high-concentration tailings stockpiling. The main conclusions are as follows:

(1) The curing agent system for tailings consolidation and stockpiling composed of slag, quicklime, cement, gypsum, and bentonite is feasible. The reaction mechanism involves synergistic interactions among these compositions, promoting the dissolution of active compositions in slag and tailings. In addition to their inherent cementitious activity, the excitation agents such as cement, quicklime, and gypsum in the curing agent also serve as alkaline excitation agents for the volcanic ash reactions of slag and tailings. The formation of calcium hydroxide and various C-S-H/C-A-H enhances the disintegration of the slag glass network. The presence of gypsum aids in the formation of AFt or AFm through reactions with active compositions. Bentonite contributes to ion exchange and volumetric expansion, reinforcing the solidifying system;

(2) The optimal proportions of each material in the curing agent are as follows: slag, 58%; quicklime, 15%; cement, 8%; gypsum, 9%; and bentonite, 10%. The proportion of industrial waste steel slag reached 58%, which reflects the low-cost nature of the curing agent. Under this condition, the modified unclassified tailings slurry exhibited a yield stress of 43.62 Pa and a viscosity coefficient of 0.67 Pa·s at a slurry concentration of 75%, and the average uniaxial compression strength values of the MUTSSs were 0.20 MPa, 0.25 MPa, and 0.34 MPa at 7 d, 14 d, and 28 d, respectively;

(3) The hydration products of the curing agent are mainly AFt, C-S-H and Ca $(OH)_2$. With the extension of the hydration time, the contents of AFt and Ca $(OH)_2$ decreased, while the content of C-S-H increased, forming a gel system and spatial structures. The tailings particles that did not fully react acted as a skeleton and supported the strength of the solidified samples. These two aspects worked together to improve the overall strength of the solidified sample and achieved effective solidification of the unclassified tailings.

Author Contributions: Conceptualization, W.W., K.L. and L.G.; methodology, W.W. and K.L.; validation, K.L. and L.G.; formal analysis, Y.L. and Y.C.; investigation, S.W. and Y.C.; data curation, Y.L. and Y.C.; writing—original draft preparation, W.W. and S.W.; writing—review and editing, L.G.; supervision, L.G.; funding acquisition, K.L. and L.G. All authors have read and agreed to the published version of the manuscript.

Funding: This research received no external funding.

Data Availability Statement: The data supporting the conclusions of this article will be made available by the authors on request.

Acknowledgments: This work was supported by the National Key R&D Program of China (2022YFC 2903905, 2021YFC2900600, 2021YFE0102900).

Conflicts of Interest: Weixiang Wang, Kun Li, Lijie Guo, Sha Wang, Yifan Chu and Yao Lu are employees of BGRIMM Technology Group. The paper reflects the views of the scientists and not the company.

References

1. Tabelin, C.B.; Park, I.; Phengsaart, T.; Jeon, S.; Villacorte-Tabelin, M.; Alonzo, D.; Yoo, K.; Ito, M.; Hiroyoshi, N. Copper and critical metals production from porphyry ores and E-wastes: A review of resource availability, processing/recycling challenges, socio-environmental aspects, and sustainability issues. *Resour. Conserv. Recycl.* **2021**, *170*, 105610. [CrossRef]
2. Liu, L.; Xin, J.; Huan, C.; Zhao, Y.J.; Fan, X.; Guo, L.J.; Song, K.-I. Effect of curing time on the mesoscopic parameters of cemented paste backfill simulated using the particle flow code technique. *Int. J. Miner. Metall. Mater.* **2021**, *28*, 590–602. [CrossRef]
3. Edraki, M.; Baumgartl, T.; Manlapig, E.; Bradshaw, D.; Franks, D.M.; Moran, C.J. Designing mine tailings for better environmental, social and economic outcomes: A review of alternative approaches. *J. Clean. Prod.* **2014**, *84*, 411–420. [CrossRef]
4. Boger, D.V. Personal perspective on paste and thickened tailings: A decade on. *Min. Technol.* **2012**, *121*, 29–36. [CrossRef]
5. Psarropoulos, P.N.; Tsompanakis, Y. Stability of tailings dams under static and seismic loading. *Can. Geotech. J.* **2008**, *45*, 663–675. [CrossRef]
6. Birch, W.J.; Dixon-Hardy, D.W.; Engels, J.M. Tailings management facility topography modelling to identify flow accumulations and pond geometry. *Min. Technol.* **2006**, *115*, 65–74. [CrossRef]
7. Cross, A.T.; Lambers, H. Young calcareous soil chronosequences as a model for ecological restoration on alkaline mine tailings. *Sci. Total Environ.* **2017**, *607*, 168–175. [CrossRef]
8. Schoenbrunn, F.; Bach, M. The Development of paste thickening and its application to the minerals industry; an industry review. *BHM Berg-Und Hüttenmännische Monatshefte* **2015**, *160*, 257–263. [CrossRef]
9. Newman, L.; Arnold, K.; Wittwer, D. Dry stack tailings design for the Rosemont Copper project. In *Tailings and Mine Waste*; CRC Press: Boca Raton, FL, USA, 2010; pp. 315–326. [CrossRef]
10. Alshawmar, F.; Fall, M. Dynamic response of thickened tailings in shaking table testing. *Int. J. Geo-Eng.* **2021**, *12*, 28. [CrossRef]
11. Wang, Z.W.; Li, F.; Mei, G.D. OpenMP Parallel Finite-Discrete Element Method for Modeling Excavation Support with Rockbolt and Grouting. *Rock Mech. Rock Eng.* **2024**, *57*, 3635–3657. [CrossRef]
12. Fourie, A. Paste and thickened tailings: Has the promise been fulfilled. In *GeoCongress 2012: State of the Art and Practice in Geotechnical Engineering*; American Society of Civil Engineers: Reston, VA, USA, 2012; pp. 4126–4135.
13. Eng, F.P.P. Paste Thickening: Considerations for Backfill vs. Tailings Management. *Eng. Min. J.* **2011**, *212*, 34–40.
14. Yang, P.Y.; Li, L. Investigation of the short-term stress distribution in stopes and drifts backfilled with cemented paste backfill. *Int. J. Min. Sci. Technol.* **2015**, *25*, 721–728.
15. Kouame, K.J.A.; Feng, Y.; Jiang, F.; Zhu, S. A Study of Technical Measures for Increasing the Roof-Contacted Ratio in Stope and Cavity Filling. *J. Mater. Sci. Res.* **2016**, *5*, 54. [CrossRef]
16. Chen, O.S.; Zhang, O.L.; Fourie, A.; Chen, X.; Qi, C.C. Experimental investigation on the strength characteristics of cement paste backfill in a similar stope model and its mechanism. *Constr. Build. Mater.* **2017**, *154*, 34–43. [CrossRef]
17. Luo, L.; Zhang, Y.; Bao, S.; Chen, T. Utilization of Iron Ore Tailings as Raw Material for Portland Cement Clinker Production. *Adv. Mater. Sci. Eng.* **2016**, *2016*, 1596047. [CrossRef]

18. Shi, X.; Wang, X.; Wang, X. Dual waste utilization in cemented paste backfill using steel slag and mine tailings and the heavy metals immobilization effects. *Powder Technol.* **2022**, *403*, 117413. [CrossRef]
19. Wang, J.; Fu, J.; Song, W.; Zhang, Y. Effect of rice husk ash (RHA) dosage on pore structural and mechanical properties of cemented paste backfill. *J. Mater. Res. Technol.* **2022**, *17*, 840–851. [CrossRef]
20. Chen, S.J.; Du, Z.W.; Zhang, Z.; Yin, D.W.; Feng, F.; Ma, J.B. Effects of red mud additions on gangue-cemented paste backfill properties. *Powder Technol.* **2020**, *367*, 833–840. [CrossRef]
21. Zhao, Y.L.; Ma, Z.Y.; Qiu, J.P.; Sun, X.G.; Gu, X.W. Experimental study on the utilization of steel slag for cemented ultra-fine tailings backfill. *Powder Technol.* **2020**, *375*, 284–291.
22. Zhao, Y.L.; Wu, P.Q.; Qiu, J.P.; Guo, Z.B.; Tian, Y.S.; Sun, X.G.; Gu, X.W. Recycling hazardous steel slag after thermal treatment to produce a binder for cemented paste backfill. *Powder Technol.* **2022**, *395*, 652–662. [CrossRef]
23. Zhao, Y.L.; Qiu, J.P.; Zhang, S.Y.; Guo, Z.B.; Wu, P.Q.; Sun, X.G.; Gu, X.W. Low carbon binder modified by calcined quarry dust for cemented paste backfill and the associated environmental assessments. *J. Environ. Manag.* **2021**, *300*, 113760. [CrossRef] [PubMed]
24. Telesca, A.; Marroccoli, M.; Calabrese, D.; Valenti, G.L.; Montagnaro, F. Flue gas desulfurization gypsum and coal fly ash as basic components of prefabricated building materials. *Waste Manag.* **2013**, *33*, 628–633. [CrossRef] [PubMed]
25. Yum, W.S.; Jeong, Y.; Song, H.; Oh, J.E. Recycling of limestone fines using $Ca(OH)_2$ and $Ba(OH)_2$ activated slag systems for eco-friendly concrete brick production. *Constr. Build. Mater.* **2018**, *185*, 275–284. [CrossRef]
26. Li, C.; Sun, H.H.; Bai, J.; Li, L.T. Innovative methodology for comprehensive utilization of iron ore tailings: Part I. The recovery of iron from iron ore tailings using magnetic separation after magnetizing roasting. *J. Hazard. Mater.* **2010**, *174*, 71–77. [CrossRef]
27. Yao, G.; Wang, Q.; Su, Y.W.; Wang, J.X.; Qiu, J.; Lyu, X.J. Mechanical activation as an innovative approach for the preparation of pozzolan from iron ore tailings. *Miner. Eng.* **2020**, *145*, 106068. [CrossRef]
28. Barati, S.; Tabatabaie Shourijeh, P.; Samani, N.; Asadi, S. Stabilization of iron ore tailings with cement and bentonite: A case study on Golgohar mine. *Bull. Eng. Geol. Environ.* **2020**, *79*, 4151–4166. [CrossRef]
29. Fall, M.; Célestin, J.C.; Han, F.S. Suitability of bentonite-paste tailings mixtures as engineering barrier material for mine waste containment facilities. *Miner. Eng.* **2009**, *22*, 840–848. [CrossRef]
30. Opiso, E.M.; Tabelin, C.B.; Maestre, C.V.; Aseniero, J.P.J.; Arima, T.; Villacorte-Tabelin, M. Utilization of Palm Oil Fuel Ash (POFA) as an Admixture for the Synthesis of a Gold Mine Tailings-Based Geopolymer Composite. *Minerals* **2023**, *13*, 232. [CrossRef]
31. Opiso, E.M.; Tabelin, C.B.; Maestre, C.V.; Aseniero, J.P.J.; Park, I.; Villacorte-Tabelin, M. Synthesis and characterization of coal fly ash and palm oil fuel ash modified artisanal and small-scale gold mine (ASGM) tailings based geopolymer using sugar mill lime sludge as Ca-based activator. *Heliyon* **2021**, *7*, e06654. [CrossRef]
32. Wang, Y.; Wu, A.X.; Zhang, L.F.; Wang, H.J.; Jin, F. Investigation of the Sedimentation Property of Backfill Material on the Basis of Rheological Test: A Case Study of Iron Tailings. *J. Chem.* **2018**, *2018*, 9530767. [CrossRef]
33. Boger, D.V. Rheology of slurries and environmental impacts in the mining industry. *Annu. Rev. Chem. Biomol.* **2013**, *4*, 239–257. [CrossRef]
34. Yang, L.; Jia, H.; Wu, A.; Jiao, H.; Chen, X.; Kou, Y.; Dong, M. Particle Aggregation and Breakage Kinetics in Cemented Paste Backfill. *Int. J. Miner. Metall. Mater.* **2023**, *220*, 172–186. [CrossRef]
35. Mizani, S.; He, X.; Simms, P. Application of lubrication theory to modeling stack geometry of high density mine tailings. *J. Non-Newton. Fluid Mech.* **2013**, *198*, 59–70. [CrossRef]
36. Yang, L.; Gao, Y.; Chen, H.; Jiao, H.; Dong, M.; Bier, T.A.; Kim, M. Three-dimensional concrete printing technology from a rheology perspective: A review. *Adv. Cem. Res.* **2024**, *126*, 72–86. [CrossRef]
37. Li, G.Y.; Zhao, X.H. Properties of concrete incorporating fly ash and ground granulated blast-furnace slag. *Cem. Concr. Comp.* **2003**, *25*, 293–299. [CrossRef]
38. Double, D.D. New developments in understanding the chemistry of cement hydration. *Phil. Trans. R. Soc. Lond. A* **1983**, *310*, 53–66.
39. Wang, B.W.; Li, Q.L.; Dong, P.B.; Gan, S.; Yang, L.; Wang, R.Z. Performance investigation of blast furnace slag based cemented paste backfill under low temperature and low atmospheric pressure. *Constr. Build. Mater.* **2023**, *363*, 129744. [CrossRef]
40. Zhang, M.G.; Li, K.Q.; Ni, W.; Zhang, S.Q.; Liu, Z.Y.; Wang, K.; Wei, X.L.; Yu, Y. Preparation of mine backfilling from steel slag-based non-clinker combined with ultra-fine tailing. *Constr. Build. Mater.* **2022**, *320*, 126248. [CrossRef]
41. Hou, Y.Q.; Yang, K.; Yin, S.H.; Yu, X.; Wang, L.M.; Yang, X.B. Enhancing the physical properties of cemented ultrafine tailings backfill (CUTB) with fiber and rice husk ash: Performance, mechanisms, and optimization. *J. Mater. Res. Technol.* **2024**, *29*, 4418–4432. [CrossRef]
42. Salami, B.A.; Ibrahim, M.; Algaifi, H.A.; Alimi, W.; Oladapo, E.A. A review on the durability performance of alkali-activated binders subjected to chloride-bearing environment. *Constr. Build. Mater.* **2022**, *317*, 125947. [CrossRef]
43. Yuan, H.; Chen, S.Y. Effects of titanium gypsum on high calcium fly ash system: Physical and microscopic properties, hydration, and Fe transformation. *J. Build. Eng.* **2023**, *78*, 107653. [CrossRef]
44. Xia, Y.; Liu, M.H.; Zhao, Y.D.; Chi, X.F.; Guo, J.Z.; Du, D.H.; Du, J.X. Hydration mechanism and phase assemblage of blended cement with iron-rich sewage sludge ash. *J. Build. Eng.* **2023**, *63*, 105579. [CrossRef]
45. Liu, X.H.; Ma, B.G.; Tan, H.B.; Gu, B.Q.; Zhang, T.; Chen, P.; Li, H.N.; Mei, J.P. Effect of aluminum sulfate on the hydration of Portland cement, tricalcium silicate and tricalcium aluminate. *Constr. Build. Mater.* **2020**, *232*, 117179. [CrossRef]

46. Li, C.B.; Ma, B.G.; Tan, H.B.; Zhang, T.; Liu, X.H.; Chen, P. Effect of triisopropanolamine on chloride binding of cement paste with ground-granulated blast furnace slag. *Constr. Build. Mater.* **2020**, *256*, 119494. [CrossRef]
47. Soliman, M.A.; Rashad, G.M.; Mahmoud, M.R. Organo-modification of montmorillonite for enhancing the adsorption efficiency of cobalt radionuclides from aqueous solutions. *Environ. Sci. Pollut. Res.* **2019**, *26*, 10398–10413. [CrossRef]
48. Kazutoshi, H.; Li, H.J.; MatSuda, K.; Takehisa, T.; Elliott, E. Mechanism of forming organic/inorganic network structures during in-situ free-radical polymerization in PNIPA-Clay nanocomposite hydrogels. *Macromolecules* **2005**, *38*, 3482.
49. Segad, M.; Jonsson, B.; Akesson, T.; Cabane, B. Ca/Na Montmorillonite: Structure, Forces and Swelling Properties. *Langmuir* **2010**, *26*, 5782–5790. [CrossRef]
50. Liu, L.; Xin, J.; Huan, C.; Qi, C.C.; Zhou, W.W.; Song, K.-I. Pore and strength characteristics of cemented paste backfill using sulphide tailings: Effect of sulphur content. *Constr. Build. Mater.* **2020**, *237*, 117452. [CrossRef]
51. Zhao, X.Y.; Yang, K.; He, X.; Wei, Z.; Zhang, J.Q. Study on proportioning experiment and performance of solid waste for underground backfilling. *Mater. Today Commun.* **2022**, *32*, 103863. [CrossRef]
52. Grabinsky, M.W.; Theriault, J.; Welch, D. An overview of paste and thickened tailings disposal on surface. In *Symposium 2002 on Environment and Mine*; Canadian Institute of Mining, Metallurgy and Petroleum: Montreal, QC, Canada, 2002; p. 533.
53. Leong, Y.K. Controlling the rheology of iron ore slurries and tailings with surface chemistry for enhanced beneficiation performance and output, reduced pumping cost and safer tailings storage in dam. *Miner. Eng.* **2021**, *166*, 106874. [CrossRef]
54. Gao, R.G.; Zhou, K.P.; Zhou, Y.L.; Yang, C. Research on the fluid characteristics of cemented backfill pipeline transportation of mineral processing tailings. *Alex Eng. J.* **2020**, *59*, 4409–4426. [CrossRef]
55. Gu, X.W.; Yang, B.H.; Li, Z.J.; Liu, B.N.; Liu, J.P.; Wang, Q.; Nehdif, M.L. Elucidating the reaction of seashell powder within fly ash cement: A focus on hydration products. *Constr. Build. Mater.* **2024**, *428*, 136331. [CrossRef]
56. Jha, A.K.; Sivapullaiah, P.V. Volume change behavior of lime treated gypseous soil—Influence of mineralogy and microstructure. *Appl. Clay Sci.* **2016**, *119*, 202–212. [CrossRef]

MDPI AG
Grosspeteranlage 5
4052 Basel
Switzerland
Tel.: +41 61 683 77 34

Minerals Editorial Office
E-mail: minerals@mdpi.com
www.mdpi.com/journal/minerals

www.ingramcontent.com/pod-product-compliance
Lightning Source LLC
Chambersburg PA
CBHW041355210926
51850CB00013B/3090